Introduction to Energy
Third Edition

Given the potential disruption of climate change, understanding energy issues and technologies is more important than ever if societies are to make informed choices on policy. Now in its third edition, *Introduction to Energy* explores the crucial issues connected to modern energy technology and its uses. Fully updated to respond to the substantial developments in the energy sector, the book expands on the relationships of energy use and climate change; of energy availability and the alleviation of world poverty; and of energy consumption and the sustainability of the lifestyles of people in the industrialized world. Directed at a broad readership, it assumes no prior technical expertise and avoids complex mathematical formulations, continuing to provide a standard for introductory energy courses. It is also a useful supplementary text for programs in public policy, business law, and resource economics.

Edward S. Cassedy was Professor Emeritus at the Department of Electrical Engineering, Polytechnic University, Brooklyn, New York. As well as the previous editions of this book, he is also the author of *Prospects for Sustainable Energy: A Critical Assessment* (Cambridge University Press, 2000).

Peter Z. Grossman is a professor of Economics at Butler University, Indiana. He has published more than 200 works, including an analytic history of American energy policy, *U.S. Energy Policy and the Pursuit of Failure* (Cambridge University Press, 2013), which was endorsed by Nobel Laureate Thomas Schelling. He is also principal author of *The End of a Natural Monopoly: Deregulation and Competition in the Electric Power Industry* (Taylor & Francis, 2003). His scholarly articles have appeared in such journals as *Energy Policy*, the *Cato Journal*, and the *Journal of Public Policy*.

"Cassedy and Grossman's *Introduction to Energy* is a fascinating and eminently readable account of the sources of energy available to our society and a dispassionate accounting of the costs and benefits of each energy source. In this updated third edition, the authors hit all the right notes, and cover important developments in renewables as well as nuclear and fossil fuels since the second edition was published. New sections are crisply and clearly written; of special note are the sections on solar, wind, and the German *energiewende* (energy transition) in the final chapter. This book should be read by students and policymakers who wish to be able to think critically about the energy issues confronting our global society in the twenty-first century." – *Louis Schwartzkopf, Minnesota State University*

"This updated edition of the classic text by Cassedy and Grossman is essential reading for understanding the modern energy landscape in all its economic, technological, and political dimensions. It covers virtually everything, from the basics of energy supply and demand, to the changing aspects of transportation fuels and power generation, to the prospects for a transition toward renewable energy sources. This new edition offers fresh and insightful discussions of contemporary developments and concerns relating to peak oil, fracking, next generation nuclear reactors, solar and wind, energy poverty, the resource curse, the smart grid, and climate change. *Introduction to Energy* should become a staple of any college curriculum on the subject." – *Tyler Priest, University of Iowa*

"Energy spans disciplines from physics to engineering to biology to economics and public policy. The topic can overwhelm a beginning student. *Introduction to Energy* by Cassedy and Grossman gives beginners just what they need to get started. The authors take us from the timeless principles of science to – in this latest edition – such timely topics as climate change and energy transitions. This book is a classic introduction to energy, once again intelligently updated for today's students." – *Michael Giberson, Texas Tech University*

"Now in its third edition, *Introduction to Energy* is a truly impressive book. Cassedy and Grossman integrate and leverage their respective engineering and economics expertise with great effectiveness. The result is a work that will inform experts and non-experts alike. It is also an example of what a true interdisciplinary effort can accomplish. Not least of which are integrated policy implications and potential prescriptions that do not work at cross purposes. In accomplishing this, the authors do us a great service – one that leads to the avoidance of unintended and negative policy consequences." – *Jim Granato, University of Houston*

From reviews of previous editions:

". . . serves as an excellent introduction to energy . . . well written and directed at a broad readership. It provides valuable assessments of a variety of energy-related issues. Above all, the book is readable by the technical and nontechnical person." – *Energy Sources*

". . . an extremely useful book for virtually anyone who wishes to develop a working knowledge of energy, from the various ways it is produced, to the social, environmental, and economic implications associated with its production and consumption." – *Palaios*

". . . more than achieves its aims as a multi-disciplinary textbook for a range of undergraduate students." – *The Times Higher Education Supplement*

Introduction to Energy

Resources, Technology, and Society

Third Edition

EDWARD S. CASSEDY
Polytechnic University, New York, USA

PETER Z. GROSSMAN
Butler University, Indiana, USA

CAMBRIDGE
UNIVERSITY PRESS

University Printing House, Cambridge CB2 8BS, United Kingdom

One Liberty Plaza, 20th Floor, New York, NY 10006, USA

477 Williamstown Road, Port Melbourne, VIC 3207, Australia

4843/24, 2nd Floor, Ansari Road, Daryaganj, Delhi – 110002, India

79 Anson Road, #06–04/06, Singapore 079906

Cambridge University Press is part of the University of Cambridge.

It furthers the University's mission by disseminating knowledge in the pursuit of
education, learning, and research at the highest international levels of excellence.

www.cambridge.org
Information on this title: www.cambridge.org/9781107605046
DOI: 10.1017/9780511997600

© Cambridge University Press 1990, 1998, Edward S. Cassedy and Peter Z. Grossman 2017

First published 1990
Reprinted 1993
Second edition 1998
Third edition 2017

Printed in the United Kingdom by TJ International Ltd. Padstow Cornwall

A catalogue record for this publication is available from the British Library.

Library of Congress Cataloging-in-Publication Data
Names: Cassedy, Edward S., author. | Grossman, Peter Z., 1948– author.
Title: Introduction to energy : resources, technology, and society / Edward S. Cassedy, Polytechnic
University, New York; Peter Z. Grossman, Butler University, Indiana.
Description: Third edition. | New York : Cambridge University Press, 2017. | Includes bibliographical
references and index.
Identifiers: LCCN 2016049332 | ISBN 9781107605046 (pbk.)
Subjects: LCSH: Power resources. | Power (Mechanics)
Classification: LCC TJ163.2 .C4 2017 | DDC 333.79–dc23
LC record available at https://lccn.loc.gov/2016049332

ISBN 978-1-107-60504-6 Paperback

Contents

Preface to the Third Edition (Grossman)

Edward S. "Ned" Cassedy (1927–2017): An Appreciation

I was not supposed to be co-author of this edition of *Introduction to Energy*. When Cambridge University Press asked Ned Cassedy and me to do this edition, I was in the midst of writing my own book. So I told Ned to go on without me and take my name off the title page.

It seemed right in any case. This was always Ned's book in my mind. He had had the idea, and he had written the original proposal, and he had asked me to join him in writing it when it seemed likely he would have a contract.

He showed extraordinary generosity to me in asking and even more so when we started writing. I was at that time a journalist; he an academic engineer. Early on in our collaboration, he wrote something and I rewrote it. He read what I had written and said, "For this book, your style is right. Thanks."

No, Ned. Thank you.

I was flattered. Then again, nothing was more flattering than to be asked to be his co-author.

I had gotten to know Ned in 1979 when I asked for someone at the Polytechnic Institute of New York who knew electric power technology to read a manuscript I had written for Scholastic Press. The book was for 10–14-year-olds, about electric power and electric power failures.

He provided an expert read but, more important to me, we became friends and he became my first academic mentor. I was an assistant professor of journalism where he was a full professor of electrical engineering, but he helped guide me toward a more scholarly appreciation of the things I had been writing about as a journalist. It was not just energy: I was a business/economics/financial reporter and wanted to go more deeply into things, economics especially. Ned was self-taught in economics but knew a great deal more than I did. He gave me encouragement. He told me I needed to keep at it; which I did. In 1992, I received my Ph.D. in economics.

Before we wrote *Introduction to Energy*, he suggested we create a course together on energy and society for a new Science, Technology, and Society program that was starting up at Polytechnic. The course went well; he was very tolerant of my (sometimes poor) attempts to explain some economic concepts with respect to energy, telling me only after class that I had the concept just a bit off the mark – when in fact I had confused the class and he had to (gently) clear things up. Still, Ned felt our approach was exactly the kind of interdisciplinary collaboration needed for a course of this kind and he said we should write a textbook.

Soon after, we were given the very welcome news that Cambridge University Press would publish the text he had proposed. We wrote the book and it turned out very well, not just in our view but in Cambridge's as well. We were asked to produce a second edition in 1998, and then a few years ago we were asked to do it again.

Ned took on the task but, for health reasons, he could not continue. I was grieved to hear he was no longer able to do the work. When I was asked to finish, I could not refuse. In February 2017, I was saddened to learn that Ned had died even as I was correcting the proofs of the book. I deeply regretted that he would never be able to hold this edition in his hands.

But as I approach completion I must point out that Ned and I had a disagreement about an important issue: the role of nuclear power in developing a low-carbon electricity system. We had agreed that climate change is a serious matter, but he stressed that nuclear power was a dangerous – ultimately failed – option, and we needed to focus on solar and wind.

Ned and I were in agreement back in the 1980s when we wrote in the first edition of this book that nuclear power was too costly and potentially dangerous. My feeling since has changed. As I have written elsewhere, nuclear power was rushed to market before it was ready and as a consequence a lot of problems that arose should not have.

But I believe the current generation of nuclear power designs is greatly improved and the need for them important. Wind and solar, which Ned argued is the near future, I see as limited contributors to our electricity system. Replacing one large nuclear power plant with a wind installation would require an area the size of a major city; two or three would mean an area the size of Rhode Island. These renewables will become more important when, or if, low-cost electricity storage is developed as well. For now, and probably for years to come, I believe solar and wind will be small contributors, mostly kept in business by government largesse.

As I am the one to finish the book, my views on this are the ones in the text, but I have added other readings for those who want to get Ned's side of the argument. In fact, I feel this is in the spirit we started with back in the 1980s. We wanted above all to stimulate thought, ideas, *debate*. I hope I have been true to that idea.

So here we are; the third edition of *Introduction to Energy*.

In many ways, it's my version, and I should take the blame for any omissions or errors.

But to my mind, it's Ned's book.
Always.

***

Of course much of this book is still Dr. Edward S. Cassedy's and he should be credited for a good deal of what follows. In our first and second editions, I had primary responsibility for Chapters 1, 4, 9, and Appendix B. I also contributed to some of the philosophic/social/economic materials in Chapters 2, 5, 6, 10, and 11. Ned wrote all of Chapters 3, 7, 8, Appendix A, and more than half of Chapters 2, 5, 6, 10, and 11, as well as most of Appendix C.

But the last edition came out in 1998; 19 years means that a lot needed to be updated and often changed. Here is what is different in the third edition:

I have left Chapters 3, 7, and Appendix A virtually unchanged; the laws of thermodynamics have not been amended or repealed no matter the composition of Congress or the occupant in the White House. I did change Chapters 1, 2, 4, 5, 6, 8,

9, 10, 11, Appendix B, and Appendix C. Much of this was simply updating. But Chapter 1 was fundamentally changed; it had been largely about the 1970s energy crises, which were part of the life experiences of 1980s undergraduates – those who had been born in the late 1960s or early 70s – but it was no more a compelling reality to today's students than the Franco-Prussian War.

Chapter 2 has a fair amount from the earlier versions but the chapter used to be an endorsement, at least tacitly, of M.K. Hubbert's "peak oil" arguments. Now the chapter explains why Hubbert was largely wrong and leaves uncertain the amount of resources in the ground – though it is clearly a great deal more than Hubbert expected. There is a discussion also of "fracking" and its impact on resource estimates.

Chapter 4 is updated and now includes sections on the "smart grid," as well as the "rebound effect." These ideas were not current as of the second edition (1998), but have inspired large literatures since.

Chapter 5 begins now with a discussion of "energy poverty," a concept related (but not identical) to economic poverty. Also, now removed is the discussion of spaceship and lifeboat ethics. Added instead is a section on the "resource curse," or why it is not necessarily a benefit for a country to have large reserves of energy (or other natural) resources. This section brings in the importance to economic development of political and social institutions.

Chapter 6 is also largely updated but downplays somewhat the importance of coal, especially to the United States, and it considers whether natural gas can be a so-called bridge fuel – a low carbon dioxide but very abundant resource that can replace coal and provide the bulk of the United State's (and a good bit of the world's) electricity until renewables are more commercially viable. The new chapter removes the section on eco-centric ethics and instead expands issues connected to climate change, including a section on the "precautionary principle." The section on acid rain now has an extended discussion of the cap-and-trade program for sulfur dioxide. There is also information on (noncarbon) air pollution in developing countries, particularly China, and the toll it has taken on human health.

Chapter 7 is largely unchanged but does include a discussion of the Fukushima disaster of 2011.

Chapter 8 still has much of Dr. Cassedy's original discussion on the problems of long-term storage of nuclear waste and of the possibility of weapons proliferation. But instead of a close look, and critique, of breeder reactors (as in the second edition), the second half of the chapter examines the next generations of nuclear reactors and provides a discussion of why some environmentalists rest their hopes for carbon dioxide mitigation on these new nukes.

Chapter 9 previously compared the costs of coal-fired, gas-fired, and nuclear electricity; now it expands the consideration of electric power economics by including the costs of solar, wind, and other renewables.

Chapter 10 follows the older version until the last part, which in the second edition was a discussion of hard versus soft paths of development. In its place there is a section on the concept, of compelling interest these days, of energy justice.

Chapter 11 was formerly a look at what we called the "paradigms" of short-, medium-, and long-term technologies (solar thermal, synfuels, and nuclear fusion), but Dr. Cassedy had started to turn the focus toward the prospects for carbon-free solar and wind. I extended that line by changing the title to "Energy Transition" and the question of whether solar and wind could replace coal, nuclear, and gas in electric power generation (and in other applications) in the near future. The chapter, in addition to explications (Dr. Cassedy's) of the technological characteristics of wind and solar (especially photovoltaics), now considers the crucial issues of "grid parity" and energy/power density. There is also a discussion of how a transition is likely to be managed – rapidly by government mandate or slowly mainly through market forces. The German *energiewende* is treated as an ambiguous example of the transition process.

Appendix B is extensively updated as the literature on cost–benefit analysis has grown and many of the points in the chapter needed to be revised and clarified.

Much is unchanged (though all updated) from Appendix C, but some of the technologies are much closer to commercialization than the second edition had noted (tertiary enhanced oil recovery, for example), and others have generated little interest in recent years and so have been cut (magnetohydrodynamics, for example, though it is in a footnote to Chapter 3). Perhaps the most extensive changes are in the discussion of liquid and gaseous biofuels, especially given the controversy of the US ethanol mandate.

Our basic idea remains, however: we wanted to write a book that will provide an understanding of energy issues to all educated readers, and will generate discussion and debate in the classroom on a most crucial aspect of modern life: energy.

Peter Z. Grossman

PART I

ENERGY RESOURCES AND TECHNOLOGY

Introduction

The End of the Oil Age?

Thanksgiving Day, 2005.

That was the day, according to Princeton University geoscientist Kenneth S. Deffeyes, when the output of oil exceeded new additions to petroleum reserves – when world petroleum production necessarily reached its limit and would soon fall – when even with rising prices world output of oil would inexorably decline. Oil depletion would actually accelerate, declining at the same rate by which it had risen in the past to its peak, creating a "permanent oil crisis, where continued demand growth is confronted by declining supply" (Radetzki 2010).

While there may have been a bit of tongue-in-cheek ironic humor with respect to the date, Deffeyes was quite serious about the basic point. According to his 2010 book *When Oil Peaked*, world oil production had run up against natural limits. Deffeyes was not alone in that opinion. "Peak oil" had many believers, including a bipartisan Peak Oil Caucus in the US Congress. There were literally dozens of articles and books published in the first 12 years of the twenty-first century that carried the same message: peak oil was here and the world was about to change. This was, in fact, Deffeyes' third book on the subject in only 10 years. In his 2001 book, *Hubbert's Peak: The Impending World Oil Shortage,* he cited the prediction from 1956 of geologist M. King Hubbert that oil production in the United States would peak in the 1970s, a prediction that appeared to have been proven correct. The rest of the world, Hubbert had argued, would see a peak by the middle of the first decade of the 2000s. When that time arrived, Deffeyes argued that the petroleum age was coming to an end "never to rise again." Disaster loomed. Unless radical policies were adopted quickly, Deffeyes predicted, the "result [of peak oil] will be massive economic and social disruptions in a 21st-century world that has fueled itself for decades with cheap and plentiful energy."

By 2010, Deffeyes argued that the reckoning was upon us. He could point to more than just the prediction of a 1950s geologist as proof of oil's increasing scarcity. In 2008, the price of oil had soared to $147 per barrel (bbl), an all-time record in both nominal and real (inflation-adjusted) terms. This was the culmination of a decade-long rise in the prices of oil and of natural gas; the latter reached $10.79 per 1000 cubic feet, also a record, at about the same time as oil's price hit its all-time high. Both prices fell soon after, but only (it seemed) because of the worst economic downturn since the Great Depression of the 1930s. The only actual solution to the problem of peak oil, Deffeyes claimed, was a crash program to commercialize energy systems that relied on wind, solar, or biological sources that could be made into fuels; that is, any energy resource actually that was not based on oil or gas.

There was a definite logic to the notion of peak oil. Oil, natural gas, coal, and also uranium – energy resources – are products of the Earth that are by definition finite because the Earth itself is finite. There is only so much oil in the ground, and for the past 100 years – during what many have called the oil age (e.g. Heinberg 2003) – petroleum resources have been pulled from the ground by the billions of barrels and burned for their energy value. But there can only be so much oil. The resource has to run out . . . at some point. Why not Thanksgiving Day, 2005?

By 2014, it was clear that oil had not in fact peaked. The prediction was very obviously wrong. We had not come to the end of the oil age, had not come to the end of our ongoing dependence on oil and natural gas; we had not even reached an end to the age of cheap oil and gas, as some had put in a few years earlier (e.g. Ahmed 2010). In 5 years, natural gas prices had fallen by three-quarters; the price of oil stayed high a bit longer, but by 2015 the price had collapsed to around 30 percent of its all-time high. In real terms it had gone from an all-time record to a level it was at 30 years before. So far from a falling rate of production, there was a worldwide glut of oil and, at least in the United States, a vast over-supply of natural gas. The oil age had not ended in 2005 or 2015; the problem of peak oil seemed fantasy.

A Look Back

In reality, fears of exhaustion of energy supplies are not just a recent phenomenon. Throughout history people have had concerns over the depletion of energy resources. Sometimes the fears were justified but the result was seldom disaster; rather it was a transition from reliance on one source of energy to another.

A dramatic example was England. By the late sixteenth century, England was becoming rapidly deforested. Of course, wood had uses beyond energy – in building especially – but it was also the primary fuel of the time, and the need for fuel contributed significantly to a sense of impending crisis. Wood fuel was used in several key industries such as glassmaking, metal smelting, and extracting salt from seawater. The concerns over deforestation lasted for about a century, but in the process of allaying them, the British transitioned to a different energy resource, which in turn became the leading fuel of the Industrial Revolution (Nef 1977).

Great Britain had an alternative to wood: coal. Coal was abundant, in many places coal seams reached the surface, and coal had been used as a fuel in Britain to some extent since Roman times. But it also had drawbacks that limited its applications. Above all, when it burned, it emitted dirty smoke. This led to such bad air pollution in London that, as long ago as the 1300s, burning of some varieties of coal was banned. Also, coal made a poor industrial fuel as it left impurities on the final product.

The need to find an alternative to wood, however, led to two notable technological advances. First, the *reverberatory furnace* was developed. It was essentially a hollowed sphere, but its shape caused the heat to reflect back onto itself – to reverberate –

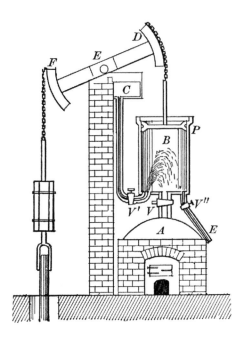

Figure 1.1   The Newcomen steam engine used to pump water out of deep underground coal mines in eighteenth-century England.
Source: Courtesy of Babcock & Wilcox Co.

raising the temperature and resulting in a cleaner burn (by the standards of the day) of the coal fuel. With the furnace, coal could replace wood for most industrial purposes, and by the end of the 1600s it had.

Much of the coal used in the 1600s was easy to extract from the surface. As more of it was used, however, it had to be mined underground. Great Britain is a wet country and underground mines quickly fill with water. But in the early 1700s a technological solution to mine flooding was found. The Newcomen steam engine (see Fig. 1.1) pumped water out of the mines, allowing expansion of coal extraction from deeper mines.

Coal seemed abundant, but even in the midst of plenty there were fears of depletion. The nineteenth-century British economist William Stanley Jevons (1865) predicted the exhaustion of British coal mines over the ensuing century and with it the end of British preeminence in manufacturing. He dismissed the possibility of substituting a different fuel to replace coal; he declared that such an idea was in fact, "useless to think [of]."

But by the late 1800s, there was a resource available to replace coal in heating, one that could also provide fuel for lights: petroleum, that is, oil. But oil, too, was said by experts to be in short supply. In 1886, one geologist declared that young Americans living then would outlive American oil resources. In 1909, President William Howard Taft was persuaded that large tracts of land on which there were oil reserves needed to be put off limits to private energy companies and kept in the ground for national security reasons. The navy was concerned that in the event of war there would not be enough oil to fuel American warships. In the 1920s, the US Geological

Survey predicted domestic oil would be depleted by the end of the decade. Many of the large international oil companies concurred; they abandoned oil prospecting in the United States, regarding America's resources as largely exhausted.

In every case, the predictions were quickly proven wrong. Jevons had made two mistakes: First, he believed that the rate of the *growth* of coal consumption in the past would have to continue for Britain to remain prosperous. But trends that are not sustainable will stop, and if coal or any resource becomes more expensive to produce, people will find ways to use it more efficiently, so that the growth in coal consumption would slow. That was occurring in Great Britain by the late 1800s and on into the twentieth century. By 1913, Britain was, in fact, producing about half as much as Jevons had predicted in 1865, but economic growth continued and there was still more than enough of the resource for domestic use.

Jevons's second mistake was in believing that no alternative could ever be found that would be as practical as coal for Britain's primary fuel. But the uses for oil grew, replacing coal in many instances, most especially in transportation. The nineteenth century was the age of the coal-fired steam locomotive and the steamship. The twentieth became the age of the oil-based gasoline and diesel engines that replaced most coal-fired transportation throughout the world.

Of course, as noted, there were fears that oil would soon run out – especially in the United States. Such fears increased as the world made a slow transition from the coal age to the age of oil. But the transition to oil took place because the substitute resource oil proved not only plentiful – especially in the United States – but also easier to transport. The early twentieth century saw the first major oil strikes in Texas and Oklahoma; there were also major finds in California and in several other states. In 1930 the largest oil field ever was discovered in East Texas. Giant fields were also found outside the United States: in the Middle East, Russia, Canada, South America, and Africa.

Nevertheless, fears of oil as well as natural gas (which was an increasingly important resource itself) exhaustion persisted. At the end of World War II, the US government began to study ways of turning coal (which in fact was proving less expensive and more abundant than Jevons had ever imagined) into substitutes for oil products such as gasoline. The United States had lots of the former but were (said to be) running out of the latter; actually it was soon apparent that US oil reserves were steadily rising. Against a backdrop of plenty, Hubbert made his famous predictions about the future of US oil, gas, and coal resources (more about this in Chapter 2). It was only a decade later, however, in the late 1960s, when it seemed to many that Hubbert was right.

In the 1970s, the predictions of oil and natural gas exhaustion grew more numerous and more believable. American consumers faced a series of what were called energy crises: energy market problems that led to shortages of oil products (especially gasoline) and prices that rose quickly and dramatically. Experts of the time said the end of the oil age would happen sometime between 1980 and 2000. In 1975, the head of the US Geological Survey believed that the United States would run out of both oil and natural gas by 1995 (cited in Manne 1975). Other estimates, including one from the US Central Intelligence Agency, envisioned exhaustion of oil and gas resources throughout the world by 2000. Only the cartel of oil exporting countries (the Organization of the Petroleum Exporting Countries or OPEC) would have any oil left, but by 2000 the price would be so high that dependence on oil would destroy an industrial economy like America's.

The predictions of the 1970s and later years, however, like those of the previous 100 years, did not come to pass. Indeed, as before, soon after most predictions of resource exhaustion have been made, new large oil fields have been discovered. In the 1980s and 1990s and again after 2010, oil gluts followed predictions of scarcity. Prices thought to be ever-rising, fell. By the mid-1980s, the price of oil was lower in inflation-adjusted terms than it had been since the early 1970s, and in the 1990s it fell to its lowest level since the end of World War II. High prices for oil and gas in the 1970s and early 1980s had stimulated the search for new oil fields and many were found in places no one had looked before. And around the time peak oil was said to have been reached, new methods of extraction of oil and gas made exploitation of hard to reach resources cost effective. In fact, the economies that have suffered most from the ups and downs of the oil market have not been industrial economies such as that of the United States, but rather countries such as those of OPEC that have depended on high and consistent revenues from oil (Grossman 2013, Chapter 7).

Energy Issues of Today and Tomorrow

While peak oil is not an imminent problem, there are many ongoing vital issues related to energy we know we need to confront now and will likely continue to confront in the future. This book is about these issues, their import and possible solutions. Of course, there is no single solution, no panacea, no magic bullet, for the dilemmas related to energy today or in the years ahead. There are most certainly going to be challenges. In fact, since these issues will affect our wealth, our lifestyles, our health, and our children, ultimately they engage us all.

a. An issue that is in fact connected to the peak oil debate is *sustainability* (see Chapter 2). Nearly everyone who writes about energy has something to say about sustainability, and many would begin by saying that fossil fuels – oil, gas, coal – are, by definition, not sustainable. They will become depleted, if not on Thanksgiving Day 2005, then at some point. But what do we mean when we talk about sustainability? We will consider the definitions of sustainability in the next chapter, but the overriding question is: *what* exactly should be sustained? Should we seek resources that last forever (or at least for many hundreds or thousands of years)? Or do we seek sustainability of the ecosystem and so our primary concern should be to find energy resources and technologies that do not pollute or degrade the air, the land, or the seas? Or do we want to sustain, if not advance, the services that modern society gains from energy-using technology? Many people would say we need all three, but some argue that the protection of the ecosystem is so crucial that we must limit our use of energy and the technology it powers in order to save the Earth.

b. Concerns about peak oil, as already noted, also raise the issue of *transitions*. Just as coal succeeded wood and oil succeeded coal, at some point in the future *something* (or a few somethings) will likely succeed oil. Some have said it might, in the short term, be natural gas, because it is relatively abundant and burns more cleanly than oil or coal. Others claim it will be solar power or wind or a new generation of nuclear-powered electricity generators (Chapters 8 and 11). Proponents of alternatives with great promise, such as hydrogen or nuclear fusion, argue that those will be the successors to fossil fuels, but those technologies are not close to cost effectiveness.

But the major question for society is not only *what* will replace oil (or all fossil fuels), but also *how* will that decision be made? Should energy markets or government policymakers choose the next sources of primary energy? Or perhaps it could be some combination of government and market? But then, what should be the relationship between these and which of them, ultimately, should "decide?" And on what basis should they make their choices? As we will discuss in Chapter 11, the government of Germany has embarked on a policy of energy transition (*energiewende*) with controversial results. But in other countries there might be social, legal, historical, demographic, and topographical barriers to following Germany's lead.

c. Of course, one reason for government intervention in markets is to correct what are termed *market failures*. That is, markets do not take into account all the costs or benefits that are involved in the production and sale of any particular good. One major example of a market failure is an *externality,* especially a "negative" externality such as pollution: if the production of a good or service causes, say, air pollution, then costs will be imposed on everyone since there is no market mechanism to transfer the cost to the seller or the buyer. Instead it is borne by all users of the atmosphere. Too much of the product is produced and it is sold at too low a price (because external costs are not included). In a situation like this, there may be a role for government to require in some way that the cost be included in the price of the product – for example, through taxes or abatement requirements. In many chapters we will be considering the external costs of using energy resources and technologies. But sometimes costs are prospective (intergenerational) and difficult to measure. How then to include them is a question of great importance but also great debate.

d. One externality problem will (as it must) be featured in this book. Oil, coal, and natural gas (our fossil fuels) are burned for their energy, and the more fossil fuels are burned the more the atmospheric component carbon dioxide (CO_2) will be emitted into the atmosphere. As we will discuss in Chapter 6, CO_2 can act like the glass of a greenhouse, letting in heat but not letting nearly as much of it out. The effect will mean a warming Earth. Indeed, increased CO_2 in the atmosphere will almost certainly affect Earth's climate (and some claim it is doing so already), but by how much? Some scientists have argued that the world is already in a danger zone where climate change will cause great suffering to all life, including the possibility that life itself will be extinguished (Lovelock 2007). Others foresee a much less threatening scenario. Still, *climate change* is probably the most urgent issue related to energy production and consumption. If life itself is threatened, should we stop burning fossil fuels? Is it a moral imperative that we live more simply and lead less energy-intensive lives? We consider these philosophical questions in Chapters 4 and 6, but we will refer to the issue of climate change throughout the book.

Although burning fossil fuels adds CO_2 to the atmosphere, is climate policy really just energy policy? That is, should policymakers stipulate energy technologies and resources that are to be used or should they stipulate reducing CO_2 and other climate-altering gases and allow markets to choose how that is to be accomplished? There are arguments that energy policies – such as replacement of coal electric power generation with wind-generated electricity – do not in fact reduce CO_2 emissions nearly as much as people think.

Externality problems related to energy are not limited to the issue of climate change. Air pollution (from components of fossil-fuel combustion other than CO_2) in some cities in China, for example, is severe and can cause or exacerbate many ailments from asthma to heart disease. The means of extracting resources can have important environmental consequences as well. Extraction of oil from heavy sands in Canada, extraction of natural gas and oil through shale rock (hydraulic fracturing or "fracking"), and the extraction of rare earth metals used in components of wind generators are among the other environmental issues connected to energy resources and technologies, and ones that will be considered in various places throughout the book.

e. But even when there are potentially dangerous externalities, all choices of remedial action involve societal *trade-offs*. For example, coal smoke can cause intense local pollution (as in China), but decisions to end coal combustion could mean that people in developing countries are unable to get the energy they need to build prosperous societies. A very important problem for the twenty-first century is *energy poverty* (Chapter 5). More than one billion people have no access to electricity and another billion have access only intermittently. Energy poverty keeps people more generally impoverished; it reduces not only their wealth but their health as well.

f. It is commonly thought that developing countries that export oil or other natural resources have an advantage over countries that need to import resources. But, as we will see, that is often not the case. In fact, oil exporters in particular have become so dependent on oil revenues that little in the way of investment for broader economic development occurs. The problem of the *resource curse* (Chapter 5) is another factor that complicates solutions to poverty as well as to global environmental problems. Throughout the book, we will note the trade-offs, the conflicts, and the barriers to wise use of energy resources and technologies.

Thinking about Energy Issues

The issues noted above have societal and philosophical importance, but the causes and the potential solutions are also technological. For the individual who is not an energy expert, the question arises: How can one provide answers to technological problems without being an engineer or a scientist or an energy analyst? As we will discuss in the chapters that follow, energy technologies, though intricate at times, are not beyond the average individual's comprehension. More importantly, we will see that any analysis of energy issues must go beyond the strictly technical. Technology itself does not stand in isolation from the world. Judging its place in, and value to, society is something that everyone can, and arguably must, do, since such judgments will directly affect the way we all live.

The specific answers to the questions on energy we face now and will face tomorrow may not be easy to implement. But such answers must come from a full analysis, a sense of the social and philosophical context in which energy technology and resources are used, and a keen appreciation of what energy issues mean to the way we live and to the world we live in. This book does not provide a prescription for the future. Instead, our quest is for a *critical* appreciation of energy issues, including resources, technology, safety, economy, behavior, and so on. Only in this way are we likely, as a society, to be able to answer – equitably and clearly – the questions that will arise in the years ahead.

References

Ahmed, N.M. 2010. "The Age of Cheap Oil is Over." *New Statesman*, 12 November.

Deffeyes, K.S. 2001. *Hubbert's Peak: The Impending World Oil Shortage*. Princeton University Press, Princeton, NJ.

Deffeyes, K.S. 2010. *When Oil Peaked*. Hill and Wang, New York.

Grossman, P.Z. 2013. *U.S. Energy Policy and the Pursuit of Failure*. Cambridge University Press, Cambridge, UK.

Heinberg, R. 2003. *The Party's Over: Oil, War and the Fate of Industrial Societies*. New Society Publishers, Gabriola Island, British Columbia.

Jevons, W.S. 1865. *The Coal Question*. Macmillan and Co, London.

Lovelock, J. 2007. *The Revenge of Gaia: Earth's Climate Crisis & The Fate of Humanity*. Basic Books, New York.

Manne, A.S. 1975. "What Happens When Our Oil and Gas Run Out?" *Harvard Business Review*, 53, 123–136.

Nef, J.U. 1977. "An Early Energy Crisis and its Consequences." *Scientific American*, October, 141–151.

Radetzki, M. 2010. "Peak Oil and Other Threatening Peaks: Chimeras Without Substance." *Energy Policy*, 38, 6566–6569.

2 Energy Resources

Introduction

Natural resources may be placed in two categories: *depletable* and *renewable*. Depletable resources are those that are used up before nature can replenish them. Our major fuel resources today are fossil fuels, particularly coal, oil, and natural gas. These resources effectively can never be replaced since immense amounts of time were required to create them. They are, by definition, diminishing, and as an existential matter they could "run out" someday because they are finite. As a practical matter, however, they will never "run out," because before that time they would become so expensive as to be not worth extracting. It would be a problem if this stage of resource extraction occurred rapidly. As noted in Chapter 1, people have predicted the end of oil and natural gas and even coal. But the "end" of fossil fuels has been like the horizon. One can see it in the distance but can never quite reach it. As we write this in 2017, the "end" of fossil fuels seems a long way off.

Other resources either can be replaced or are not depletable; water power and plant materials (called *biomass*) do indeed replenish themselves, if they are allowed to, and therefore are truly renewable. Others, such as solar energy and wind power are not depletable; they are *inexhaustible*, but are commonly also included in the category of renewables. In the pre-industrial world, people used these renewable resources primarily. *Wood*, for example, was the fuel for heating and cooking, *wind* powered sailing vessels, and *water* power turned the wheels that ground grain.

But since the beginning of the industrial age, we have relied increasingly on fossil fuels. Today, 84 percent of the United States' and 81 percent of the world's energy comes from fossil fuels.[1] In recent years, scientists and businesses have given increasing consideration to renewable energy resources. But, as we see in Chapters 10 and 11, we are only beginning to utilize the potential of renewable resources.

In spite of, or more probably *because of*, our reliance on fossil fuels, the use of these resources has caused controversy since the Industrial Revolution, which began in Great Britain in the late 1700s. Questions about who controlled these resources, who benefited from their control, and what were the environmental consequences of this have been raised repeatedly. Fossil fuels have been the focus of intense and protracted controversy throughout the industrialization of Europe, the United States, and, more recently, in the industrialization of Asia.

In the United States, these disputes have raged off and on for over a century and have common threads that can be followed to the present time (Wildavsky & Tenenbaum 1981). Some of these threads are part of the broader historical fabric of this country, such as the anti-monopolist movement of the last decades of the nineteenth century, which grew into the Progressive movement of the early twentieth century. A symbol of these political struggles was John D. Rockefeller and his Standard Oil Company. Following the discovery of oil in Pennsylvania in 1859, Standard's rapid acquisition of oil production and refining operations provoked the charge that the company sought monopoly power. Whatever the truth of the charge, Rockefeller's company was following a larger trend of the era whereby consolidations of industries, for example in steel, were being formed.

The popular reaction against large business amalgamations led to, among other consequences, the passage of the Sherman Antitrust Act (1890) and the founding of the Progressive movement. Progressivism extended into the twentieth century with the trust-busting of President Theodore Roosevelt (1901–1908) and the presidential candidacy in 1924 of Robert La Follette, who campaigned on an antitrust platform.

Standard Oil was subsequently charged with antitrust violations under the Sherman Act and was broken up into separate corporations. But Standard's disintegration did not settle all of the issues that the oil trust had raised. The most crucial of these was the question of control of the extraction and use of resources. For a number of years, Standard had the power to determine how much oil and its associated products reached the market. In theory, they could always keep the amounts low with respect to demand in order to keep prices and profits high and to maintain supplies for the longer term. Standard was able to do this for a time, but in part that was because most oil production was located in western Pennsylvania, close to Standard's Cleveland refineries.

Ironically, Standard was broken up after the company had begun losing much of its market power because of discoveries far from Standard's territory – primarily in Texas. Whereas in the early 1900s, Standard controlled about 90 percent of refined petroleum products produced and sold in the United States, by 1911, the year of the Standard dissolution, its market share had already fallen by one-third. As a historian of the US oil and gas industries has noted, "[T]he arduous antitrust campaign … failed to grasp momentous industry changes that signaled a new competitive era" (Bradley 1996).

Although no one company had the power to control the supply of the country's oil any longer, it remained in the interest of the largest oil companies to maintain policies that would result in the same kind of market control that Standard had exercised. But could they? Over time, there have been some instances of collusion among oil producers. There have been many more times in which they have been accused of collusion, but the accusations went unproven.

But a curious dynamic evolved within the oil industry in the twentieth century: While most attention of regulators and the general public focused on the major international oil companies (including successor companies of Standard, such as ExxonMobil), there have been over the years literally thousands of small oil producers as well as small independent refiners and marketing companies. These independents organized into associations which by the 1930s had gained considerable political influence, especially at the state level in Texas, Oklahoma, and other oil-rich states. But the independents and the major oil companies often had very different policy goals. The independents were instrumental in the cartelization of the American oil industry – cartelization run by the Texas Railroad Commission and supported by the administration of President Franklin D. Roosevelt. The Commission controlled domestic production until 1972, keeping supplies tight in order to maintain prices high enough for small operators to be profitable (Libecap 1989). At the national level, small producers benefited from the Mandatory Oil Import Program, which from 1959 through the early 1970s, limited the quantity of imported oil – oil that was largely produced and sold by the major oil companies and was typically cheaper to produce than domestic oil.

Although a state government entity controlled the level of domestic output and the federal government partially controlled imports, no government entity controlled the *information* about proven or potential oil resources in the United States or anywhere else. Such information is of course of great importance with respect both to resource markets and to government policies. This was especially true once the United States became a net importer of oil and then suffered through the Arab oil embargo in 1973, which resulted in the quadrupling of the price of crude oil and an explosion of profits for major oil companies. Government officials (and the general public) wanted to know just how much domestic oil was actually available. Could domestic production have filled in for embargoed oil? Were the oil shortages of 1973 contrived by the oil companies simply to raise prices and profits, as some people in and out of government alleged?[2] These questions were hard to answer because most of the data needed were controlled by the oil companies themselves and thus suspect. It is not surprising therefore that much of the controversy about resources has to do with information – what is it, how do we determine it, who should control it?

It also is not surprising that throughout the twentieth century, anti-monopolist politicians and public interest groups have challenged estimates of oil (and later natural gas) reserves. The industry trade associations, the American Petroleum Institute (API) and the American Gas Association (AGA), have maintained committees for estimating supplies and production capacity. Although the US government created its own information agency in 1977, the Energy Information Administration (EIA or USEIA), statistics provided by the API/AGA have continued to be a primary source of information. The reliability and accuracy of those statistics has remained a matter of contention even into the twenty-first century.

Controversies over Fossil-Fuel Resource Estimates

How much oil (or gas or coal) was available in the 1970s? How much is available today? How much will we have in the years ahead? We must emphasize at the outset that determining what is available, and where, is no easy matter. Estimates offered by government as well as industry have been routinely challenged, and have often been proved wrong. Two groups whose origins go back to the 1800s have been especially notable in questioning resource estimates.

The first – the *consumer movement* – is rooted in the Progressive tradition. The second is the *environmental movement*, which began as the conservation movement in the late 1800s. Although the focus is the same, the emphases of the two groups are different. Consumerists are more concerned with the quality and price of products than with the structure of companies. Nevertheless, there is an implicit sense of mistrust of large corporations, including the major oil companies, underlying this activism. Ever since the turn of the twentieth century, consumerists have expressed fears about massive transfers of wealth to big business, and the political control that is perceived to go with it. In the last 40 years there have been few businesses richer than the major oil companies.

Conservationism (or *preservationism*, as it was often called historically) was distinct from Progressivism and in fact did not embrace a distinct political ideology. Like today's environmental advocates, conservationism united individuals across the political spectrum who shared a belief in the protection of the world's resources. Interestingly, although they often are advocated by the same public figures, consumerism and conservationism have exerted opposing pressures on public policies regarding extraction of energy resources. In general, conservationism advocates the safeguarding of natural resources and thus slow rates of extraction of fossil fuels. Consumerism, on the other hand, advocates maintaining low prices, which encourages increased extraction of fossil fuels in an unrestrained market. With respect to consumerism, however, free market advocacy is often inadvertent since these activists often explicitly oppose greater market freedom, which it is thought would benefit major oil companies. Still, it would appear that freer markets would be the logical outcome of their program. Some consumer groups have noted this at least tacitly and have favored outright government control of the market, to control prices and at the same time limit oil company profits. Conservationists would tend to endorse control over free markets – if not by government then by the major oil companies who might be inclined to reduce production, albeit for their own ends. It therefore has been said, "the monopolist is the conservationist's best friend," and, we might add, the consumerist's worst enemy.

Though they have what in some sense are contrary interests, consumerists and conservationists have been allied in their mistrust of the public information supplied by the oil, natural gas, and coal resource industries. Specifically, both have felt that industry-supplied data on fossil-fuel supplies are self-serving. According to this view, companies will say that oil and gas supplies are lower than they are in order to keep prices high, when such statements are in the companies' interest. Or they will declare that there is more oil or gas when they wish to encourage demand or influence favorable legislation.

Perhaps the most acute public expression of mistrust occurred during the gasoline shortage of the summer of 1979, which prompted a special government report on the practice of using industry data. In this report, consumer groups challenged the confidentiality of oil industry resource information. Individual oil companies claimed that they needed to protect proprietary resource information from their competitors. The issue raised a fundamental question for democratic government: When (and why) must a private concern make public disclosures that may be against its own interests? It is a question that has important legal and ethical implications.

Whatever the validity of industry data, it should not be forgotten that resource industries are organizations with agendas that can affect their choice of data. But then the groups that challenge industry have their own biases and predispositions. As we will see, others besides the oil companies have put forward estimates of resources that were later proven to be far from the mark.

The Estimation of Fossil-Fuel Resources

Before we look at how fossil-fuel resources are estimated as well as the extent of current estimates, it is important to understand the taxonomy of resource quantities. In general, there are two categories: *reserves* and *resources*. Reserves are the most commonly cited figure for fossil-fuel quantities since they are the most certain. Reserves are those resources known to (or are thought likely to) exist that are economically recoverable with current extraction technology. Reserves are sometimes divided into *proven reserves* (known to exist); *probable reserves* (thought likely to exist); *possible reserves*; and *unproven reserves*. All of these are part of the overall resource base, which also includes all existing known or probable resources that are not either economically or technically recoverable (or both), depicted here in Figure 2.1. The latter may become reserves when prices rise or the technology of extraction improves. There is, however, some level of resources that is ultimately unrecoverable, shown on the figure as "Remaining oil and natural gas in-place". In other words, we will never completely run out of any fossil fuel, but remaining resources will be either too expensive or too difficult to extract.

The information problem results from the fact that it is very difficult to estimate even the size of reserves, much less the entire resource base, though as the figure indicates, reserves are "more certain." In fact, there is no known method for making accurate and certain predictions of future amounts of any resource that will be available as a fuel. This has been true for oil, natural gas, and coal and is also expected to be the case for other fossil resources such as shale oil (and gas) and oil sands.

Geologists have tried various techniques for estimating these resources based on physical evidence and statistical data. The first estimates of US fossil resources were attempted by the US Geological Survey (USGS) from 1908 to 1918 using what would now be considered primitive methods. The intent of these estimates was to aid the then-emerging oil industry, but the industry reacted hostilely to the USGS's

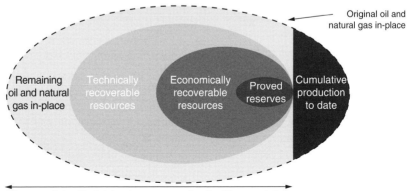

Figure 2.1 Stylized representation of oil and natural gas resource categorization.
Source: US Energy Information Administration.

conclusions. The estimates were considered by the industry to be too low and created what was then termed the *exhaustion bogey*. In retrospect it appears that the estimates were approached from what has been called the *limitationist* (or *Neo-Malthusian*)[3] viewpoint and the criticisms arose from opposing *expansionist* (or *cornucopian*) views. The limitationist geologists at the USGS believed that oil was basically running out and they feared wasting it. They wanted the government to regulate the oil companies' production rates. The expansionist companies, on the other hand, did not want any government restrictions on their extraction at that time, and sought to preserve their interests by forming their own committees for collecting data and estimating reserves.

History soon proved the geologists wrong. Their estimates were made just before large new deposits of oil were discovered in the US Southwest, and therefore took place within a period when the experience in the cycle of discovery was limited. Perspectives and the nature of debate changed radically with big strikes in Texas and Oklahoma. Glut replaced scarcity as the focus of concern, especially for the major oil companies, who worried about collapsing prices and profits in the long term. Then in the early 1930s, because of larger estimates of the resource and enormous supplies on the market, the same companies that had opposed government intervention in the oil market reversed themselves, and advocated production restrictions, either voluntarily or by regulation to change and stabilize the market in their favor. The Texas Railroad Commission and the US Congress obliged, with the former acting as regulator of an oil cartel.

Following World War II, government officials became pessimistic again and feared resource depletion. In fact, reserves in the United States rose undisputedly for the following 22 years, but exploratory drilling efforts fell off after 1957. Some critics (e.g. Commoner 1979) claimed that this drop-off of drilling effort had much to do with the growing supply of imported oil. This alternative source of cheap oil cut the incentive to drill in the United States and, since drilling is part of the process of discovery, additions to reserves slowed. Thus, according to the critics, estimates of reserves (and even of ultimately recoverable resources) were too low.

There were also questions during this period about the *finding rate* of oil, which is the rate at which oil is discovered per unit of exploratory drilling effort. Evidence was growing that during the post-war era oil was becoming more difficult to find in the heavily worked US oil basins. However, even these finding-rate statistics, which were used by geologists to extrapolate discovery rates, were not immune to interpretation to the contrary.

The technology of resource estimation has improved over the years. Today geologists are able to use computer imaging of subsurface strata and as a result are often fairly certain whether drilling in a given location will produce commercially viable quantities of energy resources. But even with such tools, it is still impossible to know just how much oil, gas or coal ultimately there is in the world. It is prohibitively expensive to prospect for such natural resources exhaustively over vast land areas by computer imaging, much less by definitive determinations such as drilling. Therefore, various broad survey methods have been tried to keep this cost of information reasonable, while at the same time providing statistics that are useful in formulation of corporate strategies as well as government resource policies.

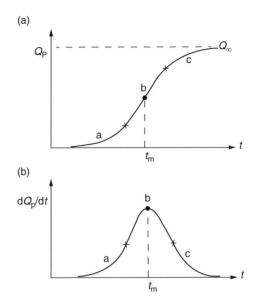

Figure 2.2 Logistics curves. (a) Cumulative production versus time. (b) Production rate versus time. a: early-time segment, b: midrange segment, c: late-time segment.

One method of resource estimation, the volumetric method, relies on surface reconnaissance (often from the air) and mapping to find geological formations that are likely to be rich in energy resources. If the reconnaissance of the Earth's surface were reliable for such identification, it would then be relatively simple to estimate the volume of the resources below these areas. However, surface reconnaissance data have turned out to be highly uncertain in the identification of resource-bearing areas. Earlier in the history of oil exploration, for example, surface identification was quite successful in locating new resources in Louisiana, but a few years later it proved quite poor in Texas.

Since cumulative production (Q_p) varies with time (t) it is described by a mathematical function of time: $Q_p(t)$, and the production rate is described by the derivative in time: $P = dQ_p/dt$. Examples of such curves (called *logistics curves*) may be seen in Figure 2.2, and are appropriately used for large regions to determine the cycle of exploration of an exhaustible natural resource over time. These curves are interpreted in three major segments: the early-time segment, the midrange segment, and the late segment. During the early segment (a), when the resource is just beginning to be extracted, the rise in the amount extracted each year is a fixed fraction of the cumulative amount extracted to date – that is, the growth is compounded according to its magnitude at a particular point in time. As a result, during this period the rate curve and the cumulative curve are exponential in time. Were it not for the finite nature of the resource, exponential growth could continue, if there was a demand for it.

Segment (b) of the S-shaped curve is the earliest indication of the finiteness of the resource. Here the curve in its midrange has reached its maximum slope versus time, and the rate curve passes through a peak (at $t = t_m$) and thereafter drops. This peak in the *rate* of production is a significant signal that the resource will become increasingly scarce.

The late-time segment (c) is the period when the cumulative production (Fig. 2.2a) is reaching the limit of the resource: Q_∞ (*ultimate cumulative production* or *ultimate recoverable resource*). Here, as the cumulative curve approaches its asymptotic limit (Q_∞), the rate curve (Fig. 2.2b) falls to zero (zero production). Note also that the total area under the production *rate* curve must equal the ultimate cumulative production (Q_∞).

This description of the logistics curve thus shows its appropriateness for describing the time history of the use of a *finite* resource. Actually, it is but one of many uses of this classic curve. In biology, for example, the growth of a species population (e.g. protozoa) in a limited environment is known to follow the S-shaped curve. In areas of human activity, such as the historic adoption of new technologies (e.g. steamships), the S-shaped curve may be observed in the pattern of development over time. (The correlation of the curve with respect to technological adoption will be explored in Chapter 10.)

Now, in order to understand how the logistics curve is currently used to estimate oil, natural gas, and coal resources, we must consider two more curves: one for *cumulative discovered reserves* and another for *remaining proven reserves*. First, we should explain what these terms mean and define their relationship to the production curves above. In terms of formulas:

$$Q_p(t) = \text{cumulative production}$$

$$Q_d(t) = \text{cumulative discovered reserves}$$

$$Q_r(t) = Q_d(t) - Q_p(t) = \text{remaining proven reserves(or simply } reserves)$$

where *cumulative discoveries* are simply assumed to precede production with the same relative shape versus time, as follows:

$$Q_d(t) = Q_p(t + \Delta t)$$

with

$$\Delta t = \text{the time lag between discovery and production}$$

Also:

$$
\begin{aligned}
Q_\infty &= Q_d(\infty) \\
&= Q_p(\infty) \\
&= \text{ultimate recoverable resource or ultimate cumulative production}
\end{aligned}
$$

Figure 2.3 shows the use of all three curves to describe the time history of a particular resource. Note that the vertical axes of the cumulative (and reserve) curves will be labeled in the appropriate units for the particular resource: billions of barrels (BBL) of oil, billions of metric tons (Bmt) of coal, or trillions of cubic feet (TCF) of natural gas. The rates of extraction, on the other hand, should be expressed as measures per unit of time: billions of barrels of oil per year (BBL/year), billions of metric tons of coal per year (Bmt/year), and trillions of cubic feet of natural gas per year (TCF/year). Special note should also be made of the fact that *reserves* ($Q_r(t)$) vary with time. The popular impression is often that reserves are a fixed amount of resources held in reserve for later use, and thus the use in the oil and gas industry of the reserve/production ratio:

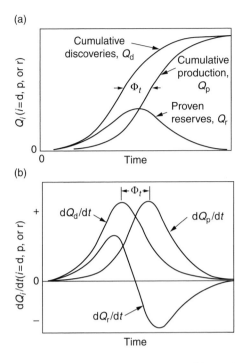

Figure 2.3 Time histories of resource use. (a) Cumulative quantities versus time; (b) quantity rates versus time. *Source:* Adapted from Penner & Icerman (1974).

$$\frac{R}{P} = \frac{Q_r}{dQ_p/dt} \text{ [year]}$$

This might seem to give the impression of a static prediction, whereas its strict interpretation is: "If this year's reserves (R) were to be depleted at the rate of this year's production (P), those reserves would be exhausted in R/P years." Reserves can and usually will, vary – due to changes of either discovery or production, and the R/P ratio can remain constant for decades, or it may even rise as new resource areas are discovered and extracted.

Since information is updated continuously with respect to reserves, any given logistics curve may be inaccurate from one year to the next. No one knows what proven reserves will be in a year or 10 or 20 years, and of course, no one can accurately predict the size of the ultimate recoverable resource (Q_∞) for any one of them. Nevertheless, a logistics model is a useful way of conceptualizing the ultimate quantity of a resource because it provides a clear (albeit idealized) construct of the production and consumption of a depletable resource.

It should also be noted that the logistics curves and R/P ratios refer to what are called *conventional resources*. Some resources exist that are termed *unconventional*, when they cannot be extracted for either economic or technical (or both) reasons. Technical breakthroughs and price changes could cause an unconventional resource to become "conventional" (in the sense of becoming a part of the proved reserve base), and as a result raise the level of both reserves and the ultimate recoverable resource. Some resources

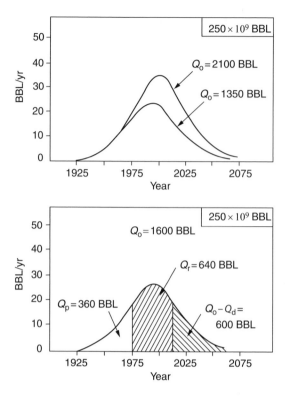

Figure 2.4 World oil: projected extraction for (a) Q_∞ estimates of 2100 BBL and 1350 BBL. Adapted from Hubbert (1973). (b) Q_∞ estimate of 1600 BBL. Adapted from Hubert (1973) and Wilson (1977).

labeled unconventional in the past have in recent years become "conventional" but some unconventional resources may never be technically or economically recoverable.

We can see, then, that judgments must come into play at two levels: first as to the size of the total resource in the ground (*in place*, that is, whether it is economically recoverable or not) and second as to what fraction of it is economically recoverable. (Indeed, even with respect to a producing oil or gas well, the ground is never drained of 100 percent of its resources.) When these judgments are combined with the social and economic pressures associated with fossil fuels, controversies ensue. New estimates are accepted or disputed according to whether they support or contradict the current point of view of a given interest group. An optimistic or pessimistic outlook on the part of the estimator can even color the predictions. The various predictions arising from these controversies will usually be represented by *production rate* curves, where the size of the areas under the curves corresponds to various estimates of Q_∞. This production rate is, of course, the most easily corroborated of these resource measures.

Figure 2.4 illustrates how such variations in estimated Q_∞ show up on resource production curves. These curves in Figure 2.4a were proposed by the controversial geologist M.K. Hubbert (Chapter 1) and illustrate the range of earlier estimates of ultimate, recoverable, world-oil resources (Q_∞). Hubbert is cited as the person who in the post-war era of rising reserves first argued that the United States faced imminent limitations of oil and natural gas resources. His prediction (made in the 1950s) of a peak for US oil production sometime around 1970 seemed in the 1970s to have been remarkably accurate. His estimate for world oil (made in 1969) was that the peak of production would occur during the first decade of the twenty-first century. That prediction influenced the peak oil movement and was the basis for Kenneth Deffeyes' assertion that world peak oil had passed in 2005 (see Chapter 1).

Figure 2.4b shows an alternative to Hubbert's logistics model, although it falls within the range Hubbert had proposed. The effect of these different estimates of total resources on production rates, as forecast by the logistics curves, is interesting. The first feature is that the asymptotic time to exhaustion is *not* proportional to the ultimate size of the resource. Hubbert's two world-oil curves, for example, represent a variation of over 50 percent in the estimated resource, but the (asymptotic) times to exhaustion were nowhere near that ratio. The logistics curve predicts that for an increase in the estimated Q_∞ there is an *increase* in the level of production *as well as* an extension in time. The reasons for an expected rise in production, as well as an extension in time, have to do with the effects of perceived scarcity or abundance on the demand and price of the resource (see Chapter 4).

Note that it is possible to divide the total area under the production curve (Q_∞) into various cumulative quantities. Figure 2.4b, for example, gives the world oil production rate for the year 1975, and therefore a vertical line is drawn under the curves at that year. The area under the (rate) curve to the left of that year's line must be the cumulative production (Q_p) *up to* that year. Of the remaining area under the curve, the cross-hatched area marked Q_r (Fig. 2.4b) represents the cumulative reserves (Q_r) that were estimated *as of* the year 1975. Furthermore, the sum of the two areas Q_p and Q_r should represent the cumulative discoveries (Q_d) up to that year, since:

$$Q_d = Q_p + Q_r$$

The last segment represents the remaining resources that are as yet undiscovered. Starting logically from the cumulative-discovery area (Q_d), the remaining area under the production rate curve to the right of the Q_p and Q_r areas must be $Q_\infty - Q_d$, the *remaining undiscovered resources.* Part of these remaining resources is the reasonably certain "inferred reserves," meaning that they will come from extensions of existing fields or result from more complete recovery from existing fields. The remainder after that, however, is merely informed postulates about the geology outside of known fields. It represented the undiscovered oil in the United States in the year 1975. This uncertain quantity was the object of controversial judgments concerning the size of the remaining resource and the fraction that will be economically recoverable.

United States Oil Resources

Was Hubbert Right?

Looking at Figure 2.5a, it would seem that the answer would have to be "yes." Domestic oil production leading up to 1980 is shown in the figure, with the caption showing estimates of *reserves* (Q_r) and *total resources* (Q_∞) using the

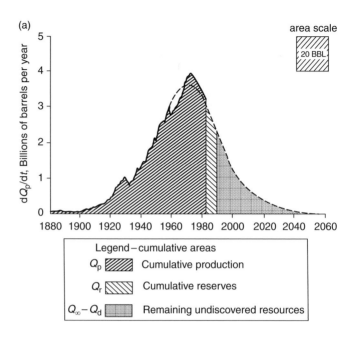

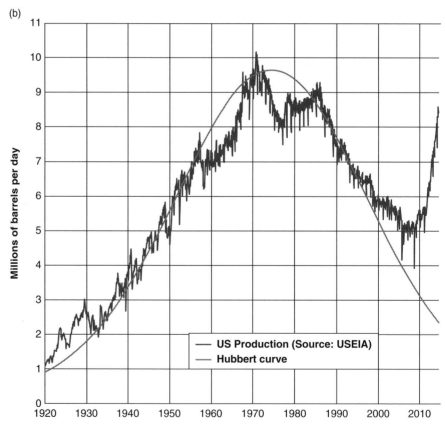

Figure 2.5   US oil: (a) production rate versus year, 1880–2040; production characteristics: $Q_\infty = 170$ BBL, $R = Q_r = 22$ BBL, $P = dQ_p/dt = 2.4$ BBL/year; $Q_\infty - Q_d = 41$ BBL. (b) Production rate versus year and Hubbert's logistics curve. *Source:* US Production USEIA (1994).

production data up to that year. The production curve was based on the data collected up to that year and the Q and P production estimates shown are also based on those data. Note Hubbert's conclusion that Q_∞ for US oil was 170 BBL.

But around the time Hubbert was making his forecast, his numbers were challenged. A.D. Zapp, a USGS geologist, published an estimate for Q_∞ of 590 BBL for US oil resources (Zapp 1962). Hubbert had already made a lower estimate and in 1969 published a figure of 165 BBL. In any event, for the next several decades, Zapp's optimism was not borne out. Subsequent estimates tended to confirm Hubbert's. Part of the difference rested on the assumption of the rate at which new oil would be discovered (the *finding rate*). Zapp had assumed a constant finding rate, suggesting that oil discoveries would continue at the same rate of previous years. Hubbert, however, assumed that the finding rate would decline, which for the most part it did in the United States through the end of the twentieth century.

Hubbert's viewpoint that the rate of production of domestic oil resources had peaked was generally accepted. However, it, too, turned out to be wrong. As Figure 2.5b shows, after several decades in which Hubbert's curve seemed to fit perfectly US oil production and US reserves, reserves rose and the actual production curve changed.

Hubbert erred for two reasons: First, he did not take full account of market dynamics. Indeed, his model posits an idea that to an economist is nonsensical. That is, the demand will keep rising regardless of price and that supply will keep falling. Of course, at first US consumers will replace domestic oil with foreign oil and so there will be a stable market equilibrium. But since world oil would peak in the early years of the twenty-first century, according to Hubbert, eventually there would not be sufficient supply to meet ever-rising demand. The result, as shown in Figure 2.6 would be a gap that will never be filled. The implications of this are, first, that no producers will try to extract unconventional resources (regardless of the price they might receive). And second, that consumers will not seek substitutes for petroleum as the price rises. In fact, the idea of a supply–demand gap has always been, according to one renowned economist, "intellectually bankrupt" (Alchian 1975), but is one that has been widely touted, especially in the 1970s.

Hubbert also did not account for technical change. Because of various technological improvements in oil extraction, unconventional resources and hard to reach conventional resources have been added to the reserve base (by 2015), while the price of oil was relatively high, making many resources previously too costly to extract, economically viable. As Figure 2.7 shows, after falling for many years, reserves have risen from around 25 BBL to over 37 BBL. The R/P ratio is still only about 10 years because more US crude oil is being utilized as well as discovered. But then the R/P ratio for US oil has been in the range of 8–15 years since the 1920s.

Hubbert made several slightly different estimates for ultimately recoverable oil in the United States, and he qualified his initial numbers after the discovery of oil in Alaska by limiting his pessimistic estimates to the lower 48 states. Nevertheless, he put Q_∞ at approximately 170 BBL and $Q_\infty - Q_d$ at roughly 40 BBL.

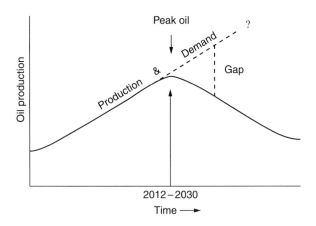

Figure 2.6   Supply–demand "gap."

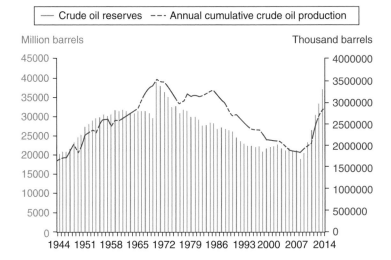

Figure 2.7   US oil reserves.
Source: Energy Information Administration (in "U.S. Oil Reserves, Resources and Unlimited Future Supply," by Jude Clement, 4/2/15 at http://www.forbes.com).

Moreover, as the United States approached its peak, he anticipated that daily production would fall to 1.5 million bbl/day. Neither of these have proven accurate. By 2014, US reserves had risen by 36.5 BBL – an amount almost as large as had been thought in 1980 by Hubbert and his followers as the total remaining resources of oil in the United States. Since 1980, US consumption of *domestic* oil has totaled 87.7 BBL. By 2014 daily production averaged 8.7 million bbl/day and in 2015 reached more than 9.4 million bbl/day, the highest rate of production since the 1970 "peak."

As always, current estimates for $Q_\infty - Q_d$ are disputed. In 2013, the EIA put "technically recoverable resources" at 238 BBL or an implied supply that would last 66 years at current rates of production.

It should be noted that the EIA expresses uncertainty about any long-term projections. The agency has calculated three scenarios: the resources estimated for the "Reference case" are the ones above. In the "High" resource scenario, there would be 430 BBL of recoverable oil in the United States (including Alaska), while in the "Low" scenario $Q_\infty - Q_d$ equals 209 BBL.

Some other estimates are lower; others far higher. An energy researcher from Rice University has been quoted to the effect that there might be as many as 2 *trillion* barrels of recoverable oil in the United States, or about 575 years at current rates of production! That is an extreme view and even if correct may include large quantities that would be too difficult or costly to recover. Most analysts believe there is far less than that. Although the specter of "peak oil" has receded, it is important to emphasize that resources *are* finite. We simply do not know when we will see definitive signs that resource depletion is approaching – in the United States or the world as a whole – but it does not seem likely to be soon.

United States Natural Gas Resources

Hubbert's estimate of US natural gas resources was more quickly shown to have missed the mark, but in the 1970s it also seemed an accurate picture of that resource. As Figure 2.8 shows, Hubbert had identified an early peak in the late 1970s, but by the late 1980s, production of natural gas in the United States began to diverge from Hubbert's curve. By 2011 actual production was greater than it had been in the early 1970s and was trending upward.

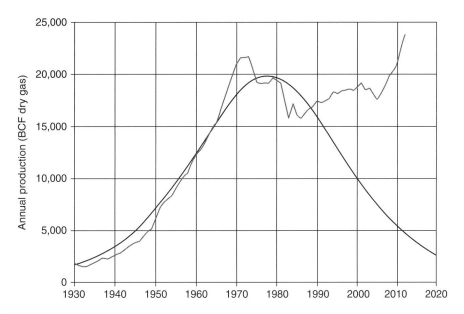

Figure 2.8   US natural gas production versus Hubbert's prediction.
Source: USEIA (2015).

According to current estimates from the EIA, remaining US natural gas resources will last throughout the twenty-first century. Overall, the stock of natural gas has been reduced by about one-tenth since we began extracting it – at least according to the government information service. Their view is, as of 2015:

$$Q_\infty \sim 3438\,\text{TCF}, R = Q_r = 338\,\text{TCF};$$
$$P = \mathrm{d}Q_p/\mathrm{d}t = 25.7\,\text{TCF}/\text{year}; Q_\infty - Q_d = 2266\,\text{TCF}$$

But there is again considerable debate as to just how large is America's natural gas resource. As shown above, the EIA puts the total technically recoverable remaining natural gas at 2266 TCF or about 87 years of supply at current levels of production. But estimates range from 1800 TCF to 3900 TCF, the latter an R/P of about 140 years. Note that prices of natural gas have been low and thus consumption has increased as industries that require large amounts of energy, such as electric companies, have switched from coal to natural gas – if they have been able to.[4]

An interesting feature of the gas production curve is its relationship to changes in the regulatory environment. The process started in the 1930s when Congress placed natural gas pipelines that crossed state lines under the jurisdiction of the Federal Power Commission (FPC). Initially, the FPC's mandate concerned construction of pipelines and the price for natural gas that pipeline owners could charge local gas distribution companies (Libert 1956). Local prices charged to consumers were already regulated by state public utility commissions. But ambiguity in the language of the Natural Gas Act of 1938 left it unclear just how extensive the FPC's jurisdiction was over natural gas pricing. Some in government and outside it wanted to see the FPC set the price for natural gas from the wellhead to the pipeline operator as well as the price from the pipeline to the distributors.

After years of wrangling over this question, in 1954 the Supreme Court put the FPC in charge of setting natural gas prices for all phases of its transit from wellhead to distributor (as long as it was destined for interstate commerce).[5] This proved problematic since, as economic theory would tell us, setting prices at any given level means that if the cost of production (or transmission) rises or if demand rises, the higher costs cannot be automatically passed on to consumers. Consequently, the fixed price may create a disincentive to seek new supply, and thereby set the stage for shortages in future years.

Unsurprisingly, after 17 years of price regulation, there were indeed supply shortages and continual debate over the future of natural gas – in terms of both price and supply. A leading official of the Carter administration in fact said in effect that the US natural gas supply was quickly disappearing and could not be counted on for the future.[6] But was the drop in supply due to a physical limitation or to a political one? That could only be tested by price deregulation. Would market prices create incentives for more production?

Finally, in 1978, after an extraordinary contentious debate in Congress, the Natural Gas Policy Act was passed, in which deregulation of natural gas prices was to take place in stages through the 1980s. In the meantime, the FPC was eliminated and replaced in 1977 by the Federal Energy Regulatory Commission (FERC), which was to control pricing and oversee the deregulation process. Deregulation of natural gas pricing was, for the most part, made complete only in the year 2000.

As Figure 2.8 shows, by the early 1990s, after most of the remaining controls had been eliminated, there was a clear increase in supply, but by 2008 there were also record high prices. Once more there were predictions of the exhaustion of natural gas supplies. As one scholar wrote in 2007, "Production [of natural gas], both in the U.S. and Canada, is now in terminal decline" (Korpela 2007).

But Then It Was No Longer in Decline

United States producers had long known that gas resources were locked up in Devonian shale in geopressurized formations that were *not* extractable economically by technology up to the 2000s. Indeed, such unconventional gas resources, although known to exist, had been incompletely explored and estimates of their size were largely unconfirmed hypotheses. Still, these gas deposits held the promise of an abundant alternative fossil fuel, which emitted far less air pollution than coal.

Just about the time natural gas was supposed to be in "terminal decline," energy entrepreneurs began using a combination of technologies to extract the shale gas deposits cost effectively: first, they developed the ability to drill horizontally as well as vertically (called *directional drilling*), thus reaching several natural formations with a single drill bit. Second, they utilized a method of breaking through rock that was actually developed in the 1940s and 1950s called hydraulic fracturing or *fracking*.[7] This technique uses water and chemicals, which are injected under high pressure into underground shale rock formations, so as to open up fissures in the rock layers and allow gas to escape. These fissures commonly run parallel to the Earth's surface and can be split apart for lengths of several thousand feet underground by means of directional drilling. Gas collection is made through perforated pipes, which connect to a pipe rising up from the gas-rich layer of shale deep underground to the wellhead at the surface (see Fig. 2.9).

Note in the figure that there are many ways natural gas appears: It is often "associated" with oil wells; a product found in coal beds; locked up in shale rock and in "tight" sands; and of course found in conventional natural gas wells.

Large-scale *shale gas* production has begun; in 2013, for example, shale gas production was over 11 TCF, about 45 percent of total US natural gas production. Still, there are many more known deposits that will only be developed if environmental concerns are settled. These concerns include the possible contamination by the fracking chemicals of potable groundwater supplies. Investigations of underground fractures of bedrock *must* give assurances that these vital groundwater resources are not disturbed or contaminated. It has been reported that in some instances shale gas production has contaminated nearby groundwater. Water, it has been alleged, had been allowed to "commingle" with the fracking fluids, and natural gas had also migrated to underground aquifer water supplies.

But a 2015 study by the US Environmental Protection Agency (EPA), while noting some reported cases of contamination, could find no "widespread systematic impacts" from fracking; the report was strongly criticized by environmentalists.[8] Perhaps the real issue was summed up in the headline to a report on fracking in *The Journal of the American Medical Association*: "Evidence Slim for Determining Health Risks from Natural Gas Fracking" (Mitka 2012). Until the evidence becomes more conclusive, shale gas production is likely to be slowed.

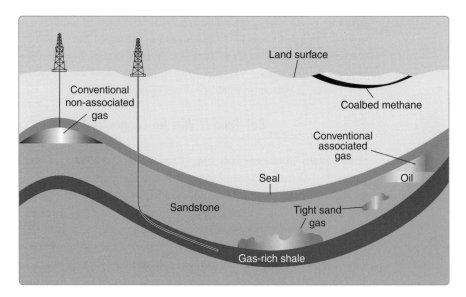

Figure 2.9 Natural gas resources.
Source: USEIA: Gas Deposit Diagram; retrieved from http://en.wikipedia.org/wiki/File:GasDepositDiagram.jpg.

However, the quantities of natural gas from shale deposits are already shifting resource estimates higher. Shale gas *basins* are known to exist throughout North America.[9] Among the most prominent are the Marcellus basin in the Northeast, the New Albany shale in the Illinois basin, the Barnett shale in the Fort Worth (Texas) basin, the Haynesville basin in Louisiana, and over a dozen others spread across North America. The quantities are thought to be immense. One formation alone, the Utica shale, is thought to contain 782 trillion cubic feet or about 30 years of US supply at current rates of consumption.

In recent years, natural gas has been used in many applications. Some are new; for example, to a limited extent, natural gas in compressed or liquid form has replaced oil as a transportation fuel. To a much greater extent natural gas has replaced coal in electricity generation. Natural gas is less polluting than coal or even oil. It contains virtually no sulfur and therefore emits negligible sulfur pollutants when burned (see Chapter 6). If environmental concerns related to production can be allayed, shale gas is likely to be a significant part of US energy supply in the years ahead.

United States Coal Resources

Unlike oil and gas, all analysts have understood that the United States possesses vast coal resources (Fig. 2.10). Hubbert's curve for the United States (Fig. 2.10a) is incorrect, but not because of any mistake related to finding or production rates. Rather, Figure 2.10a does not take account of the fact that coal is so problematic from an environmental standpoint that the United States

(a)

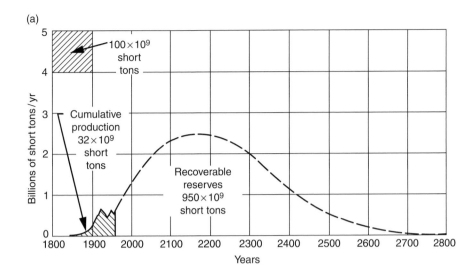

(b)

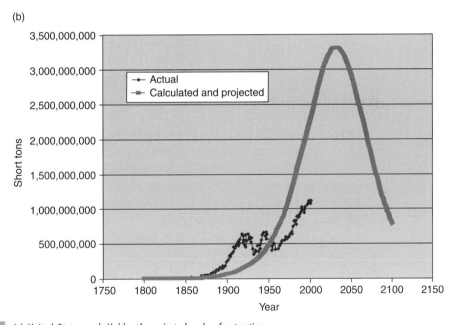

Figure 2.10 (a) United States coal: Hubbert's projected cycle of extraction.
Source: M.K. Hubbert (1956).
(b) Actual US coal production and a fitted Hubbert curve.
Sources: Vaux (2004); Data: USEIA, 2004.
Note: The smooth curve rising to a peak value over 3 Bmt is the theoretical logistics curve, while the irregularly rising curve, reaching a little over 1 Bmt at the year 2000, is from actual data. In the last 15 years the curve has flattened out and production levels have been roughly constant.

extracts far less than it could (and could consume much more than it does) – and that rate is actually falling.

From appearances, the logistics curve for coal is currently in that *early time segment* where growth would be expected to be exponential in time. This resource is so large, compared to present demands, that there are no concerns, or controversies, over scarcity. It is worth noting that such logistics curves are customarily drawn assuming that only 50 percent recovery is economic; in other words, the total *resource in place* is about twice as large as the amount assumed to be recoverable. The *proven recoverable reserve* for coal production is estimated to have been about 305 Bmt as of 2009, of which 69 Bmt had already been produced and consumed (Milici et al. 2013). The total remaining US resource is estimated to be 434 Bmt. In other words, the EIA estimates that a further 232 Bmt can be recovered. But if 232 Bmt can be mined economically using present technology, that implies a current R/P ratio of over 200 years.

Despite this huge abundance of coal, it currently supplies less than one-fifth of US annual energy demand. Coal use is not expected to grow rapidly in the coming decades because of the limitations and constraints it places on the environment and public health. Moreover, it produces the most carbon dioxide, the major greenhouse gas, of any fossil fuel and so has been shunned in some countries that seek to mitigate the problems of climate change. The idealized logistics curve does not reflect the likely path of US coal consumption moving forward because its utilization will depend on policy more than on availability.

Do Logistics Curves Have a Future?

Hubbert's predictions were wrong, but does that mean we have seen the end of the logistics curve approach to resource estimation?

The answer is almost certainly no. Hubbert's approach was overly ambitious. He believed that he could capture all elements of geology, economics, and presumably institutional factors in a single curve that would not change substantially in the years that followed. This was extremely naïve. Still, the fact that it captured US oil production as it seemed to be evolving, and was largely "correct" in that sense for the next 30 years, demonstrated that it was useful as a way to conceptualize resource limits. As a tool to understand a given country's resources based on known reserves and the historical development of production, the logistics curve is clearly advantageous, and continues to be utilized in that way (e.g. Juvkam & Dessler 2009, Saraiva et al. 2014).

The methodology has been revised so that it can accommodate new information, that is, ways of updating the model in the light of new information. New efforts use what is sometimes called a "multicyclic model" (as opposed to the single cycle Hubbert model) or a "modified multi-Hubbert model." As Nashawi et al. (2010, p. 1788) explain, the original Hubbert model was too restrictive, but the multicyclic models allow for "continuous updating ... depending on the historical oil production trend and known oil reserves." Indeed, though the field has gone beyond the basic Hubbert model, some forms of logistics curves are still widely applied. It is even argued that to leave out logistics methodology in resource studies "is to not actually understand nonrenewable natural resource economics" (Reynolds 2014).

World Fossil-Fuel Resources

If US resource estimates are contentious and diverse, worldwide estimates are much more so. As you can see below, many analyses of oil resources suggest that oil and gas will be sufficient to meet the growing demand from the developing world, India and China in particular, for another two or three generations at least and probably more than a century. Daily oil production, already over 90 million bbl/day, is expected to exceed 100 million bbl/day in the 2020s.

But there are those who retain a belief that oil production will peak sooner rather than later. A website, "Peak Oil Barrel" (which has as a subhead "The Reported Death of Peak Oil Has Been Greatly Exaggerated"), recently predicted a world oil production peak "in 2018 or a few years later."

The recent history of world oil production and projections of oil resources may be summarized as follows:

World oil – Sources: USEIA, USGS, BP Statistical Review
Production (2014): $P = dQ_p/dt = 33.9$ BBL/year = 93 million bbl/day
Reserves (2014): $R = Q_r = 1700.1$ BBL
Ratio: $R/P = 1700.1/33.9 = 50$ years

But What Are the Ultimately Recoverable Resources, Q_∞?

For both oil and natural gas the ultimately recoverable resources depend on what is counted. The world has consumed about 950 BBL of oil. Add that to proven reserves of 1700 BBL gives 2650 BBL. The USGS has estimated (as of 2012) that 565.3 BBL technically recoverable conventional resources remain to be discovered, giving a Q_∞ of 3215.3 BBL. But if one includes such resources as oil sands, shale oil, and other resources listed previously as unrecoverable cost effectively, the number rises to perhaps as high as 5000 BBL (that is, 5 trillion barrels), with an R/P ratio of about 120 years. If history is to be the guide, there will probably turn out to be more than even this high estimate.

World *natural gas* estimates are even more varied. In 2014, proven world reserves of natural gas stood at 6600 TCF.

World natural gas – Sources: USEIA, BP Statistical Review, Holz et al. (2015)
Production (2013): $P = dQ_p/dt = 119$ TCF/year
Reserves (2013): $R = Q_r = 6557$ TCF
Ratio: $R/P = 6557/119 = 55$ years

But what is the correct number for remaining undiscovered resources? The USGS estimated that 5605 TCF of conventional resources remain to be discovered for a total of 12,205 TCF – that is, Q_∞. But could it be higher – perhaps as much as *nine times* as high? According to the International Energy Agency (IEA) Q_∞ for conventional gas alone is over 16,000 TCF. Include shale gas and other unconventional resources and the number soars. The IEA estimates total gas resources at 28,500 TCF and, if true, an R/P of 239 years. Other estimates are higher still. Aguilera et al. (2014) put total gas resources at over 50,000 TCF and an R/P at 571 years.

These estimates do not include *methane hydrates*, which are essentially methane molecules encased in water ice. This resource is found primarily in oceans but also in permafrost and under ice caps. The quantity of methane – the main constituent of natural gas – is unknown, but estimates of methane in hydrates range to over a million TCF, although the generally accepted estimate is 3000 trillion cubic meters or 107,000 TCF – enough at current usage rate to last for 1000 years (Chong et al. 2016). As of this writing, however, these remain unrecoverable resources and cannot be extracted cost effectively. In fact, technical issues remain to be solved if they are ever to be utilized.

The size of world coal resources is also uncertain but they are generally believed to be immense.

World coal – Sources: USEIA, World Coal Association, World Energy Council
Production: (2012): $P = dQ_p/dt = 7.42$ Bmt
Reserves: (2012): $R = Q_r = 887$ Bmt
Ratio: $R/P = 887/7.42 = 119$ years

Remaining undiscovered resources:

$$Q_\infty - Q_d \approx 17,000 \text{ Bmt}$$

Source: German Federal Institute for Geosciences and Natural Resources (2010)

Obviously, the size of this estimate is staggering, but remember that this is an estimate of remaining resources, not a measure of how much coal can be mined cost effectively with current technology. Clearly a good bit of this resource will never be produced. In fact, given the high costs coal imposes on the environment as noted (and see Chapter 6), a large amount of coal that could be produced cost effectively may be purposely left in the ground. That is, we may never be *running out* of coal but we may well choose to *move away* from coal long before the supplies are actually depleted.

United States Renewable Resources

Unlike fossil-fuel resources, which lie essentially hidden in underground wells or mines, renewable resources can be more clearly identified and estimated. Moreover, we know where they are likely to be developed if and when they become technologically and economically practical. One renewable resource is *hydropower* – long utilized, a resource that provides an important contribution to electric power generation. (This will be included in the discussion of conventional energy technologies in the next chapter.)

The largest renewable resource of all is *solar energy*. Each day the Earth receives from the Sun over 15,000 times the energy production of all sorts by humans. However, at present a very small fraction of this solar energy is converted directly into the production of useful energy. Moreover, because of the seasonal variations in the amount of sunlight, it would be difficult in many regions of the world to

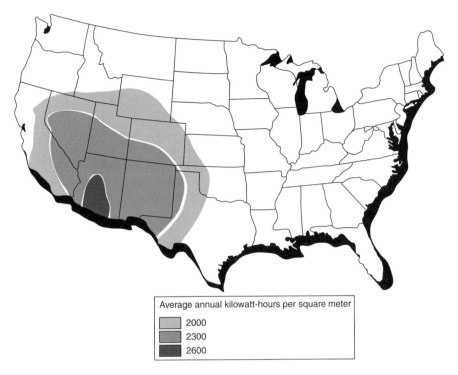

Average annual kilowatt-hours per square meter

- 2000
- 2300
- 2600

Figure 2.11　Solar energy resources in the United States.
Source: Electric Power Research Institute, Palo Alto, CA.

harness solar resources effectively. As Figure 2.11 shows, the southwestern United States has the highest potential for solar utilization in the country. That is, the average total solar energy falling on a square meter during a year is higher in Arizona and parts of neighboring states than it is in the rest of the United States.

Wind resources are distributed not only according to the *strength* of the winds in a particular region but also the *persistence* or consistency of the winds. The very highest average speeds are in mountainous areas and in some coastal regions. The most practical places to erect wind generators, however, may be in the open plains, where there are few obstructions and where the land can also be used for farming. Figure 2.12 shows that the upper Great Plains region (the Dakotas, Iowa, Minnesota, Kansas, Oklahoma and Nebraska) and the northern part of Texas already have extensive wind energy development. Wind generators have proven more efficient in these areas than offshore or in any of the coastal regions.

Biomass energy resources should be produced in land areas where crops or trees can readily be grown for use in making fuels or burned directly for their heating value. The use of such land, however, should preferably not compete with food crops or with lumber uses of wood. Also, the energy crops and wood will best be grown with a minimum of irrigation and fertilizer in order for the resulting "biofuels" to be economically competitive (see Appendix C). With these considerations

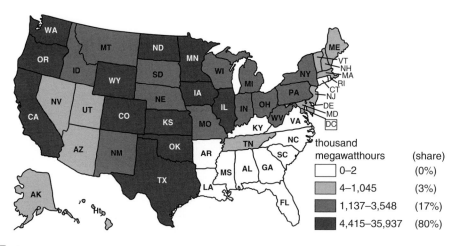

Figure 2.12 Electricity generation from wind, by state (2013).
Source: USEIA.

in mind, certain regions are more likely to become important for biomass energy resources (Figs. 2.13 and 2.14).

Biomass energy programs have been controversial. In 2007, the US Congress passed the Energy Independence and Security Act (EISA). This bill contained an expansion of a mandate created in 2005, called the Renewable Fuel Standard or RFS. According to the EISA, 36 billion gallons of the biomass fuel ethanol – in reality alcohol – was mandated to be produced and blended into transportation fuels by 2022 and continued thereafter annually. More than half of this amount was supposed to be produced from nonfood plants such as wood chips and cellulosic plants such as switch grass. Fifteen billion gallons were to be made from corn. This last objective was achieved by 2014, but it required that 40 percent of the US corn crop be used for fuel instead of food. There were claims that corn ethanol required more energy to make than it provided as a transportation fuel; that it raised food prices; that it increased air pollution and carbon dioxide emissions; and that any amount of ethanol above a blend of 10 percent ethanol to 90 percent gasoline damaged engines that were not specifically designed for ethanol fuel. The controversy continues but so does the RFS, even though cellulosic ethanol is still not commercially viable and has yet (as of 2017) to be produced in large quantities.[10]

Hydropower is an operating reality in much of the world today. The amount of energy that *hydro* contributes for human consumption is, however, limited by the water resources available, together with the right conditions for electricity generation (see Chapter 3) and meeting the limitations required for environmental protection.

For 2014, for example in the United States, hydropower supplied only 6.3 percent of the 4 trillion kilowatt hours (kWh) of *electric energy production*, and that hydro itself accounted for about 7.5 percent of the country's total *electricity*

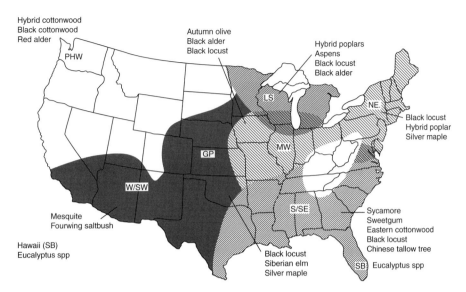

Figure 2.13 US wood resource regions for energy.
Source: den Boer et al. (1986).

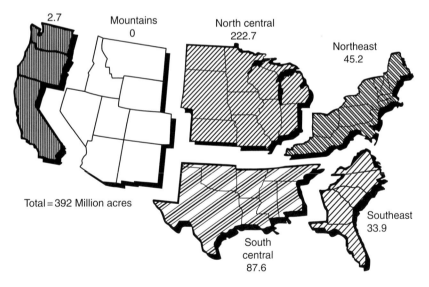

Figure 2.14 Regions for potential energy crop production, without irrigation.
Source: Courtesy of the US Department of Energy and US Department of Agriculture (1992).

generating capacity (power, in kilowatts [kW]). The greatest concentration of US hydroelectric facilities is in the Pacific Northwest, but according to a 2014 US Department of Energy (USDOE) Study that there is potential for increases in US hydrocapacity of more than 50 percent.[11]

Of course, other than hydropower generation, the renewable resources are not yet available on the mass basis of fossil fuels. The development of the technologies required to make mass economical use of the renewables is still very much work in progress, and the final outcome for any one of them is still uncertain. It cannot be stated with much certainty when success will be achieved. The research and development processes for these various technologies, and the status of each, are discussed in Chapters 10 and 11 and Appendix C.

Conflicting Views Concerning Resource Depletion: Limitation, Expansion, Sustainable Development

Thus far we have painted a generally optimistic picture of future US as well as world fossil-fuel resources: possibly more than 100 years of oil; perhaps more than two centuries of natural gas; a millennium or more of coal. But this picture must be strongly qualified. Note that in each case the R/P assumes that everything remains the same – the same levels of consumption and production. This is unrealistic.

By 2050 the world's population is expected to grow by as many as 2 billion people. More people will need more energy resources. In fact, millions of people now need more access to energy, not less. In many developing countries today, people consume little oil, little natural gas, and have little or no access to electricity. Their energy resources are mainly wood and dried animal dung, which they gather to burn in inefficient, indoor stoves, causing unhealthy indoor air pollution. They seek a better life and look to follow countries such as India and China that have the highest growth of new car sales, the greatest increases in electricity production, and have had immense increases in consumption of oil, coal, and other natural resources. China's annual consumption of oil, for example, increased by over 85 percent from 2003 to 2013, India's by more than 50 percent. Overall, world oil consumption rose by around 12 percent during that period.

Of course it is not inevitable that consumption of any resource will rise. In the United States and most of the developed world consumption of some major energy resources has been falling. For the 2003–2013 period, US consumption of oil fell by 6 percent, while Japan's was down by more than 18 percent. Consumption of coal has risen on a worldwide basis but has fallen in the United States. All of this is to say that R/P ratios are suggestive of what *could* happen but will almost certainly not tell us what *will* happen with respect to either production or consumption of any energy resource.

In fact, there is an extraordinarily wide range of forecasts – a continuum ranging from a confidence in ever-rising world prosperity fueled by our current mix of energy resources on the one hand, to fear of catastrophe on the other. The catastrophe will result from either unplanned-for resource "peaks" or environmental disasters, or both. In other words, with respect to energy resources in the future, there is little agreement on *what* to do and even on *whether* there really is a problem in the first place.

These views can be described as falling into a range of what, as we have already noted, are termed *expansionist* (or *cornucopian*) and *limitationist* (or *neo-Malthusian*) perspectives. Optimism about the potential of technology underlies the expansionist approach (see, among others, Kahn et al. 1976, Simon 1981, Bryce 2014, Epstein 2014). This view holds that technological development is key to economic growth, and that in general resources should be used at whatever rates are required for economic growth. Expansionists are not necessarily blind to the possibility of resource depletion and other resource problems, but they are not worried about the outcome. Wealthier societies are more resilient societies and can better adapt to resource and pollution challenges. As one expansionist put it, "In effect, technology keeps creating new resources" (quoted in Simon 1981).

The limitationist view stresses that natural resources are finite and that we must deal now with the reality that they will, if not run out, at least prove insufficient to meet growing world demand. Furthermore, rapid extraction would be bad even if there were no danger of depletion. It causes pollution and despoils the land; it risks climate change and threatens all life by destroying the ecosystem itself. Limitationists are basically pessimistic about the ability of technology to solve these problems quickly and, in fact, are pessimistic generally about what lies ahead for the world because of resource-related problems. Some see truly ominous possibilities for the future if we do not take drastic steps *now* to limit resource extraction.

The limitationist position is exemplified by the work of Donnella and Dennis Meadows (1974, 1992), whose books on the "limits to growth" demonstrated through computer simulations why industrial society might be headed for collapse. Not surprisingly, most limitationists advocate policies that limit resource extraction, population, and even economic growth.

If such arguments have validity, it is hard from a moral perspective to argue against limitationist policy prescriptions. How could we continue to commit acts that contribute to the collapse of civilization or the destruction of the ecosystem? Such destruction clearly must be considered a great evil and action to forestall or prevent it good. Yet it also must be noted that to take actions to forestall these projected disasters would have immediate consequences that are not beneficial. There are, as economic theory tells us, always *trade-offs*. For example, limiting growth might deprive poor people of material goods; government restrictions on consumption would limit individual freedom. However, these deprivations and restrictions would not be based on an evil that exists, but rather on one that is only hypothesized.

Indeed, there have been many criticisms of limitationist arguments. Critics (Cole et al. 1973, Simon 1981) have noted that the Meadows's simulations used questionable assumptions; when these assumptions had changed, the forecasts for the future became far less perilous. Moreover, many of the catastrophes predicted by the "Limits" books and by works of other limitationists, such as Paul Ehrlich or Kenneth Deffeyes, simply have not come to pass.

There is a long history of failed predictions of catastrophe (Bailey 2015). Perhaps the most famous historical example was the catastrophe theory of Thomas Malthus

(1766–1834), a British philosopher and economist. In 1799, he predicted that population growth would outstrip agricultural production and mass starvation would follow. His argument was in part mathematical: population would expand at a *geometric rate*, he maintained, while agricultural production would expand only *arithmetically*. Malthus had good reasons for his arguments, but they were not borne out because he had overestimated population growth and had underestimated the productive potential of agriculture. In other words he had underestimated technology. But Malthus is forever remembered for his failed catastrophist predictions and many who make similar bleak forecasts have been labeled *neo-Malthusian* as a consequence.

Limitationist and expansionist positions embody the ambivalence that industrial society has shown for the technology that created it. The expansionist belief in a *technical fix* for problems that beset us is deeply rooted in the experience of the twentieth century. We have seen technological progress at work and believe in its potential. But at the same time, many see technology itself as the cause of our problems not the solution to them. Although a machine may seem morally neutral, its existence is inseparable from its uses. The limitationist view reminds us that employment of technology creates moral and social problems: pollution, population pressures, resource depletion, and so on. The limitationist contends that we need to gain control of technology before we are destroyed by its effects.

The expansionist view is optimistic on both the future of technology and the capacity of the free market economic system to adjust itself to cope with problems unrelated to near-term profits and losses (Simon 1981). Technology in the context of the free market becomes not merely a more efficient approach, but a way to better the lot of all humanity (Friedman 1962). Like the limitationist argument, the expansionist approach is clearly a moral position, but the focus is changed. It seeks to maximize both economic growth and liberty simultaneously, now and in the future, although it is based on an assumption – not a certainty – that the future will find a way to take care of itself.

The limitationist view emphasizes that the solution to current problems comes from *limits* to extraction and development; the expansionist sees the solution in *growth*. These positions might seem to preclude one another, that is, we can either have growth or limits but not both. Yet an important concept, introduced in the 1980s, has held some promise of reconciling limits and growth. The concept revolves around the idea of *sustainability* and has been termed "*sustainable development.*"

This idea was given international prominence in 1984 at the Global Possible Conference of the World Resources Institute and again at the 1987 meeting of the World Commission on Environment and Development (Brundtland Commission 1987). The underlying idea was, according to the 1987 document, "development [should] meet the needs of the present without compromising the ability of future generations to meet their own needs." Economic growth was possible, but an economy should exist "in equilibrium with the earth's resources and its natural ecosystem."

But just what does "sustainability" mean? The notion has been adopted by many, but its meaning varies greatly. One study published in 2007 counted "some three hundred definitions of 'sustainability' and 'sustainable development'" (Johnston et al. 2007). Because it is so difficult to define precisely, interpretations are as far apart as ever, especially with regard to achieving sustainability through policy.

Consider the idea of sustainability from the perspective of energy: what does it mean to have sustainable energy? As noted in Chapter 1, there is an overriding question: what do we want to sustain?

Sustainability with respect to *resources* could begin with the following guiding principles: "Renewable resources [should be used] at rates less than or equal to the natural rate at which they can be regenerated," while nonrenewables could be depleted but with "optimal efficiency" (Pearce & Turner 1990). These principles suggest policies that would emphasize more efficient technologies, careful management of nonrenewable resources, and the development of renewables – in all likelihood with government support at least for research and development.

Energy sustainability with respect to the world's *ecosystem* would likely require movement away from depletable resources such as fossil fuels and an effort to encourage renewable energy technologies. These would, from this point of view, need to be employed not only in the developed world but among developing nations as well. It is in these countries that we see the most rapid growth in the consumption of fossil fuels and also (as we discuss in Chapters 5 and 6) the greatest increases in pollution. This not only creates polluted cities but also increases greenhouse gases. Ultimately, the goal of energy sustainability with regard to the ecosystem would be: reduce fossil-fuel consumption to zero.

Energy sustainability could also mean sustainability of the *services* energy provides (Jaccard 2005). It should not matter to one's lifestyle, happiness, utility, etc. if electricity is supplied by wind or natural gas as long as it is supplied. Energy is part of a system and it is the system that we look to retain (and improve) indefinitely. Of course, it matters to the environment which form of energy is chosen, and this concept entails a view that energy production should be environmentally "benign." This does not mean abandonment of fossil fuels, at least in the near term, but rather it entails finding ways to reduce carbon dioxide emissions and other pollutants in using fossil resources. In time, several generations perhaps, the world can reduce fossil-fuel use significantly, but in the meantime people would not have to sacrifice the energy services of modern technological society.

In general, energy sustainability is like any form of sustainability: difficult to define. Rather it is argued that it can be directed to three basic goals: Low pollution, use of resources that will be many decades before depletion or are basically non-depletable, and no "significant" social injustices in production and consumption.

To many people fossil fuels fail on all counts. They pollute – both in production and consumption. It is unclear how close oil especially is to depletion, but it seems an oil economy involves the risk of high economic costs along with the environmental costs, growing over time. Moreover, fossil fuels are produced in some countries where social injustice is prevalent, and even in more egalitarian societies costs of various kinds may fall disproportionately on the poor.

It would seem that greater emphasis on renewable energy technologies and resources would be the answer, the natural outgrowth of the sustainability concept. It must be kept in mind, however, that the greater the emphasis on renewable resources, the greater the uncertainty inherent in the program. Renewables are not yet available to replace fossil fuels on a mass scale; nor is it known when or if these technologies will become widely available and affordable. Therefore, policies that force society to become dependent on renewable resources in the future might be a costly mistake. But then continued reliance on fossil fuels has costs that are known, that cross borders, and that may grow in the future.

In 2005, Sheik Ahmed Zaki Yamani, the former Saudi Arabian oil minister said, "The Stone Age didn't end for lack of stone, and the oil age will end long before the world runs out of oil." In one sense he is quite correct. The world is not "running out" of oil or any other of the fossil fuels in the common understanding of the term. The wells will not soon be completely dry. But there are two important differences between the transition of thousands of years ago and today. First, the Stone Age ended because of exogenous technological change (the smelting of metals), not because it became either more difficult or costly to gather stones, nor did stone gathering risk environmental damage to the Earth. Unlike stones, further extraction of oil will mean eventually greater scarcity and higher costs, and it may also mean more environmental damage adding to those costs. Second, because of that environmental damage, society – possibly even the world acting in concert – may *choose* to transition away from oil and all other fossil fuels, and seek resources so that the energy system can be sustained in a way that is truly beneficial and environmentally benign.

References

Aguilera, R.F., R.D. Ripple, and R. Aguilera. 2014. "Link between Endowments, Economics and Environment in Conventional and Unconventional Gas Reservoirs." *Fuel*, 126, 224–238.

Alchian, A.A. 1975. "An Introduction to Confusion," in *No Time to Confuse: A Critique of the Final Report of the Energy Policy Project of the Ford Foundation, a Time to Choose America's Energy Future*. Institute for Contemporary Studies, San Francisco.

Bailey, R. 2015. *The End of Doom: Environmental Renewal in the Twenty-first Century*. St Martin's Press, New York.

Boer, W.K. et al. (eds.). 1986. *Advances in Solar Energy, Volume 3*. American Solar Energy Society, Boulder, CO.

BP (annual), *BP Statistical Review of World Energy*. http://www.bp.com/en/global/corporate/energy-economics/statistical-review-of-world-energy.html.

Bradley, R.L. Jr. 1996. *Oil, Gas & Government*. Roman & Littlefield, Lanham, MD.

Brundtland, G.H. 1987. *Our Common Future. The World Commission on Environment and Development*, Oxford University Press, Oxford, UK.

Bryce, R. 2008. *Gusher of Lies: The Dangerous Delusions of "Energy Independence."* Public Affairs, New York.

Bryce, R. 2014. *Smaller Faster Lighter Denser Cheaper*. Public Affairs, New York.

Chong, Z.R., S.H.B. Yang, P. Babu, P. Linga, and X.-S. Li. 2016. "Review of Natural Gas Hydrates as an Energy Resource: Prospects and Challenges." *Applied Energy*, 162 (1), 1633–1652.

Cole, H.S.D., C. Freeman, M. Iohoda, and K.L. Davitt. 1973. *Models of Doom: A Critique of Limits to Growth*. Universe, New York.

Commoner, B. 1979. *The Politics of Energy*. Knopf, New York.

Deffeyes, K.S. 2009 (Re-Issue), Copyright 2001. *Hubbert's Peak: The Impending World Oil Shortage*. Princeton University Press, Princeton, NJ.

Epstein, A. 2014. *The Moral Case for Fossil Fuels*. Penguin, New York.

Friedman, M. 1962. *Capitalism and Freedom*. University of Chicago Press, Chicago.

Grossman, P.Z. 2013. *U.S. Energy Policy and the Pursuit of Failure*. Cambridge University Press, Cambridge, UK.

Holz, F., P.M. Richter, and R. Egging. 2015. "A Global Perspective on the Future of Natural Gas: Resources, Trade, and Climate Constraints." *Review of Environmental Economics and Policy*, 9 (1).

Hubbert, M.K. 1949. "Energy from Fossil Fuels." *Science*, 109 (2823), 103–109.

Hubbert, M.K. 1956. *Nuclear Energy and the Fossil Fuels*. Publication No. 95, Shell Development Company, Houston, TX.

Hubbert, M.K. 1969. "Energy Resources," in *Resources and Man*. National Academy of Sciences-National Research Council; W.H. Freeman, San Francisco, CA.

Hubbert, M.K. 1973. "Survey of World Energy Resources." *Canadian Mining and Metallurgy Bulletin*, 66, 37–53, July. (Also reprinted in M.G. Morgan (ed.) 1975. *Energy and Man*. IEEE Press, New York.)

Jaccard, M. 2005. *Sustainable Fossil Fuels*. Cambridge University Press, Cambridge, UK.

Johnston, P., M. Everard, D. Santillo, and K.-H. Robert. 2007. "Reclaiming the Definition of Sustainability." *Environmental Science and Pollution Research*, 14 (1), 60–66.

Juvkam, H.C. and A.J. Dessler. 2009. "Using the Hubbert Equation to Estimate Oil Reserves." *World Oil*, 230 (4).

Kahn, H., W. Brown, and L. Martel. 1976. *The Next 200 Years: A Scenario for America and the World*. Morrow, New York.

Korpela, S. 2007. "Oil and Natural Gas Depletion and Our Future." *Resilience*, http://www.resilience.org/print/2007-07-21/oil-and-natural-gas-depletion-and-our-future.

Libecap, G.D. 1989. "The Political Economy of Crude Oil Cartelization in the United States, 1933–1972." *Journal of Economic History*, 49 (4), 803–832.

Libert, D.J. 1956. "Legislative History of the Natural Gas Act." *Georgetown Law Journal*, 44 (4), 695–723.

Meadows, D.H., D.L. Meadows, J. Randers, and W.W. Behrens. 1974. *The Limits to Growth*. Signet, New York.

Meadows, D.H. and D.L. Meadows. 1992. *Beyond the Limits*. Chelsea Green Publishers, Burlington, VT.

Milici, R.C., R.M. Flores, and G.D. Stricker. 2013. "Coal Resources, Reserves and Peak Coal Production in the United States." *International Journal of Coal Geology*, 113, 109–115.

Mitka, M. 2012. "Rigorous Evidence Slim for Determining the Health Risks from Fracking." *Journal of the American Medical Association*, 307 (20), 2135–2136.

Nashawi, I.S., A. Malallah, and M. Al-Bisharah. 2010. "Forecasting World Crude Oil Production Using Multicyclic Hubbert Model." *Energy & Fuels*, 24, 1788–1800

Pearce, D.W. and R.K. Turner. 1990. *Economics of Natural Resources and the Environment*. Johns Hopkins University Press, Baltimore, MD.

Penner, S.S and L. Icerman. 1974. *Energy, Volume I: Demands, Resources and Policy*. Addison-Wesley, Reading, PA.

Reynolds, D.B. 2014. "World Oil Production Trend: Comparing Hubbert Multi-Cycle Curves." *Ecological Economics*, 98, 62–71.

Saraiva, T.A., A. Szklo, A.F.P. Lucena, and M.F. Chavez-Rodriguez. 2014. "Forecasting Brazil's Crude Oil Production Using a Multi-Hubbert Model Variant." Institute of Development Studies, Brighton, UK, www.academia.edu/12876494/Who_Drives_Climate-relevant_Policies_in_Brazil.

Simon, J.L. 1981. *The Ultimate Resource*. Princeton University Press, Princeton, NJ.

Stock, J.H. 2015. *The Renewable Fuel Standard: A Path Forward*. Columbia University, Center on Global Energy Policy, New York.

United States Environmental Protection Agency. 2015. *Assessment of the Potential Impacts of Hydraulic Fracturing for Oil and Gas on Drinking Water Resources*. Office of Research and Development, Washington, D.C.

Vaux, G. 2004. *The Peak in U.S. Coal Production*. The Wilderness Publications, www.FromTheWilderness.com.

Wildavsky, A. and E. Tennenbaum. 1981. *The Politics of Mistrust*. Sage, Beverly Hills, CA.

Wilson, C.L. project director. 1977. *Energy: Global Prospects 1985-2000. Report of the Workshop on Alternative Energy Strategies*. McGraw-Hill, New York.

Zapp, A.D. 1962. *Future Petroleum Producing Capacity of the United States*. US Geological Survey Bulletin 1142-H.

Zhang, J., S. Palmer, and D. Pimentel. 2012. "Energy Production from Corn." *Environment, Development and Sustainability*, 14, 221–231.

NOTES

1 Data from the World Bank, world development indicators: http://wdi.worldbank.org/table/3.6.

2 Allegations of this sort have been made several times during the last 50 years – in fact, whenever oil prices spike. But investigations by the Federal Trade Commission (FTC) have never substantiated these sorts of allegations (Grossman 2013).

3 Named for Thomas Malthus, one of the first economists, who in 1799 predicted widespread famine in Great Britain because, he reasoned, population would rise more quickly than food production. His prediction, though logical in light of history, proved wrong in almost every respect. Discussed in the last section of this chapter.

4 Some parts of the country do not have a pipeline system that can transport natural gas from the wellhead and so cannot easily switch from coal to natural gas.

5 *Phillips Petroleum Co. v. Wisconsin*, 347 U.S. 672 (1954).

6 John F. O'Leary, Deputy Secretary of Energy.

7 George Mitchell, head of Mitchell Energy Co., has been considered the "father of frack-ing" (according to the *Economist* magazine); he began experimenting with hydraulic fracturing and horizontal drilling in 1981.

8 A study of water contamination in Wyoming thought by some to be due to fracking was shown instead to be caused by bacteria. (Associated Press story, November 10, 2016)

9 A map of US shale basins (lower 48 states) can be found at www.lib.utexas.edu/maps/united_states/united_states-shale_plays-2011.pdf.

10 There is a large literature on biofuels and the RFS. See for example, Bryce (2008), Zhang et al. (2012), Stock (2015).

11 The study, "New Stream-Reach Development: A Comprehensive Assessment of Hydro-power Energy Potential in the United States." Prepared by the Oak Ridge National Laboratory for the USDOE.

3 Conventional Conversion of Energy

Energy Conversion

To power industrial society, energy resources have to be converted into useful work. In fact, the technical wonders of the modern world depend on just a few of what we will call *conventional* energy conversion technologies, the principles of which have been known for a century or more. Most of these technologies are *heat engines*; the major exceptions are *hydropower* and *wind power*. Hydropower (water power) and wind power have long been sources of energy, going back to ancient times. Hydroelectric plants of various sizes currently provide roughly 7.5 percent of the electricity generating capacity in the United States. Modern wind turbines have experienced a great surge of development in recent decades (see Chapter 11) and may supply a significant fraction of worldwide electricity in the years ahead. Some other alternative energy technologies, such as solar-photovoltaic (PV) conversion, are also not of the heat-engine type. Photovoltaics, with help of policies to encourage it, has begun to achieve some penetration into the energy market, especially for home rooftops and for locales that have predictable sunlight most days, such as deserts.

Most energy technologies, however, have been based primarily on conversion of *thermal heat* energy to useful purposes. These technologies have evolved greatly since the start of the Industrial Revolution. This concentration on thermal conversion seems likely to continue into the future. Not only have there been huge investments in these current technologies, but some prospective alternatives, notably synthetic fuels (*synfuels*) are themselves also based on thermal conversion. Even *nuclear fission* technology today is used as the heat source to drive power plant turbines; in other words, they are nuclear-fueled *heat engines* driving the electricity generators (Chapter 7). These differ from fossil-fuel fired boilers only as to the source of the heat. The same is likely to be the case for *nuclear fusion*, if a working fusion reactor can be achieved (see Chapter 10 and Appendix C). Only a major technological innovation, such as a breakthrough in solar-direct conversion or low-cost grid-scale electric storage (see Chapters 10, 11 and Appendix C) would change this historic reliance on thermal conversion.

Conventional Heat Conversion

The category of conventional heat engines includes two particularly important varieties: the *steam power plant*, by far the major means of generating electricity today, and the *internal combustion engine*, the principal power source for transportation in our industrial society. Indeed, two somewhat different forms of internal combustion engines – the gasoline and diesel engines – today power about nine-tenths of all passenger travel and about one-third of all freight transport in the United States. The diesel engine has also found a place in small-scale electricity generation in addition to its automotive use. Another form of internal combustion engine, the *gas turbine*, now figures prominently in electricity generation as well as in aircraft propulsion.

But most electricity is still primarily generated by *external* combustion steam engines. And we will therefore begin our discussion of conventional energy technologies with a consideration of steam engines and of the steam electric power plant.

Steam Engines

Steam engines operate on the *Rankine cycle* (after W.J.M. Rankine, 1820–1872) and follow thermodynamic paths that exhibit changes in pressure and volume of the steam as the machine goes through its cycle of operation. In part, the paths of the Rankine cycle are determined by the particular characteristics of the *working substance* as it changes from water to steam or from steam to water. (See Appendix A for a review of thermodynamic principles.)

A classic example of the Rankine cycle is the reciprocating steam engine, which powered factory machinery in the nineteenth century and is similar to those engines used on railroad locomotives and steamships from the early 1800s through the mid-twentieth century. In Figure 3.1, we can see a piston inside a cylinder that is connected to a flywheel by a drive shaft. There are also inlet and exhaust valves on the cylinder. The basic action of the engine is probably familiar. First, the steam enters from the right into the cylinder. As the steam expands, the piston is driven horizontally to the left. But after the steam thrust, the motion imparted to the flywheel forces the piston back to the right. In the process, the spent steam is

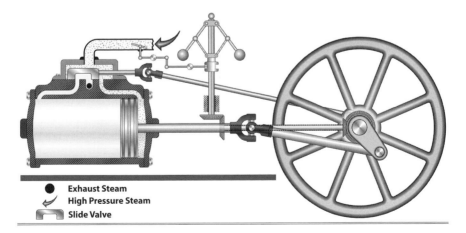

● Exhaust Steam
↙ High Pressure Steam
⊓ Slide Valve

Figure 3.1   Reciprocating steam engine.
Source: Fouad A. Saad/Shutterstock.

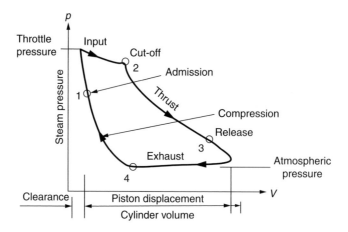

Figure 3.2 Pressure–volume (p–V) diagram of Rankine cycle in the reciprocating steam engine.
Source: Adapted from F.T. Morse (1947), *Power Plant Engineering*, Van Nostrand, New York.

expelled through the *exhaust valve*. The motion of the piston and the flywheel also control the slidevalve that regulates steam *input*, as well as the exhaust. As long as the steam is fed to the piston chamber, the reciprocating motion will continue. The reciprocating steam engine is an example of an *open* thermodynamic cycle, in that the *working substance* (steam) is exhausted from the system after its use.

The changes that take place in the pressure and volume of the steam as the cycle of reciprocating motion takes place can be described on a pressure–volume (p–V) diagram. A path on a p–V diagram describes the changes in pressure that accompany changes in volume of the working substance (in this instance, steam) of a heat engine (see Appendix A).

We can also consider the p–V diagram in terms of the positioning of the valves that control steam input and exhaust. In Figure 3.2, we see the four Rankine cycle p–V paths:

Path 1–2: steam (heat) input
Path 2–3: steam expansion (*thrust*)
Path 3–4: steam rejection (*exhaust*)
Path 4–1: compression

where:

Point 1. *Admission* of steam starts as the input valve is opened; the exhaust valve remains closed.
Point 2. *Cutoff* of the input valve occurs and thus both ports are closed for path 2–3.
Point 3. *Release* as the exhaust valve opens while the input valve remains closed.
Point 4. *Compression* starts as the exhaust valve closes; again, both valves are closed.

Modern steam plants no longer use reciprocating engines to deliver mechanical work. Rather they use *steam turbines* that usually operate in a *closed* thermodynamic cycle. In closed cycle operation, the working substance is constantly

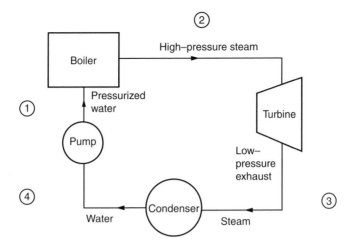

Figure 3.3   Schematic diagram of a steam-turbine power plant.

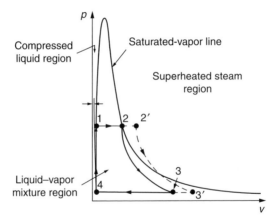

Figure 3.4 p–V diagram of a closed-cycle steam plant.
Source: Adapted from Faires (1948).

recycled without any exchange of its mass. In terms of its thermodynamic description, the p–V paths are simply retraced as the cycle is repeated over and over again, as the working substance itself is circulated over and over again. A schematic diagram of the closed cycle for a modern steam-turbine power plant is shown in Figure 3.3. The water–steam working substance is shown there to be circulating around a physically closed path: from the *boiler* to the *turbine* to the *condenser* to the *pump* and back to the boiler again. In contrast, in the reciprocating engine, the steam is discharged at the end of each cycle.

The thermodynamically *closed cycle* of the steam plant is depicted on the p–V diagram in Figure 3.4. Starting at point 1, water under pressure is fed into the boiler, where heating changes it to steam. The boiler output (point 2) is in fact high-pressure, high-temperature steam that is fed to the turbine. The steam expands through the turbine, causing it to spin, and yield useful work output. After the temperature and pressure drop from expansion, the steam exits the turbine (point 3) and is fed into the condenser. There the steam's latent heat of vaporization (the energy required to turn water to steam) is given up to the cooling water, without any exchange of mass from the working fluid to the cooling water. At the condenser the working fluid returns to water – point 4 in the cycle – at essentially atmospheric pressure. A pump then raises the water pressure to that in the boiler, and the cycle repeats, starting again from point 1. This cycle of operation is a *closed* Rankine cycle.

The *p–V* paths on Figure 3.4 for the closed Rankine cycle are similar, but not identical, to those for the open reciprocating steam engine (Fig. 3.2). The heat input path (1–2) of both is a constant-pressure process that results from the liquid-to-vapor transition. Over most of this *p–V* path, the boiler is adding the necessary heat to change the *phase* of the working substance from liquid to vapor. The *p–V* expansion path (2–3), taking place within the turbine, is similar to the thrust path of the reciprocating engine. The exhaust path (3–4) is a constant-pressure process for the closed cycle, as it was for the open-cycle steam engine. Finally, the compression (path 4–1) in the closed Rankine cycle takes place at essentially constant volume through the action of the separate pump.

In the closed Rankine cycle, the working substance is typically brought to a *superheated* condition by continuing to add heat beyond the point where normally water turns to steam. The key is that water under pressure boils at higher temperatures. For example, at sea level, water turns to steam at 100 °C (Celsius) or 212 °F (Fahrenheit). But at a pressure equal to 20 times the sea-level atmospheric pressure (called 20 *bar*), water boils at more than 200 °C (over 400 °F) and thus the steam produced is called superheated.

When this is done (in superheater coils in the boiler), the heat-input path is extended to point 2′ in Figure 3.4 and is then in a superheated state. The path extension 2–2′ is merely a horizontal line, since the superheated extension is still at constant pressure. However, the steam temperature rises in this superheated extension, which results in an increase in the thermal efficiency. That is, the ratio of energy available for useful work to input energy increases (see Appendix A).

The thermal efficiency of the modern steam turbine plant is often enhanced by other means. For example, some of the steam may be diverted back to the boiler before it passes through the turbine. This steam is then reheated and sent back to the turbine, where it reenters near the point where it has been diverted. *Reheating* can improve efficiency by as much as 3 percent. (See Appendix A for further discussion of thermal efficiency.)

Steam Power Plants

In a steam power plant, we can see the physical realization of the Rankine steam cycle. The entire process of energy conversion is depicted in Figure 3.5, starting with the fuel input (coal) at the top and ending with the electric output (the high-voltage transformer) on the right.

If we examine the components of the thermal core (Fig. 3.5), beginning with the boiler, we see:

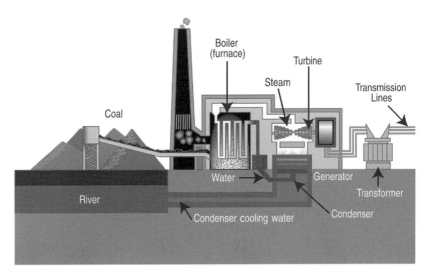

Figure 3.5   A fossil-fuel fired generating station: cross-sectional drawing.
Source: Courtesy of the New York State Electric & Gas Corporation.

1. *Fuel loading* – a conveyor carries the coal across the top and drops it into a *storage bunker.*
2. *Fuel input* – a *traveling grate* moves the coal into the furnace, where it combines with *a forced draft* of heated air to achieve combustion in the *fire box.*
3. *Water input/steam output* – boiler tubing in the furnace fire box contains the working substance, which enters as water and leaves as steam on its way to the steam turbine.

The combustion of fuel – together with the transfer of its heat – in a modern steam plant takes place at high temperatures. As we note in Appendix A, the temperature at which the input heat is supplied has a critical effect on the thermodynamic efficiency (that is, the amount of energy *output* from a given energy *input*) of the cycle. The higher the input temperature, the higher will be the efficiency that can be attained. The highest temperature in the fossil-fired power plant is at the flame itself, typically around 1650 °C. The hot combustion gases in the furnace chamber in the immediate vicinity of the flame, however, range from 1150 to 1260 °C. These combustion gases actually transfer the heat into the (pressurized) working fluid in the boiler tubes. But as this transfer takes place there is a considerable temperature drop; the working fluid will only be at around 250 °C (~500 °F) for a conventional steam power plant.

Tracing the remainder of the closed cycle of the plant for Figure 3.5, we find:

4. *Steam drum and superheater* feeding the turbine.
5. The *turbine* itself.
6. The *condenser* directly below the turbine.
7. *Boiler feed pump* feeding the pressurized water back to the boiler.

The turbine, of course, is where heat energy is converted into mechanical work, to turn the electricity generator. A simplified example of a steam turbine is shown in

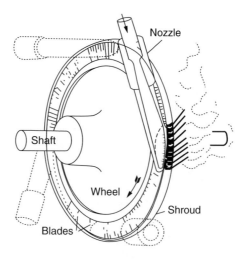

Figure 3.6   Simple impulse steam turbine.
Source: Adapted from Severns and Degler (1948) © John Wiley & Sons.

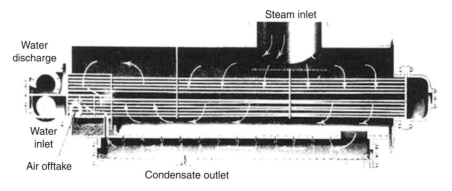

Figure 3.7   Longitudinal section of a small surface condenser.
Source: Severns and Degler (1948) © John Wiley & Sons.

Figure 3.6. Called an *impulse turbine*, it consists of a bladed wheel set on a shaft. Essentially, steam is forced out through the nozzle at high pressure and is directed onto the blades of the turbine wheel. The impulse of the jetting steam turns the wheel, which turns the shaft for useful work. In a second type of turbine, the *reaction* turbine, steam enters through passages in stationary blades and turns other sets of blades as it builds up velocity and pressure. A turbine in any given power plant can be a *composite* of both types, using both impulse and reaction processes, or it can be a simple impulse or a simple reaction turbine, depending on its design.

The next major component in the closed Rankine cycle is the *condenser* (see Fig. 3.7). The condenser shown is an early version of a modern *surface condenser*, in which the steam and condensate do not come into direct contact with the cooling water, but rather undergo heat transfer only. In the condenser, steam enters through the top and passes over metal tubes carrying the cooling water. Upon contact with the cool tubes, the steam turns to water, which drops to the bottom of the condenser and flows through the outlet. The condensing of the steam is accompanied by a pressure drop, which creates a vacuum and thereby reduces the *back pressure* into

Figure 3.8   Power plant cooling towers.
Source: ProjectB, iStock / Getty Images Plus.
Note: A photo of this particular power configuration has been included because the cooling tower depicted here has become the symbol of the contemporary power plant and particularly the nuclear power plant. But it actually represents only one part of thermal plant operation and is not distinctive of nuclear plants only.

the turbine exhaust. (See Appendix A for a discussion of the thermodynamics of the Rankine cycle.) The heat of condensation from the steam is transferred almost entirely to the cooling water circulating in the tubes and is carried away.

The final major component in the closed Rankine cycle is the *boiler feed pump*, which returns condensed water to the boiler at the *back pressure* of the steam in the boiler. In its simplest concept the feedwater pump merely raises the pressure of the liquid condensate to match the boiler's back pressure. However, the pressures required by modern boilers can be quite high – in a range of more than 100 bar – for the feed pump to overcome.

The condenser not only transfers heat from steam into the cooling water, it also transports waste heat to the outside environment. If we return to Figure 3.5, we see at the bottom how the loop of cooling water, with the aid of the circulating pump passes the water through the condenser. When the cooling water is discharged, its temperature can be sufficiently high to raise the temperature of the body of water it drains into. In the early days of condenser-equipped power plants (late nineteenth and early twentieth centuries), there was little concern about the discharge of waste heat. However, with the development of large-scale plants, biologists discovered that temperature changes could have noticeable effects on aquatic life. Spurred by the environmental movement of the past few decades, governments have established regulations on thermal pollution, limiting the permissible temperature increase of a body of water, be it a river, a bay, or a lake. These regulations sometimes require electric utilities to construct cooling ponds or towers to reduce the temperature of the

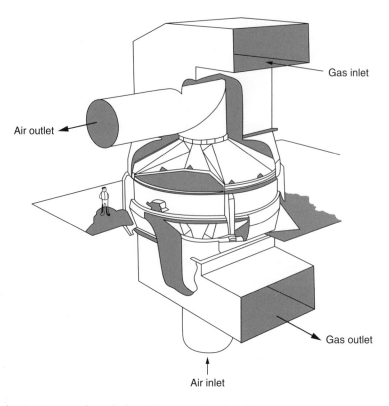

Gas inlet

Air outlet

Gas outlet

Air inlet

Figure 3.9   Air preheating in power plants: Rothemuhle regenerative air preheater.
Source: Courtesy of Babcock & Wilcox Co.

cooling water being discharged. At some sites, particularly in arid regions, *cooling towers* (see Figure 3.8) are needed, regardless of regulations, in order to recycle the coolant back to the condenser.

The operating principle of cooling ponds or towers is the same, namely partial evaporation, a commonly known cooling mechanism. In cooling ponds or towers, a portion of the water evaporates, taking heat from the remaining water and thereby cooling it. Cooling in a pond is achieved through evaporation by exposing a large surface area of a shallow body of water to the air, thus maximizing the ratio of cooling area to volume of water. Even more effective cooling is achieved in a tower (Figure 3.8), in which water descends in sprays or sheets over baffles through an upward draft of air. The air draft can be achieved through a natural convective flow or can be forced with fans.

One last component of the thermal plant is worth noting – the *air preheater*. The purpose of a preheater is to improve the plant's thermal efficiency by recovering some of the heat exhausted to the stack after combustion. This partial recovery of heat is accomplished through a heat exchange from the hot stack gases to the incoming flow of air. (The two streams of gas are physically separate.) Figure 3.9 illustrates the actual equipment used in a large-scale power plant.

Since air preheating is *external* to the closed Rankine cycle, it should *not* be thought of strictly as an improvement in the Rankine cycle's efficiency, as is the case with the reheating of steam. However, air preheating eliminates part of the need for heat to be gained from combustion. That means less fuel can supply the same heat input to the working substance, and so it serves in the nature of an efficiency improvement.

Hydroelectric Generation

Although we will focus most of our attention in this chapter on heat engines, we should not entirely overlook water power. The *conversion of falling or flowing water to useful work* is an ancient technology. Before the development of the steam engine, water wheels were used as engines and powered much of the early industrialization of Europe. In the United States, the availability of water power determined the locations of the early factories, such as the textile mills in New England.

Even though hydrocapacity grew in absolute terms during the nineteenth century, the huge demand for mechanical power brought about by the Industrial Revolution in America was far beyond the capacities of small hydropower sites that were developed in the Northeast. Consequently, coal, used with the newly invented steam engine, supplied industrial power increasingly, beginning in the nineteenth century. Later in the century, coal-powered electric plants and the electricity they produced supplied an increasing proportion of the energy burden. Electricity generation, powered by firing coal, grew rapidly during the twentieth century (usage doubled every decade for more than five decades) and soon supplied most of the power needed by industry, as well as a great deal of the energy used for residential and commercial purposes.

Much of electric power generation required thermal-type plants. Today, about four-fifths of electrical energy is generated by thermal plants (fossil-fuel fired and nuclear). About two-thirds of the remainder is generated by water power. Most hydroelectric development in the United States has either utilized sites of tremendous power potential, such as the Niagara River or Hoover Dam, or has been part of large regional development schemes, as in the Tennessee Valley or Pacific Northwest watersheds. In the expansion of electric power earlier in the twentieth century, not only were smaller water power sites ignored for hydroelectric development, but in many cases existing small-dam facilities were closed down in favor of the larger fossil-fuel fired plants.

But actually, these small sites, if added together, could provide a great deal of power. This potential remains undeveloped. In some regions less than half the potential hydrogenerating capacity has been tapped. As noted in Chapter 2, according to a study by the DOE, the United States could add 65 GW of electric power; each gigawatt being approximately equal to the output capacity of a single, present-day thermal power plant. In other words, full development of hydropower in the United States could replace 65 coal or nuclear power plants. A similar situation exists in other parts of the world, particularly in developing nations (see Chapter 5).

Prior to the energy crisis era of the 1970s, the conventional wisdom of the electric power industry was that hydropower had already been fully developed in the United States. What was meant, of course, was that the *large* hydrosites had been developed. However, with rapidly escalating nuclear plant construction costs, and controversies over the environmental consequences of coal and the safety of nuclear power plants, small hydrosites came to be viewed in an entirely different light, and a number of small hydroprojects were undertaken, but potential has remained for considerably more.

Hydroelectric Basics

Hydroelectric generation is based on the principle of the use of *potential energy*. Natural processes of evaporation and rain place volumes of water at various altitudes above sea level. Bodies of water at elevations above some point of use can provide energy for useful work by falling water. As the water falls (by gravity) from the higher elevation, the potential energy of the water is transformed into kinetic energy (see Appendix A for a discussion of the different forms of energy). This kinetic energy is then converted into mechanical form by a hydraulic turbine. The conversion to kinetic energy may occur gradually over a downhill slope or over a sharp drop but the overall principle is the same. This general principle is illustrated in Figure 3.10.

Hydro facilities are grouped according to their *head* (designated by "h" in the figure), which is the altitude of the fall before the water does its work. At a *high-head plant*, the drop is in the range of 500 feet (150 m) or more; a *low-head plant* usually has a drop under 65 feet (20 m); those in between are termed *medium-head plants*. A clear example of a high-head plant is the Hoover Dam on the Colorado River in Nevada (Fig. 3.11).

Hydroelectric plants, like other types of generating plants, are also grouped according to the size of their electric power output. The Hoover Dam is both a high-head and a *large-scale* plant, having an electric power output capacity of 1344 MW, comparable to the capacity of a major coal-fired or nuclear power plant. The much

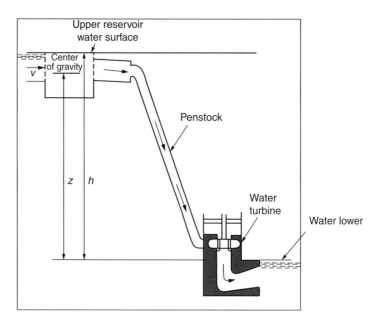

Figure 3.10   Principles of hydroplant operation.
Source: Adapted from Davis and Sorenson (1969).

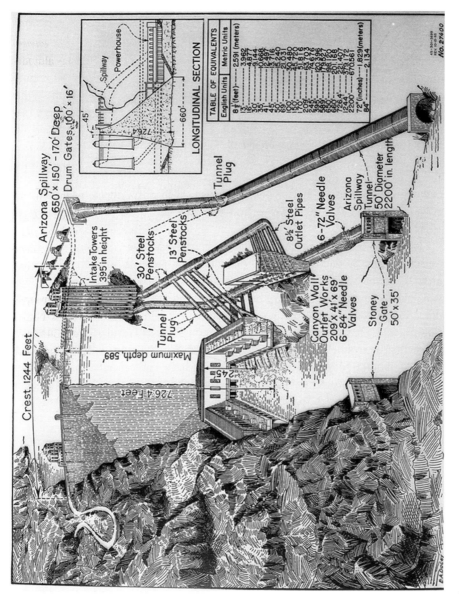

Figure 3.11 The Hoover Dam: a high-head hydroelectric plant.
Source: Bureau of Reclamation (https://www.nps.gov/nr/twhp/www/ps/lessons/140hooverdam/140visual1.htm).

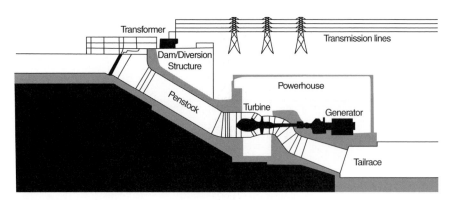

Figure 3.12   Low-head hydroelectric generation.
Source: Courtesy of City of Seattle, Light Department.

more enormous James Bay Complex in Quebec, if fully developed, could have an output capacity of 27,000 MW – and currently has a very large capacity of 16,527 MW.

Large-scale output is not achieved using high heads alone. Rather, high output capacity is the product of the head (height of the reservoir measured in feet or meters) and the rate of water flow (gallons or liters per second). Accordingly, large-scale output power can be achieved with a low-head plant having a large volume of flow. An example is the Robert Moses hydrostation on the St. Lawrence River (not shown), which has a generating capacity of 800 MW, but a head of only 30 feet.

Small-scale hydroelectric plants typically operate in the range of a few kilowatts (kW) to about 5 MW. Most small hydroplants have low heads, such as the dam shown in Figure 3.12. The flow of the river together with the pressure of the head behind the dam supply the power. The output obtained for the normal volume of flow is called the *run of the river.* The hydroplant for the dam in Figure 3.12 operates essentially on the flow volume of the run of the river, with virtually no head.

Conversion to Electricity

The energy conversion of modern hydropower and steam power to electricity is based on the same concept. Energy in the form of either falling water or steam turns a turbine that runs an electric generator. Since most electricity is generated in thermal plants, we focus here on the conversion of thermal energy to electricity. To generate electricity in a steam power plant, a turbine is connected to an electric generator by a common rotation shaft (see the center shaft in Figure 3.13). The turbine is the *prime mover*, meaning that it supplies the mechanical work to rotate the shaft of the electric generator. (In the case of hydrogeneration, the *hydraulic* turbine performs the same task.)

Electricity is generated according to the principles of *electromagnetic induction*, by which an electric voltage is induced in a coil of wire by a moving or rotating magnetic field (see Appendix A). Two types of electricity can be generated – direct current (*d.c.*)

Figure 3.13   A large steam turbogenerator.
Source: Copyright "Siemens Pressebild", obtained with the friendly permission of Siemens Germany by
Christian Kuhna.

or alternating current (*a.c.*). But since virtually all centrally generated electric power is
a.c., we will focus on the conventional a.c. generator as depicted in Figure 3.14.

Essentially, an a.c. generator has two major parts: a *stator* and a *rotor*. A stator
(Fig. 3.14a) is a fixed structure that holds the coils of wire in which voltages will be
induced. The stator structure is made of laminated steel, which channels the magnetic
flux within the machine. Iron (or steel) concentrates and guides the magnetic flux of
the rotor poles as they pass from one stator coil to the next at the periphery of
rotation. The purpose of the laminations is to reduce magnetically induced *eddy
currents*, which cause electrical losses (see Tarboux 1946).

The rotor (Fig. 3.14b), which is driven by the turbine, contains electromagnets. At
the periphery of the rotor are magnetic *poles* that are magnetized by *field coils* placed
just inside the radius of the pole faces. This particular example is a *four-pole
machine*, in which two pairs of north–south (N–S) poles are created by winding
field coils on the rotor (shown as single turns in Fig. 3.14b). The field coils receive
electric current from a separate d.c. source, which can be produced by a d.c.
generator on the same shaft as the main a.c. generator.

As the magnetic poles turn on the rotor past the coils embedded in the stator, the
a.c. voltage in each coil alternates in polarity, positive and negative. A smooth
alternation of polarity ideally traces out a sine-wave shape in time (see Appendix
A), with a repetition rate of 60 times per second – 60 Hz (*hertz*, a measure of
frequency). If a current flows from the stator coil, it too has a smooth waveform
with a repetition rate of 60 Hz. These two waves together determine the power

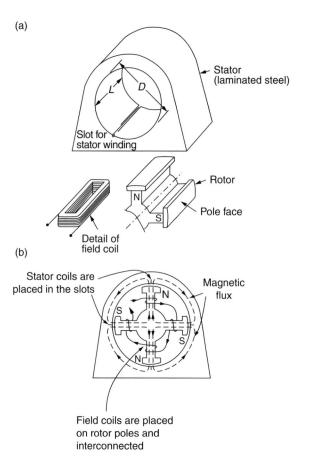

(a)

(b)

Figure 3.14 Alternating current generator. (a) Disassembled parts. (b) An assembled four-pole machine.
Source: Adapted from Elgerd (1978).

output, because the power flow of a generator at any instant equals the product of voltage times current (and is measured in watts, see Appendix A). Of course, the power output will then also alternate as a sine wave. The *average,* over time, of a generator's power output waveform represents the useful energy the generator delivers. That is, the energy that can be converted back into mechanical work by an electric motor or can be converted back into heat for customers connected to the power grid.

But if we look at these alternating waveforms of voltage and current, we might surmise that the power also will vary in time. In order to avoid having a fluctuating supply of power, electric systems are usually operated on a *three-phase basis,* resulting in three distinct power waveforms (P_a, P_b, and P_c), each one staggered equally in time from the others as shown in Figure 3.15. These voltages are applied to circuits having three conductors (in addition to a possible ground connection). The three-phase operation makes the *sum* of the power outputs of *all three* phases *constant*, not fluctuating (Fig. 3.15). A constant electric output is a major advantage of the multiphase a.c. power system.

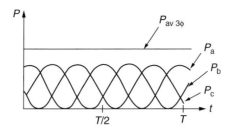

Figure 3.15   Three-phase power waveforms versus time.

Figure 3.16   High-tension transmission line.
Source: pjhpix, iStock/Getty Images Plus.

After the three-phase power is generated, the voltage is raised for transmission. Electrical transformers change low voltage to high voltage and reduce it again at the point of use (see Appendix A for a discussion of the transforming of a.c. voltages and currents). High voltages are the best way to transmit electricity over long-distances because they reduce the electrical losses in the transmission line conductors. Electrical losses are dramatically minimized by using high-tension (that is, high voltage) transmission (see Fig. 3.16), because the electric currents are reduced inversely in proportion with the voltage and electrical losses vary as the square of the magnitude of the current.

Electric power demand fluctuates over daily cycles, rising to peaks when there is activity during the daytime and falling to a minimum at night as activity goes down.

Electric generating systems are therefore designed for two major phases of operation – *baseload* and *peak load*. Baseload service supplies that minimum level of power demand that continues night and day. Peak-load service helps to supply the maximum power needed by the system during the short period of the day when it is demanded. Electric utility companies have traditionally been required to supply power at whatever level was demanded and at the lowest possible cost – although as we will see in later chapters the lowest cost principle has been altered in order to encourage the use of nonfossil-fired electric generation.

Peak-load demand is best supplied by generators with relatively low capital costs but high running costs (such as the gas turbines described later in this chapter). Since the high running costs are incurred over only a small fraction of the time, they are more than offset by the low capital costs (see Chapter 9).

The Efficiency of Steam Electric Power Plants

As we have indicated, a typical late twentieth-century steam power plant (schematically shown in Fig. 3.17) did not convert all, or even close to all, of its energy input to electricity. In the figure, useful output – electricity – was only 35 energy units, out of input equal to 100 units. Therefore, the plant efficiency (the ratio of useful energy output to the required fuel input energy) was only 35 percent. A small amount of power (~2 percent) was needed for plant operation, such as the boiler feed pump, but the rest of the energy was lost as waste heat – some at practically every step along the way. Stack losses, for example, accounted for 12 units, and boiler radiation and

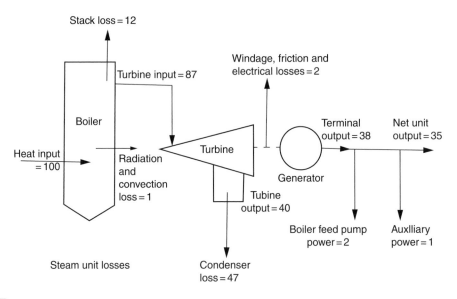

Figure 3.17 Steam power plant: input/output energies.
Source: Adapted from W.D. Marsh (1980), *Economics of Electric Utility Power Generation*, Clarendon Press, New York.

convection loss took away 1 unit. The largest loss, however, was from condenser waste heat – 47 units – more energy than was delivered as output. Yet, even as we knew from thermodynamics that most of these losses were unavoidable (see Appendix A), some technological improvements have been achieved in recent years.

Thirty-five percent efficiency had for decades been representative of conventional steam power plants, without prospects for dramatic improvement. But plant efficiencies can be, and have been, improved markedly in recent years, by raising both the pressure and the temperature of the boiler. Once pressure passes 220 bar (that is, pressure of 220 atmospheres) and temperature 374 °C, it is said that the working fluid has reached the *critical* point. Raising both, to 221 bar (or higher) and 374 °C+, the system is said to be *supercritical*. At this juncture the water does not boil, there is no phase change from water to steam. In fact, supercritical systems operate typically at between 230 and 250 bar, and temperatures of more than 540 °C (over 1000 °F). At supercritical levels, the thermal efficiency is over 40 percent. Some plants have exceeded the supercritical range. Termed *ultra-supercritical* (USC), they operate at pressures as high as 300 bar and temperatures of over 600 °C – with average efficiency rising above 45 percent (see Beér 2007, Tan et al. 2012). Even higher efficiency improvements appear possible with technologies under development, until the limits of thermodynamics are approached.

Higher efficiencies have also been achieved with *combined-cycle* natural gas power plants (CCG) and through *cogeneration* (in which the heat is utilized concurrently to generate electricity and to provide heat for other purposes). The former (discussed in detail below) have thermal efficiencies of close to 60 percent while the latter can reach efficiencies of close to 80 percent. More exotic electric power generation technologies such as *magnetohydrodynamics* (MHD) also have the potential for high efficiencies.[1]

Internal Combustion

Most modern transportation uses a different heat engine technology from the one employed in electric generation. Automobiles, trucks, and aircraft use forms of *internal combustion* engines. Automobile and truck engines are mainly of two types: the gasoline engine, operating on the Otto cycle (named after N.A. Otto, 1832–1891) and the diesel engine, burning a lower petroleum distillate and operating on the Diesel cycle (named after R. Diesel, 1858–1913). While Otto-cycle engines are used almost exclusively in powering cars and trucks, diesel engines also run small generating facilities and most present-day railroad locomotives.

Jet aircraft use a third type of internal combustion machine, the *combustion* or *gas turbine*. Gas turbines, however, are also used as prime movers for electric generators in specific and important applications. Most notably, they provide extra power during times of the day when electric demand reaches peak load. A gas turbine has special advantages for these applications. First, it can be started quickly. (A boiler needs considerable time for steam buildup; turbine combustion does not.) Second, although it burns high-cost fuels, such as petroleum distillates or natural gas, the turbine equipment itself is a much less costly investment than a steam generator or a hydroelectric dam. Since it is operated for small periods of time throughout the year, a gas turbine uses little fuel in total and so it is more cost effective for peak-load service when both capital and operating costs are considered.

The term *internal* combustion refers to the fact that the fuel is burned *within* the engine. In steam-cycle conversion, the fuel is burned externally, inside a separate furnace. The heat of combustion is then transferred into the *working substance* (usually steam) and the working substance does the work for useful output. In internal combustion engines, the combustion takes place within the working machine and the hot combustion gases then make up the working substance, delivering the useful work. Also, unlike the modern steam-cycle operations, internal combustion technologies are all thermodynamically *open cycle* operations. In gasoline and diesel engines and gas turbines, the combustion gases are exhausted from the machine once they have delivered their work.

Otto and Diesel Cycles

Almost all gasoline and diesel engines operate with systems of pistons, and the thermodynamic cycles follow paths similar to those of the open-cycle steam engine. The Otto cycle for conventional gasoline engines follows the thermodynamic paths indicated on Figure 3.18. The cycle requires four *strokes* (one-way motions). One of these strokes, however, is not strictly a path in the thermodynamic cycle, but rather a fuel suction stroke (0–1). Also, one of the mechanical strokes combines two thermodynamic processes. The thermodynamic cycle therefore consists of the paths 1–2–3–4, as indicated on the *p–V* diagram.

The four strokes of the Otto cycle are:

Stroke 0–1. During the *suction stroke*, the fuel is drawn into the cylinder through the open intake valve as the piston moves downward.

Stroke 1–2. During the *compression stroke*, the fuel mixture is compressed as the piston moves up and both valves close. This stroke, a path in the thermodynamic cycle, is ideally an *isentropic* compression process (see Appendix A).

Stroke 2–3–4. The *power stroke* consists of two thermodynamic paths (2–3 and 3–4). The first, path 2–3, is the heat input process, where the compressed fuel mixture is ignited by a spark at point 2. Path 2–3 is nearly a constant volume process; the combustion process raises the pressure while the piston starts its motion from minimum volume. Expansion process 3–4 delivers work to the engine crankshaft. It starts once maximum pressure (point 3) has been reached.

Stoke 4–0. During the *exhaust stroke*, the heat is rejected, corresponding to the heat rejection stroke of the steam engine. Just as the steam engine ejects

(a)

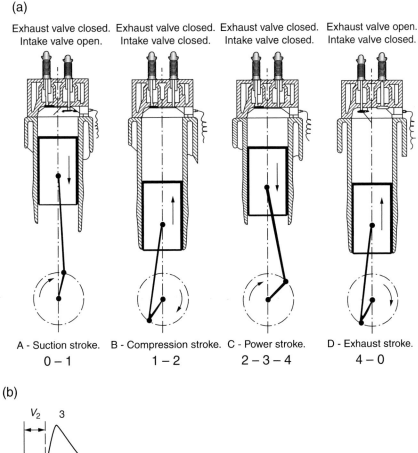

(b)

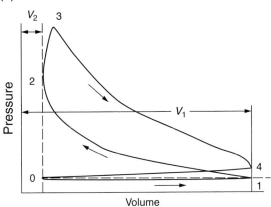

Figure 3.18 Otto-cycle four-stroke internal combustion engine. (a) Cycle of operation. (b) Indicator *p–V* diagram. *Source:* Adapted from Severns and Degler (1948).

steam in this stroke, the gasoline engine ejects the products of combustion (gases) from the cylinder through an exhaust valve.

Whereas Otto-cycle engines use electrical ignition to initiate combustion in the cylinders, Diesel-cycle engines do *not*. Ignition is achieved in the diesel engine when the compression stroke raises the fuel mixture temperature sufficiently high for

(a)

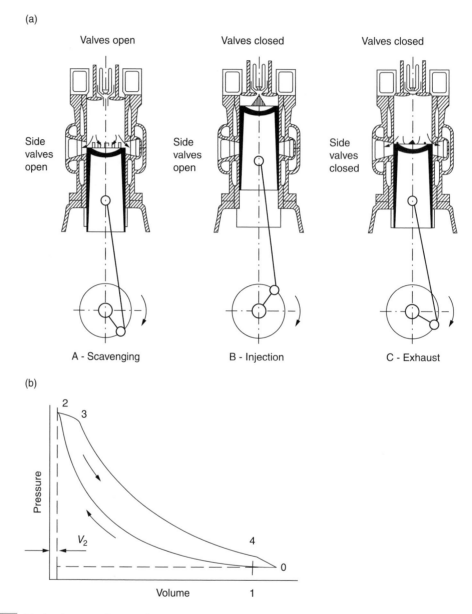

Figure 3.19 Diesel-cycle two-stroke internal combustion engine. (a) Cycle of operation. (b) Indicator p–V diagram. *Source:* Adapted from Severns and Degler (1948).

spontaneous ignition (*auto ignition*) to take place. Also unlike the Otto cycle, the Diesel heat input path is a constant-pressure process, rather than constant volume (see Figure 3.19).

The Diesel cycle shown in Figure 3.19 is for a two-stroke machine, meaning that the entire thermodynamic cycle is completed in two strokes (up and down each) of the piston. (Four-stroke diesel operation is also possible.) In the two-stroke operation, four thermodynamic processes are incorporated:

Stroke 0–1–2

Path 0–1. With the intake top valve open, compressed air forces products of combustion of the previous cycle out of open exhaust valves on the side, called the *scavenging* operation.

Path 1–2. During the *compression* process, which is ideally *isentropic* (see Appendix A), ignition temperature is reached at point 2. Both valves are closed.

Stroke 2–3–4–0

Path 2–3. *Injection* of heat takes place. Ideally, this is a constant-pressure thermodynamic process (position B in Fig. 3.19a), with all valves closed.

Path 3–4. The *expansion* process delivers work. Both valves are closed.

Path 4–0. During the *exhaust* operation, the side valves open at the end of the expansion stroke and heat rejection is accomplished.

Figure 3.20 shows the major parts of an automobile engine. Whereas the engine shown is a gasoline engine, many of the parts are similar to diesel engines. The key difference is that after the fuel is injected into the cylinder, a spark plug ignites the fuel, whereas in a diesel engine compression alone causes ignition.

The Diesel cycle holds several advantages over the Otto cycle. First, it requires no separate electrical ignition system and no carburetor for fuel injection system (see Fig. 3.20), as does a gasoline (Otto-cycle) engine. Second, the Diesel cycle can be operated at temperatures and pressures yielding higher ideal thermal efficiencies than the Otto cycle. Finally, because it is a lower petroleum distillate than gasoline, diesel fuel is cheaper to make than gasoline.[2] Earlier disadvantages of the diesel engine, such as special starting procedures and excessive engine weights, have been overcome in recent years. As a result, the diesel engine, long a standby for trucks, has been more widely adopted for automobile use.

Thermal efficiencies of internal combustion engines ideally can be higher than those of the external combustion (steam-cycle) plants. In part this is due to the elimination of some of the losses, such as the stack and radiation losses of steam plants. It is also possible to achieve higher working-substance temperatures in internal combustion processes, which has implications for higher thermal efficiencies (see Appendix A).

A typical gasoline engine, however, is far from the ideal. The average gasoline-powered automobile, as indicated in Figure 3.21, has a thermal efficiency less than 30 percent. This measure for automotive engines is called *brake efficiency*, since it is for an engine at a constant rotational speed (rpm) working against a friction brake. This value (~30 percent) is in the midrange of efficiencies for various gasoline and diesel engines (Fig. 3.21). Some can achieve efficiencies as high as 44 percent, but each brake efficiency shown applies to one operating speed and power output only. Even though brake efficiency measures do not reflect the range of performance of driving demands, they can be good measures nonetheless of the basic technical capabilities of engine designs.

Because the working substance in internal combustion engines is not retained within the system, as it is in the steam-turbine power plant, the opportunities for

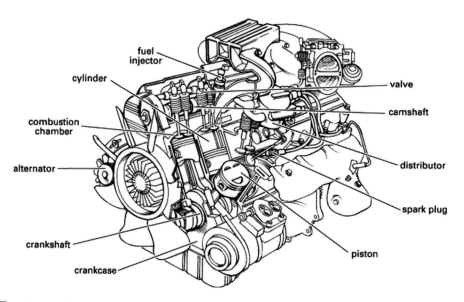

Figure 3.20 A schematic drawing of an automobile engine. Note that (as is true of most current models) this engine has a fuel injection system to feed the fuel into the cylinders. The system monitors the correct fuel mixture electronically. In older models one might find a carburetor, which mechanically mixed fuel with air in the proper proportion for efficient combustion.
Source: By permission. From Merriam-Webster's Learner's® Dictionary © 2017 by Merriam-Webster, Inc. (http://www.learnersdictionary.com).

reusing some of the waste heat for improvement in thermal efficiency are limited for automotive engines. Attempts to improve the basic thermodynamic cycles of these engines have focused on the *compression ratio*, which is the ratio of cylinder volume *before* compression to that *after* compression. In general, as this ratio increases, so does the available energy per cycle and therefore the thermal efficiency also increases. As we can see in Figure 3.21, at high compression ratios, Otto- and Diesel-cycle engines have had efficiencies of 60–70 percent in recent decades.

Although we see thermal efficiencies increasing with compression ratios, they do so with diminishing returns as the ratio approaches 20:1. Most practical engines have operated in the midrange of this curve on either side of 10:1. One of the reasons for an upper limit in the ratio for gasoline engines has been *pre-ignition* (familiarly known as *knock* or *ping*) – or *auto-ignition* as in a diesel engine. What happens in the piston is that higher compression ratios induce higher temperatures, which can create auto-ignition or pre-ignition. Lead additives in gasoline were formerly used to inhibit pre-ignition, but were phased out to reduce air pollution.

There are other ways besides thermal efficiency to measure the performance of automotive engines, ways that better reflect real-life driving conditions. The most important is *fuel economy*. Fuel economy measures, such as *miles per gallon* (mpg), are overall indicators of how a given vehicle performs its designed task.

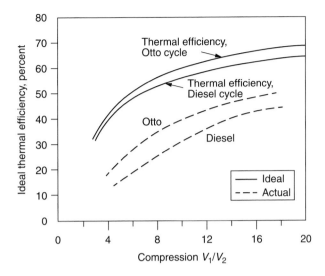

Figure 3.21 Internal combustion engine: thermal efficiency versus compression ratio.
Source: Adapted from Severns and Degler (1948) and Judge (1972).

However, this measure is merely an indicator of the fuel cost to get the vehicle to its destination and is averaged over ranges of speeds, accelerations, and road conditions. More specific measures exist for trucks in terms of *ton-miles per gallon,* which indicate the fuel costs of transporting given weights of freight over given distances. These data again reflect an average of speed, road conditions, and moderate driving habits, and thus have limitations on the extremes of the driving public. Nevertheless, measures such as miles per gallon or ton-miles per gallon tell a driver more about the fuel input requirement than does a measure of thermal efficiency.

Automotive fuel economy, in fact, involves more than maximizing the thermal efficiency of the engine (see Horton and Compton 1984). After engine efficiency, weight is the principal factor for lowering fuel requirements. With every reduction in the overall weight of the vehicle (Fig. 3.22), significant improvements in fuel economy can be achieved. This occurs because a lower inertial weight requires less power to achieve the same performance on acceleration. Weight reductions in automobiles have been achieved (and more are possible) through the use of lighter materials such as cast aluminum, plastics, and low-alloy steels. The average US automobile manufactured in 1985 weighed about 2700 pounds, compared with 3800 in 1975, which according to Figure 3.22 would imply a 12 percent gain in fuel economy for highway driving conditions. But by 2012 the average car was 3977 pounds but average fuel economy, 23.5 mpg, was higher than it had been in 1985 not just 1975, suggesting that improved efficiency was due to other factors besides weight.

Aerodynamic design can also affect fuel economy by reducing the drag of air resistance. Air resistance cannot, of course, ever be eliminated entirely. A practical limit for drag reduction currently appears to be about 25 percent below present

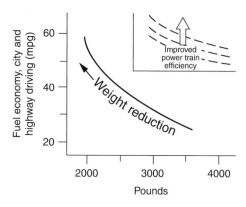

Figure 3.22   Fuel economy improvements.
Source: Adapted from Horton and Compton (1984).

designs. Since air drag is only part of the total *rolling losses* of an automobile, the resulting improvements in fuel economy are only a fraction of the percentage reduction in air drag: a 10 percent reduction in drag, in fact, results in only about a 2 percent improvement in mpg.

Further improvements have been achieved through automotive power-train designs, which include combustion and exhaust innovations as well as transmission and gearing developments (see Horton and Compton 1984). Electronic controls for optimizing engine performance are also yielding measurable gains. As with aero-dynamic improvement, each gain cannot be expected to be dramatic; rather, indi-vidual innovations will probably add only small percentage improvements to gasoline mileage, although, as we will see in Chapter 4, taken together these gains can be sizable.

Whatever the potential for improved technology, it is important to recognize that fuel economy data are based on average driving conditions and that vari-ations around this average will be substantial. Indeed, individual driving habits and personal tastes in choice of models exert an enormous influence on automo-bile fuel economy. It is fair to say that such subjectively determined factors are at least as important as engineering design in questions of national energy policy. (We will consider human behavior in all sectors of energy usage in Chapter 4.)

Gas Turbines

The third and final internal combustion engine is the *combustion* or *gas turbine*, in which the thrust of expanding combustion gases, instead of steam, drives the turbine blades, and thus converts thermal energy to work (Fig. 3.23a). As with other engines, we can follow the paths of the thermodynamic cycle, here called the *Brayton cycle* (see the *p–V* diagram at the corner of Fig. 3.23a). Starting at point a, the air input is compressed to point b. The heat input is supplied in a separate combustion chamber, fed by the compressed air through a nozzle, in a constant pressure process (path b–c). This is followed by expansion (path c–d). The expansion, in this case, is of the hot, high-pressure combustion gases (Fig. 3.23b).

This process is significantly different from the process that drives the steam-cycle turbine. First, in the gas turbine, the hot combustion gases themselves are the working substance; they enter and expand directly through the blades of the turbine, in contrast to the steam turbine where the working substance derives its heat from the combustion gases in a (separate) furnace. Second, the Brayton cycle for the gas turbine is an *open cycle* – unlike the steam-cycle turbine – since the hot gases simply exhaust to the atmosphere after doing their work. Finally, the working substance of the gas turbine does not undergo a phase change as it traverses its thermodynamic paths, as was the case for steam as the working substance in steam turbines.

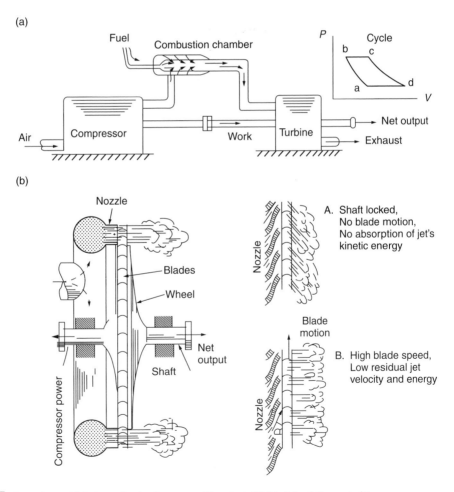

Figure 3.23 Gas turbines. (a) Basic combustion in a gas turbine plant. (b) Principle of the gas turbine.
Source: Adapted from Morse (1947).

Gas turbine cycles have *brake efficiencies* as high as 30 percent. Better utilization of the petroleum derivative or natural gas fuel can be achieved by operating the gas turbine in a *combined cycle* with a steam-turbine system. That is, the hot exhaust gases from the gas-turbine system are used to heat (or preheat) steam for the steam cycle. Thus, the gas-turbine waste heat is utilized in a combined thermodynamic cycle, resulting in a higher thermal efficiency overall for the combined system. Combined-cycle operation yields thermal efficiencies approaching 60 percent and many plants that use this method are in operation today.

Conclusion

At this point, we reemphasize the importance of these *conventional conversion* technologies in present-day energy systems. As we will see in the next chapter, they account for *virtually all* of our transportation and electric generation, and well over half of our consumption of energy resources generally. The technologies discussed in this chapter are the main reasons we need and desire energy resources. They are the principal engines of the modern industrial world and it will take a *technological revolution* or a long-term technological *evolution* to supplant them.

References

Beér, J.M. 2007. "High Efficiency Electric Power Generation: The Environmental Role." *Progress in Energy and Combustion Science*, 33 (2), 107–134.

Davis, C.V. and K.E. Sorenson. 1969. *Handbook of Applied Hydraulics*, 3rd edition. McGraw-Hill, New York.

Elgerd, O.I. 1978. *Basic Electric Power Engineering*. Addison-Wesley, Reading, MA.

Faires, V.M. 1948. *Heat Engines*. MacMillan, New York.

Horton, E.J. and W.D. Compton. 1984. "Technological Trends in Automobiles." *Science*, 225, 587–593.

Judge, A.W. 1972. *Automobile Engines*. Robert Bentley, Cambridge, MA.

Morse, F.T. 1947. *Elements of Applied Energy*. Van Nostrand, New York.

Severns, W.H. and H.E. Degler. 1948. *Steam, Air and Gas Power*. Wiley, New York.

Tan, X. et al. 2012. "Supercritical and Ultrasupercritical Coal-fired Power Generation." *Business and Public Administration Studies*, 7(1).

Tarboux, T.G. 1946. *Electric Power Equipment*. McGraw-Hill, New York

Woodruff, E.B. and L.B. Lammers. 1977. *Steam Plant Operation*. McGraw-Hill, New York.

NOTES

1 Magnetohydrodynamics (MHD) is a means of *direct conversion* of the thermal energy of fossil fuels to electricity that omits the intermediate step of conversion into mechanical

energy. MHD offers the prospect of higher thermal efficiencies for power plants, and it is one of several possible advanced fossil-fuel technologies that can lower air pollution. Although MHD is a novel concept, it is based on classical principles of thermodynamics and electricity generation. The MHD *thermodynamic cycle* is quite similar to that of the combustion turbine. But there are some important practical considerations. High conversion efficiencies require high temperatures at the start of the cycle and this, in turn, requires materials capable of withstanding temperatures of several thousand degrees. These materials have been underdeveloped for long-lived service, but must be improved if continuous operation of more than a few thousand hours is to be achieved. Prototype MHD generators were built in the United States and elsewhere starting in the 1960s. But there is little in the way of development in the world at this writing in 2016.

2 In the United States, the price of diesel fuel at the pump is often higher than that of gasoline. There is an extra 6 cents per gallon federal tax as well as state taxes that can raise diesel prices by as much as 63¢/gallon.

4 The Demand for Energy

Introduction

Modern industrial society uses enormous quantities of energy. During the course of one recent year, 2012, the United States alone consumed 95.06 quadrillion British thermal units (BTU) of energy. The 95,058,000,000,000,000 BTU (termed 95 *quads*) was energy derived from resources such as oil and coal (termed *primary* energy) and electricity, the major form of *secondary* energy. Secondary energy is an alternate form of energy that is derived from the use of primary resources. To put this in perspective, 95 quads is the energy in more than 2.3 Bmt of oil, or about 15.7 BBL, which is 660 billion gallons. The world as a whole consumed about 525 quads in 2012. In the United States, per capita consumption was 270 million BTU, while world per capita consumption was much lower, on average about 76 million BTU; in developing countries of Africa and south Asia, consumption per capita was one-twelfth that of the United States, while the poorest of the poor such as Eritrea consumed about *one–fiftieth* the energy per capita of the United States.

As individuals, we are all consumers of energy – part of a market that is both national and international in scope. Each individual's demand for energy probably differs from that of another individual, and each society's aggregate demand will be somewhat different from the demand of another society. Yet we all – individually and together – demand certain quantities of energy, in the form of fuels (primary) and electricity (secondary) from energy markets. How much we demand and what factors affect our demand decisions are important questions. The answers will tell us not only how much we use, but also whether we can continue to use as much as we do. On these questions hang the continuation of the lifestyles we have grown accustomed to as well as the economic health of both developed and developing nations.

These questions have especially important implications for the more distant future. Obviously, if we use less now, we will have more for the next generations. But here we must ask: What will happen to the economy if we use less? Can we cut consumption now without great sacrifice, without radically altering the way of life people have grown to expect? And if we can cut, by how much and by what means can we do so? Or, in essence, we can reduce the issue to a single controversial question: Do we really *need* what we *consume*?

Energy Demand: Review of the Market Model

This chapter will focus on the demand for *energy*. From an economist's standpoint, the demand for energy is no different in its basic structure from the demand for any other commodity or group of commodities. Indeed, *consumer demand* is a fundamental component of economics per se – a concept inseparable from economic descriptions of a market economy. Demand is different from wants or desires in that it reflects actual consumption intentions. In fact, demand is sometimes described as "buying plans." That is, we may want a Ferrari that costs $200,000 but we may demand a Chevy for one-tenth the price that fits within our budget.

There is not simply an *idea* of demand in economic theory. Demand has the status of economic *law*. The law of demand is a relationship between the quantity of any commodity (any good) demanded by consumers and the commodity's price. Essentially, the law states that the quantity demanded of a commodity will be inversely related to its price, or simply, the higher the price, the fewer units consumers will demand. There is a corresponding law of supply, which asserts that producers will supply an increasing amount of a good as the price rises. Although these laws admit possible exceptions and require some amplification, they have substantial empirical verification and follow common sense. How do you induce people to buy more of a product? Reduce the price: hold a sale.

There is an important qualification to these laws. That is, they apply assuming that *all other things are equal* (that is, remain constant, a usual assumption in economic analysis). Since a market economy is extremely complex and continually changing, the qualification becomes very large, since "*all*" must truly include all. For example, in considering demand, we must include such unquantifiable noneconomic variables as *taste*. Yesterday's hot fad may not be desirable today even at half the price.

One can talk about a market for energy, but individual energy resources and technologies exist in separate markets: there is a market for coal, another for oil, a third for solar panels, and so on. The demand for energy itself as well as for the individual resources and technologies will be affected by various nonprice elements, notably taste, income, and prices of related goods. Goods may be interchangeable or mutually dependent, and thus the price of one may affect the demand for the other. We can readily imagine how the demand for a commodity might be affected by the price of an obvious substitute – oil and natural gas for example. But demand for a commodity may also be affected by prices of not so obvious substitutes; for instance, the demand for energy may be affected by the cost of labor or new capital. That is, it may be economically preferable to replace an energy-intensive industrial process with one that is more labor intensive or that uses more efficient capital goods.

Demand can be thought of as a *schedule*. At any specific time – a demand schedule must be limited to a given time frame – the price of a commodity, gasoline for example, will call forth a demand for a certain quantity from consumers. At $2.50 per gallon, US consumers might demand 2 billion gallons per week. However, at $4, they might demand only 1.5 billion gallons, while at $1.50, they would purchase 2.5 billion. It is important to keep in mind that, in this specified time period, *all of these*

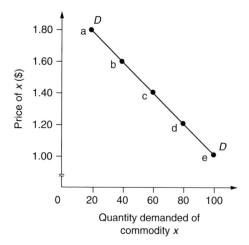

Figure 4.1   A demand curve.

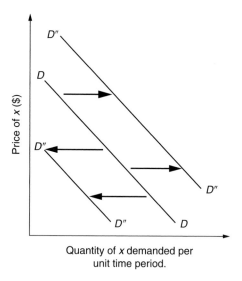

Figure 4.2   A change in demand. The shift from D to D' results from an increase in population. The shift to D'' represents a population decrease.
Source: Adapted from R.L. Miller, 1982, *Intermediate Microeconomics*, McGraw-Hill, New York.

possible price/quantity relationships are true simultaneously. A shift in price means that the *quantity demanded* will change. But the basic structure of demand, as depicted by the schedule, has not changed. A consumer demand schedule is usually depicted by economists as a downward sloping curve (Fig. 4.1). Though the curve shown here is linear, the shape of the curve and the steepness of its slope may vary from market to market or even within the same market from time to time. Still, in energy markets, the demand for individual resources, as well as the aggregate demand for energy, can always be depicted by some such curve.

A change in demand occurs for nonprice reasons such as those noted above, and the phrase "*change in demand*" means strictly that the *curve itself has shifted* (or, put another way, there is a new schedule). Price changes will then have a different impact on quantity. Consider, for instance, an increase in population due to immigration, another nonprice determinant of demand (Fig. 4.2). We can see that the curve shifts to the right. Now, according to this new schedule, at *every price*, more of the good is being demanded.

The shape and slope of any demand curve are important features not only because they depict the schedule of quantities and prices, but also because they suggest the responsiveness of demand to changes in price – what is called the *price elasticity of demand*. So a steep slope would suggest that prices have to change a great deal before there is a significant change in the quantity demanded (see Fig. 4.3a). Indeed, it is possible to imagine a vertical curve (Fig. 4.3b), where a certain amount of a good will be demanded regardless of price. For short periods of time, demand for certain energy resources has seemed very *inelastic*, that is, it exhibited a nearly vertical curve. This occurred with respect to gasoline after prices rose dramatically in the 1970s. In rural areas of the United States where gasoline was needed to run farm machinery, there was little change in the quantity of gasoline demanded despite the price rise. However, it is crucial to keep in mind that a *short-run* curve can have a vastly different shape and slope from a *long-run* curve for the same commodity.

Although the slope depicts the elasticity of a demand curve, the visual impression can be misleading. A *steep* slope is usually a sign of unresponsive, or *inelastic*,

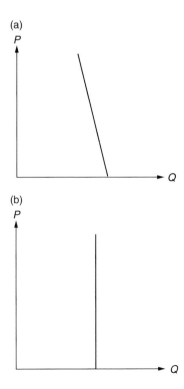

Figure 4.3   (a) A relatively inelastic demand curve (high slope). (b) A totally inelastic demand curve (vertical slope).

demand; a *flat* slope indicates a *highly elastic* demand. But elasticities can change from one part of the curve to the next. A curve could be elastic in one portion and inelastic in another portion.

The *price elasticity* (*E*) of demand can be defined as a simple ratio:

$$E = \frac{\%\text{change in quantity } (Q)}{\%\text{change in price } (P)} \text{ or}$$

$$E = \frac{[dQ/Q]}{[dP/P]},$$

where dQ is an incremental change in quantity and dP is the corresponding incremental change in price. (In general the elasticity is stated as an absolute value $|E|$, since with respect to demand, price and quantity are negatively related and thus elasticities would be negative.)

Take the example of $E = 1$. It means that for a percentage change in price (dP/P) there will be an identical percentage change in the quantity (dQ/Q) demanded. This is called *unit elasticity*. If E, on the other hand, is less than 1, the percentage change in quantity demanded will be smaller than the percentage change in price – or relatively *inelastic*. A vertical curve would have an *elasticity of zero*, for example, where no change in quantity demanded occurs regardless of price (as in Fig. 4.3b). Where E is greater than 1 (very elastic), then changes in quantity demanded will be greater in percentage terms than price changes, meaning that the market is very responsive to price changes.

How much a price change affects the quantity demanded may depend in part on whether consumers can *substitute* one good for another. The demand for energy is probably fairly inelastic even in the long term because the response to a higher price for all energy might be mainly to do without or replace energy use with labor or capital. But for individual resources where substitution is possible there may be a much greater elasticity. For example, if the price of coal rises, industrial users of coal can substitute natural gas or oil. (There are indeed some industrial furnaces that can be adapted easily to different fuels.) A rise in the price of one fuel thereby increases the demand for another fuel and the price/quantity relationship between coal and natural gas is then said to be *positive*. The sensitivity of demand for good x to price changes in good y is called the *cross-elasticity* of demand and is calculated simply as:

$$E_{xy} = \frac{\text{percentage change in } Q_x}{\text{percentage change in } P_y}$$

A corresponding mathematical formula is:

$$E_{xy} = \frac{dQ_x/Q_x}{dP_y/P_y}$$
$$= \frac{dQ_x}{dP_y} \frac{P_y}{Q_x}$$

where Q_x is the quantity of one good (x), and P_y is the price of another good (y). Therefore, if the two goods are coal and natural gas, then the greater the number of switchable industrial furnaces, the closer will E_{xy} be to unity.

It should be noted that the cross-elasticities of related goods are not always positive. In some cases, when the price of a good rises, demand for a related good falls. For instance, if the price of gasoline rises, demand for large cars will likely fall. Where the relationship is negative, the goods are called *complementary.*

Income is also an important nonprice factor that can affect demand – on both an individual level and a national one. A person may spend a certain amount of money on energy to meet basic necessities whether his or her income rises, falls, or remains constant. But if an individual's income rises, he or she will probably spend more, especially on discretionary purchases, some of which may result in greater energy consumption. For instance, the same individual might elect to spend more on energy-consuming electric appliances, a second car, or more and longer automobile trips.

The responsiveness of demand for a good to changes in income (I) – *income elasticity* of demand – can be calculated in much the same way as cross-elasticity, by the ratio:

$$E_I = \frac{\text{percentage change in } Q}{\text{percentage change in } I}$$

Here, as with all of the other measures, it is assumed that only income changes and all other things including price remain unchanged.

Such concepts of economic demand can be refined and embellished further, but the present discussion provides a sufficient framework for understanding *energy demand* as well as demand for specific resources and technologies. However, before we deal with the specifics of the energy-demand picture, we might also consider the structure of the markets in which demand for energy is observed. Our model thus far has assumed a market that is *perfectly competitive.* That means a market in which buyers and sellers are numerous and no one or group of them sets the price. The price is determined by market interactions only, and the products offered are homogeneous. Information is symmetric – everyone knows all relevant facts – and there are no costs of doing business (*transaction* costs). The model is an idealization that does not exist in the real world, but some markets, especially for homogeneous commodities such as oil and natural gas, may approximate the model in some respects – though clearly not all. With oil historically, for example, at times prices have been set in the market but at other times by governments acting individually or in concert. At still other times, the market has been controlled by collusive agreements among major producers. When individuals or groups can exercise price-setting power, the markets are said to be imperfect, defective, or to have "failed" (Grossman 2013, Chapter 2).

Market failures occur for other reasons: costs are imposed by market participants (buyers and sellers) on people outside the markets. These are spillover or *externality* problems; pollution is one example. Asymmetric information (a firm knows something that buyers do not) is another cause of market failure. Product differentiation also causes market failures. Some forms of market failure are more problematic than others; collusive agreements, for example, lead markets to inefficient allocations and unfair income distribution, while product differentiation may just be a fact or the result of clever marketing.

Often when market failures become disruptive or unfair, critics will call for government intervention in, and regulation of, the market, if not outright government ownership. Intervention in the oil market appeared to make sense in the late 1970s, when the market seemed unable to respond to rising prices; the price of oil in particular soared and demand hardly slackened. But actually, US government policies were a major reason why markets could not adjust; the prices of oil and natural gas were being set by the government (Grossman 2013). As Nobel Prize-winning economist Ronald Coase (1959) observed, one should be cautious about government imposing "special regulations" with respect to market imperfections lest the remedy prove worse than the problem it was meant to fix.

Ultimately, past experience has not and cannot answer definitely the question: does a free market in energy resources lead to the most effective allocation (i.e. does it work optimally)? Does it work equitably? Is it good policy to leave the future determination of energy supply and demand mainly to the market?

Energy Use and Economic Growth

Before we consider broad questions and policies with respect to energy markets, we need to understand the relationship between *energy consumption* (demand) and *economic growth* – for it is the pursuit of *growth* that has often justified policies about resources. There is no doubt that energy consumption is a necessary component of industrial society, and we have already suggested that there probably is a positive relationship between income growth and growth in energy consumption. In general, let us continue to assert that as income rises, energy consumption rises, when *all else remains constant*. But even with this assumption, we should consider that discretionary spending by definition may encompass a variety of options that either are energy intensive or are not.

Consider especially how an industrial or commercial firm might react to increased income. We will assume that a business tries to maximize its profitability and will choose to invest some new income toward an increase in future output. This choice conceivably could involve lessening, not increasing, the burden of energy costs. As noted above, where the choice is specifically to increase output, the producer may invest in low-energy-intensity labor or in more efficient capital. For instance, one may choose either to hire new workers or to buy a lower-energy-use capital good, such as a computer, in order to speed up the production process. But what if there are few such substitution opportunities available? What if the most profitable investment choice leads to increasing energy consumption? Then firms will likely consume more energy as they expand production. And if this is generally the case throughout the economy, then, as income grows, energy consumption will inevitably grow as well. Indeed, if the opportunity to substitute labor or capital for energy is limited, then energy *must* grow as the economy expands. Presumably, in this view, significant output growth cannot be achieved any other way.

Until the 1980s, many economists and energy analysts were convinced that increased energy consumption was a *necessary* outcome of economic growth. The consequence in the economy of using less was inevitably diminished economic activity – recession – and ultimately a lower standard of living. Superficially, history seemed to support this

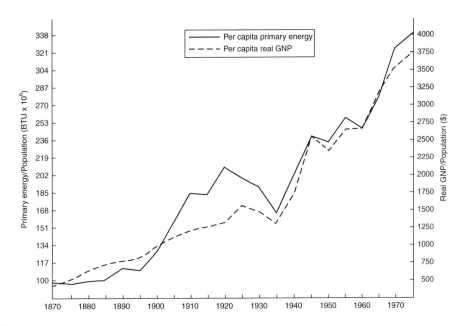

Figure 4.4   Per capita GNP and energy consumption in the United States.
Source: Schurr et al. (1979).
Note: In some of the tables and figures in this chapter, economic activity is measured in GNP (gross national product), instead of GDP. The distinction between the two is small, with GDP equal to GNP minus the net inflow of national income earned abroad

contention. In the United States, from the mid-1930s on, energy use in general rose not only in absolute terms, but on a per capita basis as well. At the same time, economic growth, as measured by the total output of goods and services (gross domestic product, or GDP), grew as well on a per capita and on an absolute basis. But remarkably (as Fig. 4.4 shows), GDP appeared to grow at a fixed ratio to energy consumption. Even during the few dips in growth both seemed to follow in step.

Actually, Figure 4.4 shows that prior to the 1930s there was little correlation between the growth in energy consumption (per capita) and the growth of per capita income. But beginning around 1935, just before World War II, per capita energy consumption and per capita income started to move in tandem. This lock-step continued for almost 40 years, until about 1975. Indeed, looking at the plot it is hard to escape the conclusion that a strong correlation existed between GDP growth and the growth in energy use.

But the leveling of per capita energy consumption after 1975 came about from the rapidly rising price of energy aided by public programs to improve energy efficiency and promote conservation, which were started during this period. These policies, along with market conditions, refuted the previous pessimistic outlook that implied that US society had only two choices: either use more energy or live less well.

The old scenario of increasing energy consumption as a condition of economic growth ignored the possibility of technical advances that raised output from less energy input, and also it denied the standard economic assumptions about the price mechanism of energy markets. That is, did the price mechanism actually function or was energy consumption a prisoner of income alone? Some free market economists, who might have been expected to have more faith in the price mechanism and to have envisioned more opportunities for technical advance and substitution, supported the bleak *either-use-more-or-live-less-well* scenario. In a 1972 study by the Chase Manhattan Bank, for example, forecasters could think of very few options – one was less television watching – for energy savings along with GDP growth.

However, such a view did seem plausible in light of the pattern shown in Figure 4.4. The only evidence that ran counter to it was comparative data between US consumption and that of most other industrialized countries. Though per capita GDP of Western European countries was comparable to per capita GDP of the United States, per capita energy consumption in Europe was typically lower, often by one-third or more. Yet such comparative data, although suggestive, have limitations; no two countries are exactly alike. The United States, which is larger and had a greater amount of heavy industry during that period, could not be compared exactly to, say, France or the Netherlands.

But the conclusion that locked the GDP/energy relationship together seemed to ignore what was actually taking place in the market. *Real* prices (that is, adjusted for inflation) of energy, on the whole, actually declined through much of the period from 1945 to 1972. Oil in particular exhibited real-price declines and was 27 percent cheaper in 1970 than it had been in 1950. Large hydroelectric projects lowered the real cost of electric power as well, and also created vast power surpluses in parts of the United States. Consequently, utilities offered incentives to users to consume as much electricity as possible: a pricing system that lowered the cost per kilowatt hour (kWh, see Appendix A) as customer use increased. What this suggests is that energy consumption actually was following the path classical economics would have predicted *–based on price not income.* Indeed, it would imply that the lockstep growth path of GDP and energy was a mirage of sorts, induced by *lower* prices for energy; in fact, if *real* energy prices had simply stayed constant, it is likely that demand would not have grown as rapidly and a gap between GDP and energy growth would have developed. As matters stood, falling real prices encouraged consumption and discouraged technical innovation for greater efficiency; there was no particular incentive to reduce energy intensiveness of production when the relative price of that factor of production was declining. On the contrary, much growth was fueled by energy-intensive, labor-saving technological change.

The fall in energy prices went largely unnoticed. And when energy prices did begin to rise in real terms in the early 1970s – showing up in both oil price jumps and rising electric rates – energy markets seemed to be income, not price, driven. The quantity of energy demanded fell, to be sure; per capita energy consumption in the United States, as measured in BTUs, plunged from 351 million BTU/year in 1973 (the year in which the first oil shock took place) to 340 million in 1974 and 327 million in 1975, and total energy consumption fell from 74.3 to 70.5 quadrillion BTU. But GDP fell too, and the US economy experienced a sharp recession. This suggested that GDP growth and energy use were not simply correlated, they were causally linked.

By 1976 GDP growth resumed and so did total energy consumption. This only strengthened the belief that a causal connection existed between energy use and economic growth. With energy usage growing again, fears of future shortages and depletion of scarce resources prompted pessimistic conjectures that Americans needed to accept voluntarily a decline in their standard of living to protect and preserve the world's resources.[1] When prices soared even higher in the late 1970s than they had earlier in the decade, consumers did not cut demand at first. Instead, consumption *rose*, GDP *rose*, and, because energy was a factor in virtually all modern production, the general price level *rose*. In other words, the energy market displayed a nearly *inelastic demand* response (Kaplan 1983) and the energy price shocks were absorbed through higher output prices. By 1979, per capita energy consumption had risen to 359 million BTU and total national demand rose 12 percent above 1975 levels to 80.9 quads.

By 1980, pessimism about future energy consumption was at its greatest. As Table 4.1 shows, studies in the mid and late 1970s, by experts from a variety of public and private organizations, forecast rising demand. Most forecasters projected energy demand in the United States to reach between 80 and 100 quads by 1980 (which it did not quite reach) and as high as 116 quads by 1985. For the year 2000, absent a scaling back of economic growth, nearly all forecasts had total consumption over 100 quads, one suggesting a massive 196 quads – the latter suggesting that annual energy demand would nearly triple during the last quarter of the century. Economic recessions in the United States in 1980 and 1982, however, made the higher end of these forecasts unlikely, well before the final tally was ever recorded. Energy consumption fell, as people had come to expect with a falling GDP.

Then, for the next 5 years, total annual consumption fell below 80 quads, and it soon became clear that most of the estimates were off, not just because of recession but because they were fundamentally flawed. Forecasters had apparently taken as given the inelasticity of the demand curve for energy, it seems, based on a rationale rooted more in psychology than economics. According to this view, the key to reducing consumption was a psychological change in our outlooks and lifestyles toward a willingness to make do with less. But it was assumed that people would not change their behavior merely because of the *price* of energy. They would not drive smaller cars, would not turn down the heat in winter or turn off the air conditioner in summer, or alter their lifestyles in any other fashion. Indeed, there had been few signs in the late 1970s that Americans were willing to make the changes that they subsequently did.

But energy prices were having an impact on life that became more apparent as time passed. In fact, there was some evidence of the impact of the price mechanism from the mid-1970s and beyond. Although overall use and per capita use were increasing, there was a sign that Americans were using energy more efficiently as well. From 1975 to 1979, though overall consumption was rising (by 12 percent), energy consumption per dollar of output had fallen by 6 percent to under 14,000 BTU/per $GDP. That is, we were producing much more but using proportionately *less* energy.

Actually, though this drop suggested greater energy efficiency, it is understandable that forecasters did not necessarily find it a proof of a trend. In fact, greater improvement in the $/GDP ratio could have signaled – and in part *did* signal – a shift from a

Table 4.1 Forecasts of US gross energy consumption: 1980, 1985, 2000, and 2010 (in quadrillion BTU)				
Study	1980	1985	2000	2010
Landsberg et al. (1963)	79.19	89.23	135.16	–
US Department of the Interior (1972)	96.00	116.60	19190	–
Ford Foundation Energy Policy Project (1974)	–	103.50	163.40	–
Historical growth	100.00	115.00	187.00	–
Technical fix	88.00	92.00	123.00	–
Zero energy growth	85.13	88.00	100.00	–
Project Independence Blueprint (1974)				
Base case	86.30	102.92	147.00	–
Conversion	82.20	94.16	120.00	–
US National Energy Outlook (1976)				
Low growth	80.23	90.72	–	–
High growth	85.40	105.64	–	–
United Nations (1976)				
Reference case	86.50	102.90		
Edison Electric Institute (1976)				
Moderate growth	–	–	161.00	–
Institute for Energy Analysis (1976)				
Low growth	–	–	101.40	–
High growth	–	–	125.90	–
National Energy Plan (1977)	–	97.00		
Stanford Research Institute (1977)				
Base case	–	–	143.20	–
Low growth	–	–	109.40	–
Electric Power Research Institute (1977)				
High case	–	–	196.00	–
Conservation case	–	–	146.00	–
Workshop on Alternative Energy Strategies (1977)				
Low growth	–	–	115.10	–
High growth	–	–	132.00	–
Brookhaven National Laboratory/Dale Jorgenson Associates (1978)				
Base case	–	–	138.50	–
National Research Council, National Academy of Sciences (1979)				
Low growth	–	69.25	66.73	65.10
High growth	–	94.29	144.28	191.60
Resources for the Future (1979)				
Low growth	–	80.52	95.00	–
High growth	–	89.50	145.00	–
The Global 2000 Report to the President (1980)	–			
Low growth	–	90.00	129.70	–
High growth	–	102.00	141.30	–
Ross and Williams (1981)	–			
Low growth	–	73.04	67.83	64.00
Business and usual	–	80.09	93.00	102.00
Actual consumption	76.20	74.00	99.00	97.5

From 1980 total energy consumption in the United States increased by about 30 percent.
Source: Adapted from LeBel (1982).

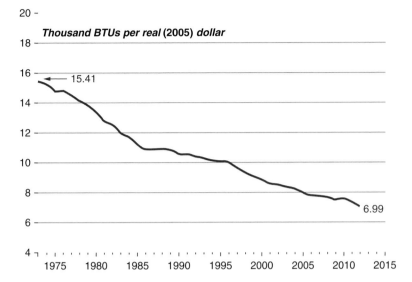

Figure 4.5 US energy consumption per real dollar of GDP, 1973–2012.
Source: USEIA 2014.

more energy-intensive manufacturing economy to one that was more service oriented. But in reality, there was a new trend toward energy efficiency in the US economy that became increasingly evident (a clear sign was the increasing fuel efficiency of new automobiles) through the first half of the 1980s. As Figure 4.5 illustrates, energy consumption per dollar of GDP decreased 21 percent from 1975 to 1985, and fell almost every year in that period and continued to fall into the 1990s and 2000s. Essentially, every dollar of GDP today requires half the energy input that was needed in 1973. This trend has strongly suggested that, given the price incentives, appreciable improvement in efficiency was both desirable and probable. It also demonstrated that no technical limitations had been at work; that is, the gains were due to people's economic behavior within the range of what was technically feasible.

Moreover as Figure 4.6 shows, consumption per capita had largely flattened out, and then had actually *fallen* from 1990 to 2011 by about 7 percent. In the meantime, per capita GDP roared ahead by 37 percent. The supposed link between per capita income and energy consumption was broken.

Why did the forecasters err? One of the main reasons was an underestimation of the impact of prices on the energy market. A strong case can be made (and has been especially by economist William Hogan [Sawhill & Cotton 1986]) that energy markets *do* respond in a fashion that is consistent with a classical market model. This has not always been apparent, because insufficient weight has been given to the importance of time for market adjustment. The crucial point is that, in an energy market, adjustments in demand take place over the long term. They may lag so far behind price changes, in fact, that at first the market appears unresponsive to price shifts, contrary to the expectations of the economic model.

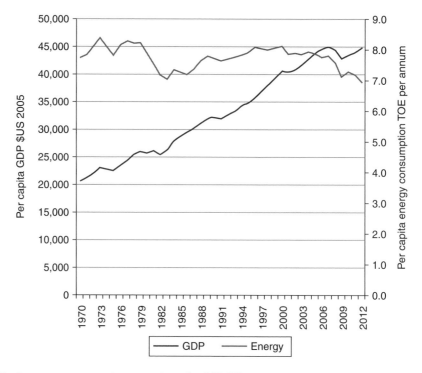

Figure 4.6 US primary energy consumption, per capita and real US GDP, per capita, 1970–2012.
Source: Energy Matters, euanmearns.com, based on BP and UN data.

This argument, in addition to fitting recent history, is logical. Energy inputs, as we have noted, are essential for much of life in an industrial society. But energy is consumed primarily by our society's machines – relatively expensive durable goods for the most part (appliances, factory equipment, automobiles, and so on). To gain greater efficiency of energy use in such an economy, machines have to be replaced. But even when energy prices rise, is there enough incentive to replace them immediately? In the short term, the answer was (and is) *no*.

Let us consider the effect of a 10 percent real increase in energy costs. In most cases, there are only a few short-term measures that can be adopted – short of doing without – that also make economic sense. There may be a few no-cost *housekeeping* measures available to individuals and businesses; for example, individuals can make sure they turn off the lights when they leave the room or, without hardship, can replace 43-watt incandescent light bulbs with 15-watt fluorescent bulbs. But the biggest gains in efficiency are *investment based*, requiring an outlay of money for new equipment or the refitting of the old. The financing cost of new equipment is often greater than the value of potential energy savings. In industry, for example, a study in the 1980s noted that energy accounted for an average of 4 percent of the factor costs of production. Thus, a 10 percent increase in energy costs would mean an increase of only 0.4 percent per unit of output. A new factory or even a new machine, on the other hand, can cost millions of dollars, with the cost of borrowing being far

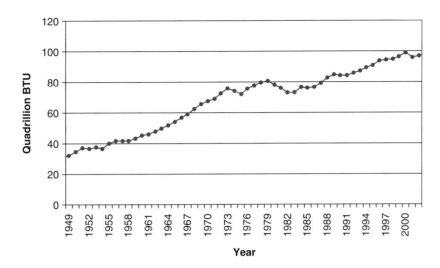

Figure 4.7 Total US energy consumption.
Source: http://wilcoxen.maxwell.insightworks.com/pages/804.html, based on data from the USEIA.

more burdensome to a business than the increase in the cost of energy. For the steel industry, for example, one economist contended that the cost of a new steel mill was "staggering" (quoted by Peck & Begg in Sawhill & Cotton [1986]). It is far easier and less expensive for the factory to raise prices a small amount in the short run to cover the increased cost. Hence, under such circumstances, prices rise while energy demand remains constant – leaving the impression of an inelastic demand curve. Individuals, to give a simple example, have little incentive to spend several hundred dollars per month to finance a new car in order to save a few dollars a month at the gasoline pump.

Still, the longer term presents a different picture. If energy prices are rising – and are expected to rise even further – then there is an incentive for business and individual consumers to buy more fuel-efficient durable goods when they are replacing them anyway – as they will over time. They, in fact, may have no choice but to buy greater fuel efficiency since it is competitively sensible for the durable goods supplier to build it into new models. And the higher energy prices go, the greater the demand for efficiency in new durables is likely to be. As a result of this pattern, over time, energy demand per dollar of GDP falls. Indeed, this lagged effect can be expected to continue even if energy prices stabilize or decline.

Total energy consumption increased steeply with minor variations into the 1970s, rose more slowly in the 1980s, and then rose still further but only incrementally from the late 1990s into the 2000s (see Fig. 4.7). In the 2010s, total energy consumption has flattened out and is far below the predictions of the 1970s. Whether it will continue to hold at this level or will resume an upward track is uncertain, especially since the prices of fuel – notably natural gas and gasoline – have fallen and are very low as of this writing. But what the rest of this chapter will argue is that further reductions in consumption – per capita and total – are entirely possible.

Efficiency and Conservation Potential

An Overview

Conservation implies extending the usefulness of resources through reduced energy consumption, and implies a reduction in energy services – less heat, less light. Strictly, it is different from *energy efficiency*, which means getting the most output from an energy input (Herring 2006). Thus far we have discussed the issue of reducing energy demand from the standpoint of increasing *efficiency*. Greater efficiency is often regarded as one means of achieving energy *conservation*. It is probably the most desirable form because it offers the potential for lower energy demand along with a continuation of expected energy services, including those that are needed for a growing economy.

Conservation also can mean the *substitution* of cheaper resources for expensive ones – a reduction in factor costs, not in total BTUs per unit of output. More importantly, we can conserve by substituting for scarce finite resources with more abundant or renewable ones. The switch from natural gas for water heating to solar heat, for example, conserves finite supplies of the natural gas while providing the same energy services.

Of course we can also do without, conserve by simply using less energy. As individuals and as a society, we might begin to shift from energy intensive activities – both work and leisure – to more labor intensive ones, for instance, orienting the economy more to specialized craft production (harking back to the society we evolved from) and away from mechanized mass production. Because we see ourselves as a free society, such change would require willing alterations in behavior and attitudes. This may only occur with education and/or because of a period of energy privation where people come to expect less and have less access to abundant sources of inexpensive energy along with a reduction in energy services. However, such possibilities remain speculative. In the realm of public policy, it is much easier and more politically acceptable for a government such as that of the United States to focus on efficiency rather than on mandating changes in attitudes and lifestyles of the public.

In view of this, most of our discussion here will focus on the potential of conservation through efficiency improvements representing *reductions* in aggregate (as well as per capita) energy consumption. These analyses generally assume the continuation of current population growth trends and attempt to factor in the impact of a growing economy. We should keep in mind that population growth will shift the demand curve to a higher aggregate level of consumption, even if per capita demand remains constant. But, as the discussion has already suggested, growth with conservation and efficiency can almost certainly continue without any significant technological breakthroughs. A study by the United Nations Foundation argued that the G8 countries (Canada, France, Germany, Italy, Japan, Russia, the United Kingdom, and the United States) could achieve an energy efficiency improvement of 2.5 percent per year from 2012 to 2030.[2] That would represent a reduction in energy consumption of over 50 percent, or, put differently, a conservation gain of that magnitude.

Table 4.2 US energy consumption by sector (quadrillion BTU)				
Year	Residential/commercial	Industrial	Transportation	Total
1970	22.1	29.6	16.1	67.8
1975	24.3	29.5	18.2	72.0
1980	26.3	32.1	19.7	78.1
1985	27.5	28.9	20.1	76.5
1990	30.5	31.8	22.4	84.6
1995	33.3	34.1	23.9	91.2
2000	37.6	34.8	26.6	99.0
2002	38.2	32.8	26.8	97.8
2004	38.8	33.6	27.9	100.3
2006	38.5	32.5	28.8	99.8
2008	40.0	31.4	28.0	99.4
2010	39.8	30.5	27.1	97.4
2012	37.1	31.0	26.2	94.5
2014	40.0	31.3	27.1	98.5

Source: USEIA 2015

In order to evaluate the prospects for conservation, it is useful to examine energy consumption in the four sectors of the economy – *commercial, residential, transportation*, and *industrial* – and examine where savings may be achieved. In the last half of the twentieth century, total energy consumption rose for all the four sectors; altogether an increase of more than double from about 42 to 99 quads in that 50-year period.

Table 4.2 shows that in just the last 30 years of the century consumption increased by 37.5 percent, but over the first 15 years of the new century, as noted, the overall level of consumption flattened, so that in total 2014 consumption was slightly lower than it had been in 2000.

There are various reasons for the changes one observes in the table. For example, the fall in overall consumption between 2008 and 2010 was to a great extent due to the major economic recession that struck in 2008 and lasted through 2009. The decline in consumption from 1980 to 1985, on the other hand, was largely the result of the rapid increases in energy prices in the late 1970s that led to greater awareness of the financial impact of energy consumption and therefore demand for more energy efficiency. Note that industrial usage has been more or less flat from 2000 to the present; this represents at least in part the sectoral shift from heavy industry to service and knowledge industries. Some of the changes were also probably the result of federal programs for energy conservation, such as the Corporate Average Fuel Economy (CAFE) standards, which have contributed to greater fuel efficiency in automobiles.

But we are looking here not so much at the past as the future. In what sectors is there the potential for further investment in energy efficiency and conservation that could lower demand by the 50 percent said to be possible? We especially want to know what can be done with *present technology* for each of the energy sectors.

The Residential and Commercial Sectors

The *residential* sector covers both single- and multi-family homes (over 100 million units in the United States). The *commercial* sector of the economy includes offices, hotels, warehouses, and retail stores in addition to service operations such as hospitals. The energy needs of the residential and commercial sectors are similar and their demands for energy parallel each other. About 60 percent of total energy demand in the combined residential/commercial sector is for space heating. Together, they also consume a significant amount of energy for lighting, water heating, air conditioning, cooking, and refrigeration (the latter two in institutional and business food services as well as in homes). Lighting, refrigeration, and most small appliances operate largely on electricity produced at centrally located generating stations. Space and water heating may be electric, but are predominately supplied by direct inputs to homes of fuels, typically oil, natural gas, butane, or propane. The combined residential/commercial sectors, taken together, used about 40 quads annually by 2014. More than half the total was residential consumption, but in 1975 residential use was 50 percent higher than commercial applications. By 2014, the difference had fallen to 15 percent.

In the wake of the energy price shocks as well as the natural gas and oil shortages of the 1970s, there was considerable interest among consumers for energy efficiency. Initially, this took the form of zero-investment options – such as shutting off lights – or mild forms of doing without – such as setting thermostats at 68 °F instead of 72 °F in winter. The turmoil in the energy markets of the 1970s also provided impetus for the wide dissemination of information to the residential/commercial sector about energy investment alternatives. Some this information was provided by the government, especially with the creation in 1977 of the US Department of Energy (DOE). Around this time also, various government programs were enacted to incentivize conservation in both the residential and the commercial sectors. President Jimmy Carter believed that efficiency would have the highest social return on investments and pushed for conservation incentives in his first major energy bill, the National Energy Act (1978). In 1987, Carter's successor Ronald Reagan signed the National Appliance Energy Conservation Act that established minimum standards for energy efficiency in household appliances.

Incentives and information focused in particular on efforts to improve insulation in houses and buildings so that space heating and air conditioning could operate with fewer losses. Consumers had a range of options, from inexpensive duct tape and caulking of existing windows to new triple-glazed windows. In some instances, people replaced heating systems to take advantage of improved efficiency of more up to date equipment. New oil and gas furnaces of that period required one-third less energy to produce the same amount of heat as the older models. Some businesses invested in more exotic equipment, such as computerized control and monitoring devices that could regulate heat and light throughout an office complex.

Heating efficiency gains were noteworthy in new housing stock. Some new homes constructed in the late 1970s and 1980s were on average 50 percent more efficient per square foot than the then-existing housing stock. The homes designed specifically for efficiency during this period turned out to be 90 percent better than the existing

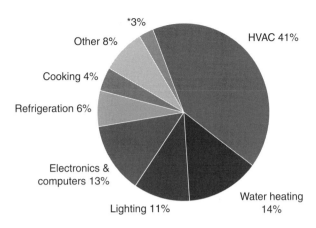

Figure 4.8   Residential buildings total energy end-use.
Source: USEIA.

average. At the same time, while these technical gains were impressive in themselves, they did not produce a dramatic fall in energy consumption by the sector, nationwide, because the turnover in the building stock was very slow and there were too many more older homes than new ones at that time to produce a dramatic difference. Even though houses and appliances had become more efficient, residential total consumption increased through the 1990s because the housing stock was increasing. But from 2000 to 2014, residential consumption was essentially flat.

Gains in the 1980s, according to some analysts, represented *easy* conservation (presumably zero or low-cost investment) (Hirsch 1987). Yet "easy" is a qualitative judgment that is not universally perceived. The most important gains were in space heating, particularly for home heating (which, as noted, made up more than 50 percent of the home energy consumption at that time). To date, much of the savings in this area have come from better insulation. Experiments have shown that experts were able to produce an average energy saving of 30 percent per unit through shell improvements alone in older homes. Those gains averaged more than 50 percent when replacement of old furnaces was included. In 1993, residential heating and cooling (HVAC) still consumed almost 58 percent of all residential energy use. By 2009, that number had dropped to 41 percent (Fig. 4.8).

The highest growth in energy consumption has been in home appliances and electronics. Residential appliances, lighting, and home electronics, which accounted for less than a quarter of residential demand, now consume more than one-third. Unlike heating, which is largely provided by primary energy sources – especially natural gas – appliances use secondary energy, namely electricity. Many home appliances are far more efficient than earlier models, but there are still many opportunities for further efficiency gains, especially in the areas of lighting and refrigeration, which consume 11 and 6 percent of residential end use, or about 2 quads for lighting and 1 quad for refrigeration.

Lighting has considerable potential for conservation. It should be noted that, in addition to residential lighting, commercial lighting consumes about 2.8 quads.

Lighting consumption has fallen because of the switch from high wattage incandescent bulbs to low-wattage fluorescents with the same amount of illumination. It was thought that consumers would be willing to pay the higher price for the latter because of the energy savings. But this transition, though called for in the energy bill of 1992, took time for two reasons: consumers resisted fluorescents because the higher initial cost proved more of a barrier than expected and also fluorescent light was considered aesthetically displeasing. Over time the price has come down, and manufacturers learned to approximate the "soft" coloration of incandescent bulbs. The mandated phase-out of high wattage incandescent light of course hastened the switch, but the use of the low-wattage bulbs is now more common.

Still, there is room for considerably more reduction in energy demand for lighting. More savings are likely when light emitting diode (LED) bulbs achieve a larger market share. As with fluorescents, the price has been a barrier to larger market penetration, but LEDs use about half as much electricity as fluorescents and are much friendlier to the environment. Fluorescents contain mercury, which can pose significant environmental dangers. LEDs have no toxic elements.

Even greater savings can be achieved with occupancy sensors that turn on lights when people enter a room and turn them off when they leave. These devices could lower consumption by an estimated 45 percent, according to the US Energy Information Administration (EIA), although some studies suggest that the amount of savings could be closer to 70 percent.[3] Because the commercial area has expanded greatly over the past few decades due to changes in the economy, a 45 percent improvement would not mean a reduction in lighting demand by that amount. But even a 30 percent reduction in overall demand through improved lighting technology would reduce energy consumption by about 1.5 quads. This means not only are consumers able to reduce their electricity bills but that society gains by foregoing the need to supply so much electricity. That cost foregone, or *opportunity cost*, would be equal to the cost of about 30 large electric power plants.

Refrigeration, which nationwide consumes about a quad of electrical energy annually (0.45 quads in the residential sector and 0.67 quads in the commercial sector), is an area where more savings could be obtained. Much has already been gained. Around 1980, the average home refrigerator consumed about 1400 kilowatt hours (kWh) of electricity per year. By 2014, the average annual electricity consumption had fallen by almost two-thirds and more efficient models consumed as little as 350 kWh/year, a reduction of more than three-quarters. In the meantime, refrigerators had a larger capacity (about 10 percent greater) and cost about 60 percent less in real terms. Much of the gain resulted from several government programs. One in the 1970s set minimum efficiency standards – standards that were amended by subsequent legislation in 1987, 1990, 1993, and 2005. The most notable program was "Energy Star", established by the US Environmental Protection Agency in 1992, essentially as a voluntary program that identified (with the "Energy Star" sign) efficient products. A product only qualified for the program by demonstrating greater efficiency than the minimum standards – in the case of refrigerators by at least 20 percent. Energy Star designation provides information to consumers not only that the appliance is efficient, but it also gives the expected

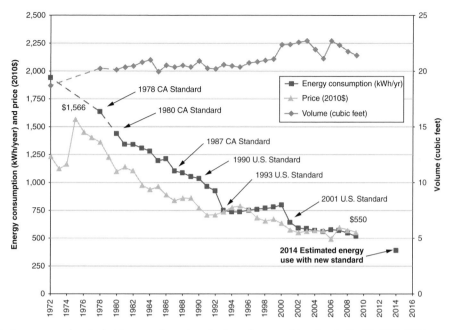

Figure 4.9 Average annual cost, electric power demand, and volume for new refrigerators in the US, 1972–2010. California (CA) was a leader in refrigerator standards in the 1970s and 1980s.
Source: Appliance Standards Awareness Project; data from Association of Home Appliance Manufacturers and the US Census Bureau.

electricity demand and probable annual electricity bill. As a result, consumers can choose knowing the longer-term cost of ownership. Energy Star, which provides specifications for many home products – e.g. heating and cooling systems, home electronics and personal computers – is credited as a major reason for the decline of energy use per household of 14 percent.

Refrigeration technology can save energy in many ways. Constant temperature is maintained through an electrically powered compressor, and the amount of electricity needed will depend on the design of the compressor and its placement, as well as on the insulation of the refrigeration unit itself. (In other words, if it is well insulated, the compressor need not run so often.) According to a study in 2008, energy savings of an additional 30 percent are possible. Other home and office appliances and machines may provide less of an opportunity for conservation gain, but in most instances there is enough room for improved energy efficiency to make the overall gains envisioned by conservationists possible.

The Transportation Sector

The transportation sector includes all means of transport for people and freight – rail, air, truck, bus, and automobile. Altogether, this sector accounted for about 27 quads or 28 percent of total demand in 2014 (Table 4.2). Energy demand in this sector had been trending upward from the 1970s through the mid-2000s but has leveled off over the past 10 years.

Most transportation in the United States uses a form of internal combustion engine, generally Otto-cycle or Diesel-cycle engines; jet aircraft use a form of combustion turbine (see Chapter 3). All three are internal power plants that use the energy of refined petroleum, and 93 percent of all transportation fuel (as of 2010) was petroleum based. Otto-cycle and Diesel-cycle vehicles, primarily trucks and private automobiles, use around three-quarters of all transportation energy. And the largest part of that category, indeed the most significant factor in the transportation sector, is the private automobile, which consumes about 10 quads of energy annually in the United States, almost half the sector total.

Conservationists have focused on the automobile, partly because of its prominence in our energy consumption patterns. But our orientation toward and dependence on internal combustion travel also carries a special danger that has worried energy analysts and government policymakers. There is, at present, limited opportunity for cost-effective substitution of fuel resources for transportation. This problem affects the transportation sector more than any other. In the electric sector, for instance, the industry will be more inclined to build more plants that burn abundant natural gas resources or use renewable sources such as hydropower if more electrical energy is demanded. In the industrial sector, manufacturers can generate process heat with gas or coal or oil or electricity, and solar-industrial heat is even a possibility.

Fewer substitution options are available with respect to transportation. Automobile engines have been designed to run on ethanol or other gasoline alternatives (e.g. *synfuels* see Appendix C). In 1980, the US Congress passed legislation to encourage the substitution of alternative fuels, mostly coal-derived synthetic oil, but also methanol and ethanol fuels. But synfuels were found to be significantly more expensive than petroleum-based technology and created various environmental problems. Nevertheless, in subsequent legislation in 1992, 2005, and most importantly 2007, as noted in Chapter 2, the Energy Independence and Security Act (EISA) increased incentives and finally (in 2007) mandated that by 2022 approximately 25 percent of automobile fuel would come from ethanol. As of 2015, the program has stalled because, as we will discuss in Chapter 10 and Appendix C, biofuel technology has not advanced to the point where large-scale replacement of petroleum fuel for biofuels is possible. More research and development is clearly required.

A second substitution possibility is battery power. All-electric vehicles have become more prominent since the beginning of the administration of President Barack Obama, which provided large incentives for consumers to purchase them. But in Chapter 10 we will explain that there remain many barriers to the widespread adoption of electric vehicles. Moreover, the programs to encourage electric cars have proved a benefit to the wealthy because most of the incentives have gone to the top 10 percent of income earners (Borenstein & Davis 2016). As of this writing, electric vehicles represented about 0.15 percent of all private automobiles in the United States (and this includes so-called plug-in hybrid vehicles that can run for a time solely on battery power but have a gasoline engine as well).

The oil supply and price shocks of the 1970s led the US government in 1975 to mandate an average fuel efficiency for new (conventionally powered) cars.[4] According to the government plan, by 1986 car makers were required to have an

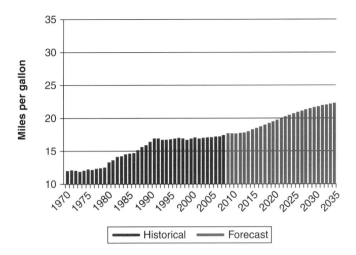

Figure 4.10 Historical and forecast values (1970–2035): Motor vehicle average mpg.
Source: Institute for 21st Century Energy, 2011.

average new fleet efficiency of 27.5 miles per gallon (mpg). However, Congress later lowered the Corporate Average Fuel Economy (CAFE) standard under pressure from the automobile industry, a step challenged by environmental groups. The political struggle has continued for 40 years. How effective have CAFE standards been? That is a matter of some controversy. A study in 2009[5] (illustrated in Figure 4.10) showed that US automakers had improved fuel efficiency over the previous 30 years by 4.1 mpg – though whether the impetus for this was the efficiency standard or changes in consumer demand stemming from the high price of gasoline is unclear. Nevertheless, the standard is continually being raised and is supposed to reach 54.5 mpg for model year 2025. As Figure 4.10 shows, the actual realized mpg is expected to only be a little over 20, and rising but still less than 25 by 2035.

It is important to recognize the magnitude of resource savings from even a small incremental improvement in fleet gas mileage. Earlier studies illustrated the national saving possible with only a *modest* increase in fuel economy. An improvement from 18 mpg to 20 mpg, for example, would mean an additional saving of 200,000 barrels (bbl) countrywide of refined petroleum per day. The further raising of the average to 26 mpg could save more than 1.5 million bbl/day. Thus far very few production models, either diesel or gasoline, have been able to maintain a 50+ mpg average. Still, if the 2025 standard is achieved, the results would be extraordinary. For example, if the number of miles driven increases by 50 percent while at the same time mileage improves to 50 mpg, then demand would fall to *less than half* of current levels, a saving of around 4 quads annually.

The Industrial Sector

Industrial consumption of energy in the United States increased steadily from 1950 to the early 1970s, after which there were two modest peaks in the late 1970s (see Fig. 4.5). Following this there was a fluctuating increase into the late 1990s, after which there was a drop off through to 2010. There have been real conservation gains in US industry, which largely account for the leveling off during the decade 1995–2005. More energy savings gains are likely going forward. However, these developments do not give the full picture of energy use in the industrial sector.

Industry consumes energy primarily for process steam (such as in paper making) and heat (such as in the smelting of metals). Also, industrial firms use electrical energy for running mechanical devices and in direct application for chemical – electrolytic – processes.

An important percentage of the energy savings has come not from conservation of any kind, but rather from a structural change in the American economy. As we have previously noted, over the past decades, and particularly since the 1970s, the percentage of output from energy-intensive manufacturing has dropped in the United States and the activity of the less energy-intensive service sector has risen substantially. (This accounts of course for some of the increase in consumption on the residential/commercial side largely from the expansion of the service economy.) In this evolutionary process, factories ceased operation and the financial resources were diverted to other endeavors. Indeed, some factories have been converted into shopping malls or otherwise adapted to the service sector. Even within the manufacturing sector itself there have been shifts away from energy-intensive manufacturing, such as steel, to capital-intensive industries such as biotechnology. More recently, there has been some expansion of energy-intensive industries because of low US natural gas and electricity prices. This is especially notable with respect to Europe and Japan, and some foreign manufacturers in industries such as chemicals and plastics have moved production to the United States, but not at so great a rate as to cause a general shift back to the industrial composition of the post-World War II American economy.

Although structural changes in the economy explain some of the energy saving, they do not explain all of it. A 1982 study, for example, found that 28.3 percent of the improvement in consumption by industry came from efficiency advances; industry was producing the same output with less energy (Peck & Beggs in Sawhill & Cotton 1986). A little over 10 percent of that improvement represented a continuation of a historical trend; there had been technical improvements in manufacturing processes that lowered the energy requirements for a given amount of output. But the remainder, 18 percent, of the savings represented a response to the rising energy costs of the 1970s energy crises – an accelerated drive for efficiency improvement as a means of keeping a crucial factor cost down.

There were important investments in energy conservation by industry during this period. New plants and equipment were more efficient and the older stock was replaced by the new. It was apparently recognized that, at any given time, a company would require a profit-enhancing rationale for making such investments. How much (and how quickly) the various industries decided to invest in conservation, of course, depended on how important a cost energy was in production. As we noted, on average, energy represents a small factor cost of production. But some industries are far more energy intensive than others, and in these, substitution of new capital for energy consumption became cost effective with the kind of price shocks the market had experienced earlier.

The cement industry provides an excellent example of the potential for, and impact of, conservation investments in an energy-intensive industry. From 1970 to 2010, the energy input to produce a ton of cement in the United States fell by almost 40 percent (Worrell et al. 2013). A significant percentage resulted from the replacement of antiquated cement plants with new ones, but also there were gains in efficiency throughout the cement-making process. However, still more improvements were available, although the potential depended to some extent on the various cement-making processes. As Worrell and Galitsky (2008) showed, there were further "energy efficiency improvement opportunities" at every stage of cement production regardless of underlying technological processes being employed.

Conservationists believe that the potential throughout all of US industry for efficiency gains is still great. A study by the National Academy of Engineering estimated that by using the most efficient technology, by the year 2020 US industry would use up to 22 percent less energy than it was using at the time of the study (2007).[6] The US DOE in 2010 set a (voluntary) goal of a 25 percent reduction in industrial energy consumption by 2020. As of 2017, it seems doubtful that the price mechanism alone would provide sufficient incentives for such gains. Prices for all fuels have been low, but one lesson from the last 50 years is that energy prices are likely to be volatile, and consequently there will be at some point market incentives for conservation. With concern about climate change, however, there is the likelihood that policy-directed incentives will continue to offer a return on industrial investment in energy efficiency, but whether they will induce manufacturers to adopt changes that would realize the full conservation potential is uncertain at best.

Electricity: The Smart Grid

A developing technological capability with respect to electricity deserves its own section since it could affect electricity usage in all sectors – even transportation if electric vehicles become more common. The concept is the *smart grid*. According to the US DOE, the idea "refers to … utility electricity delivery systems into the 21st century, using computer-based remote control and automation." More specifically, a smart grid system allows for two-way communication between electricity users and distributors, so that the former would have instantaneous information on electricity demand and prices – which would reflect dynamically the level of use. That is, a consumer would know when it would be most cost effective to engage in energy-intensive activities. Need to charge your electric car? At four o'clock in the afternoon on a hot summer's day, there is great demand on the electric system and electric power companies must typically use high-cost "peaking" generators to keep the system from an outage. But demand at four o'clock in the morning is light, and more devices – such as the electric car – could be covered by the lower-cost main baseload generator. By reducing peak demand and increasing demand at times when demand is light, electricity usage becomes more efficient, and less primary fuel is consumed. Customers' prices would change during the day and night to reflect generating costs. The two-way nature of the communications link would also mean that distributors would have real-time information on problems in the system and could react to them much more quickly than they do at present.

Most experts argue that the smart grid will be especially useful to business, although a test in the state of Washington among consumers showed that "smart" devices and dynamic pricing lowered peak demand by 15 percent (Chassin & Kiesling 2008). But even a reduction of 5 percent in peak demand, according to one analysis, could lead to the elimination of 625 peaking plants in the United States with savings of $3 billion per year (Faruqui et al. 2007). (We consider the smart grid further in Chapter 9.)

The "Rebound Effect"

There is a fundamental problem with relying on potential efficiency gains to predict the amount of likely reduction in energy consumption. It is this: Efficiency gains mean in effect that there will be more supply – unused energy so to speak – and the effect of more supply will be to lower the price. A lower price will mean that quantity demanded will increase. Thus, a 20 percent gain in efficiency is likely to mean something less than a 20 percent reduction in energy demanded. This is called "the rebound effect" or sometimes the "take-back" effect (Greening et al. 2000).

How much less? It will depend on where the efficiency is obtained. Is it in the home? In a factory? Is a more fuel-efficient car likely to be driven more, or more efficient lighting likely to lead to the purchase of more appliances? There have been many studies to test the rebound effect and they generally conclude that the effect will be between 10 to 20 percent (Herring 2006). In other words, if a new refrigerator cuts one's electric usage by 1000 kWh/year, the home owner will acquire more devices that will use somewhere between 100 and 200 kWh annually. Of course it is difficult to determine exact cause and effect. But the rebound does seem a real phenomenon and it may be especially important in the industrial sector. Indeed, it is at least possible that a reduction in energy use by 1000 kWh will lead to consumption of *more than* 1000 kWh, what is called the "backfire effect."

Consider the following: Firm A is able to cut its energy use by 50 percent. But for a firm a cut in energy use means a reduction in the cost of producing its output and so a lower price to consumers. Depending on the elasticity of demand for the product, the increase in the quantity demanded could be much greater than the percentage decrease in cost. That is, assume that the demand curve is nearly flat, so lower prices mean quantity demanded rises 10-fold. Of course that means more energy overall will be expended in making so much more of the product. In the end efficiency is gained but energy use has increased.

It should be added that the backfire effect is controversial and almost certainly not the outcome of efficiency gains for the economy as a whole. Nevertheless, incentivizing or mandating efficiency does not necessarily predict the amount of resources conserved.

Conservation and Policy

It is evident that the US economy has had the potential to conserve more energy, but important questions have remained: What would it take to realize that potential?

Would the solutions be equitable? Would a program of maximizing conservation even be desirable? Policy choices extend across a vast range, from a decision by society to let the market decide; to one that would require large-scale government intervention and control. At one end of the range, the government might only provide information; at the other end, it might be empowered to mandate conservation, control or ration supplies, and even use coercive measures to see that energy savings would be realized.

The argument in favor of letting prices alone decide would appear to have a powerful precedent on its side in the United States. In the early 1980s, the market did adjust; the price mechanism did result in a search for efficiency and substitution. If left alone, we could expect that over time there will be rising prices, including painful spikes for oil and possibly natural gas as well, which in turn would lower the quantity demanded – over time at an *elasticity* close to unity. Cost-effective technical improvements could be sought, thus lowering the energy requirements for economic growth, and new more efficient technologies researched and employed. If there is any role for government in this scenario, it might be considered educational, to help people make informed, rational market choices. For example, as in the Energy Star program, the government might help consumers understand how spending now for more efficient products will result ultimately in financial, as well as energy, savings.

Although more recent evidence has suggested that the market works better than some critics once claimed, there are nevertheless two questions that arise about the market's behavior:

> Is the market for energy resources – especially oil – free from manipulation? The Organization of Petroleum Exporting Countries (OPEC) seems to have lost the power to control the price of oil, but is that permanent or only temporary?

> Is the market farsighted? In other words, does it tend to discount important future possibilities in favor of more immediate gains?

In the 1970s, OPEC's control of the market led to a 10-fold increase in the price of oil. But the exporting countries have not been the only ones to manipulate prices and quantities of oil and natural gas over the years. In the 1920s, the major oil companies agreed to collude on oil prices (Sampson 1975). In the next decade (as noted in Chapter 2), the Texas Railroad Commission determined output levels (and hence the price as well) of domestic US oil; the Commission maintained control until the early 1970s. At that time, the federal government – the administration of President Nixon and the US Congress together – imposed price controls on all oil sold in the United States – a control that did not end until 1981 (Grossman 2013). From 1954 to the mid-1980s, prices of natural gas sold in interstate commerce were determined by the US government. These controls were mostly removed by 1989 (Bradley 1996).

Market forces took over, and shortages mostly disappeared. But even after the end of controls, there were periodic price hikes and also great, rapid swings in prices that created incentives for conservation one moment only to take them away the next. So we see conservation gaining traction with consumers in the early 1980s only to trail off in the 1990s when the real price of oil was at its lowest point since the end of

World War II. OPEC regained pricing control in the 2000s and prices rose steadily through the summer of 2008 – again incentivizing conservation – but then tumbled over 70 percent in a matter of months.

But markets are, as noted, imperfect. Thus, while government price-setting is clearly not desirable, creating and maintaining incentives for conservation might well be – especially as we look into the future. Policies that lead to lower energy consumption may also reduce pollution and particularly greenhouse gas emissions that can contribute adversely to climate change.

Private market exchange does not factor in the cost of pollution – either present or future. Pollution is called in economics a *negative externality* since a cost is imposed on society that is not included by the market in the price. Hence a tax is often imposed to bring the *social cost* of production in line with the private cost. In some cases, for example climate change, it is often difficult and controversial to impose a particular level of taxation because the future outcome is uncertain (as we will discuss further in Chapter 6). Indeed, with climate change, analyses may suggest levels of cost that can differ by orders of magnitude.

Still, a particular appeal of energy efficiency and conservation is that the greater the effort of conservation the less primary energy is consumed and therefore the lower the externality costs. Conservation produces in this context "social savings." But programs to weatherize homes, improve appliance efficiency, lower automobile gas mileage, and so on have costs. The conundrum is this: It certainly costs *society* less in terms of resources such as capital, money, and time to adopt conservation instead of new resource production. But to adopt conservation involves *private costs* to individuals and firms. Some therefore advocate government intervention in the market to create the proper incentives so that individuals will benefit more from conservation than they would if incentives came from the market alone. Alternatively, government might mandate conservation by law, as a social good. These incentives, the argument goes, should be put into effect even when energy prices are low to forestall crises of the future.

An interventionist conservation policy could take many forms. Governments could use powers of regulation, taxation, or even coercion. Today, governments offer tax credits for home and commercial conservation improvements, and provide R&D funding to produce more efficient products or expand substitution options. There are tax benefits for solar installations and for purchase of automobiles that run on battery power. Proponents of conservation have noted government subsidies over the years through tax benefits for oil and gas drilling, and have advocated equivalent subsidies to conservation.[7]

Free market proponents view any market intervention with misgivings, and they are not without foundation. Not only have energy markets proved themselves able in many respects to function efficiently over time, but more importantly, as noted, governments have often performed poorly in controlling markets. Still, the story is mixed. In the United States, government intervention in energy has been at times disastrous but at other times it has helped. The government-sponsored attempts to produce commercial quantities of substitutes for oil and gas from coal and substitute motor fuels from corn and other plant products have wasted large sums of money that could have gone for other purposes, with few positive results (see Chapter 11). But government-mandated fuel efficiency in automobiles, for example, is given credit for declines in energy demand. It is also worth noting that although the market may be said to provide greater efficiency, it may not result – especially in the short run – in greater equity. The poor may become worse off as market prices rise, for example; in the event, the social consequences of relying solely on market forces may be unacceptable.

Relying exclusively on the market may mean a policy choice of accepting economic and political problems, rather than trying to prevent them. A policy of prevention, however, may mean a loss of freedom in the economy and possibly governmental waste and mismanagement. Which course is preferable? This is one energy-related dilemma that our society faces and will have to resolve in the decades ahead.

One further consideration is useful here: Americans may or may not be able to consume less energy while preserving economic growth. But developing countries really have *no choice*; they must increase consumption to develop advanced economies. If the industrial world does not conserve, it may find itself in competition for the world's resources. As we will see in the next chapter, this has political, economic and ethical implications.

References

Borenstein, S. and L.W. Davis. 2016. "The Distributional Effects of US Clean Energy Tax Credits." *Tax Policy and the Economy*, 30 (1), 191–234.

Bradley, R.L., Jr. 1996. *Oil, Gas & Government: The U.S. Experience*, 2 volumes. Rowman & Littlefield Publishers, Lanham, MD.

Chassin, D.P. and L. Kiesling. 2008. "Decentralized Coordination through Digital Technology, Dynamic Pricing, and Customer-Driven Control: The GridWise Testbed Demonstration Project." *The Electricity Journal*, 21 (8), 51–59.

Coase, R.H. 1959. "The Federal Communications Commission." *Journal of Law and Economics*, 2, 1–40.

Faruqui, A., R. Hledik, S. Newell, and H. Pfeifenberger. 2007. "The Power of 5 Percent." *The Electricity Journal*, 20 (8), 68–77.

Galasiu, A.D. and G.R. Newsham. 2009. *Energy Savings Due to Occupancy Sensors and Personal Controls: A Pilot Field Study.* Report NRCC-51264, National Research Council Canada.

Greening, L.A., D.L. Greene, and C. Difiglio. 2000. "Energy Efficiency and Consumption – the Rebound Effect – a Survey." *Energy Policy*, 28, 389–401.

Grossman, P.Z. 2013. *U.S. Energy Policy and the Pursuit of Failure*, Cambridge University Press, Cambridge, UK.

Herring, H. 2006. "Energy Efficiency: a Critical View." *Energy*, 31 (1), 10–20.

Hirsch, R.L. 1987. "Impending United States Energy Crisis." *Science*, March 20, 1460–1470.

Kaplan, S. 1983. *Energy Economics: Quantitative Methods for Energy and Environmental Decisions*. McGraw-Hill, New York.

LeBel, P.G. 1982. *Energy Economics and Technology*. Johns Hopkins University Press, Baltimore, MD.

Sampson, A. 1975. *The Seven Sisters: The Great Oil Companies and the World They Shaped*. Viking Press, New York.

Sawhill, J.C. and R. Cotton (eds.). 1986. *Energy Conservation Successes and Failures*. The Brookings Institution, Washington, DC.

Schurr, S., J. Darmstadter, H. Perry, W. Ramsey, M. Passell, and M. Russel. 1979. *Energy in America's Future: The Choices Before Us*. Resources for the Future, Inc., Johns Hopkins University Press, Baltimore, MD.

Worrell, E. and C. Galitsky. 2008. *Energy Efficiency Improvement and Cost Saving Opportunities for Cement Making: An ENERGY STAR® Guide for Energy and Plant Managers*. Report LBNL-54036-Revision, Ernest Orlando Lawrence Berkeley National Laboratory.

Worrell, E., K. Kermeli, and C. Galitsky. 2013. *Energy Efficiency Improvement and Cost Saving Opportunities for Cement Making: An ENERGY STAR® Guide for Energy and Plant Managers*. Report, Document Number 430-R-13-009, United States Environmental Protection Agency.

NOTES

1 In the August/September 1975 issue of the journal the *Ecologist,* editors wrote, "The thesis that economic growth is nearing its end is generally accepted by all save the demented and those who simply don't want to know."

2 Report, *Realizing the Potential of Energy Efficiency* (2007), at http://www.globalproblems-globalsolutions-files.org/unf_website/PDF/realizing_potential_energy_efficiency.pdf.

3 See for example, Galasiu and Newsham (2009).

4 It was included in the Energy Policy and Conservation Act, signed into law by President Ford.

5 Institute for 21st Century Energy, November 2011, "Motor Vehicle Average Miles Per Gallon," at http://www.energyxxi.org/sites/default/files/MetricoftheMonth-NOV11Motor VehicleMPG.pdf.

6 Issue of NAE's journal, *The Bridge*, Summer 2009.

7 For many years, oil and gas drillers in the United States were allowed a tax credit called the *oil depletion allowance*. It was presumed to account for the depreciation of a natural resource property. Major oil companies lost this benefit in the 1970s although it was preserved for small producers. Conservation and alternative energy tax benefits were advanced in the 1970s, in some cases were removed in 1980s but were restored in the 1990s and 2000s.

Global Perspectives

Introduction

For the United States and the industrial world, a key question is how to sustain economic growth into the future while limiting any damage resulting from the needed consumption of energy. But for the rest of the world, energy-related questions are quite different. Developing nations need energy resources and technology in order to emerge from what in many cases is great and chronic poverty. They need to create modern economies, not just sustain them. Yet the cost of energy and the cost of growth itself have often been too great for them to pay – too great in environmental, financial, and even social terms. But the needs remain. How then do they acquire the energy resources and technologies for growth? How do they exploit what resources they do have? What policies should and can they adopt to achieve economic development?

And what policies should the industrial world adopt with respect to less-developed countries (LDCs)? Since the industrial world has most of the financial capital and the technical resources, it can play and, arguably, *must* play a major role in furthering the path of economic development in the LDCs. Their plight has both moral and political implications.

Of course developing countries differ widely, and their needs and problems differ in important ways. Some have abundant energy resources; others have little or none. Some are industrializing their economies rapidly and probably will join the ranks of the wealthy industrial world soon. Other countries, presently, have little hope of significant advancement and, in some instances, are actually becoming increasingly impoverished. Many countries have vast gaps in wealth and access to technology among their own people. Most of the world's developing nations will seek to close the gap between themselves and the industrial world, either with the help of rich nations or in conflict with them. The cost of aid to the industrial world might be great, but then the cost of conflict might be even greater. In the end, rich and poor will, perforce together, determine which path is chosen.

World Poverty and "Energy Poverty"

It has been recognized widely that consumption of energy in some form will have to grow in the LDCs if they are to grow economically. The statistics bear this out.

In 2011, according to the World Bank, more than 1 billion people lived on an income of (with the equivalent purchasing power of) US$1.25 per day or less – the level of absolute poverty established by the Bank. Another 1.2 billion people lived on an income of between $1.25 and $2 per day, still greatly impoverished. In all about 2.2 billion people, about 30 percent of the world's population, lived on $2 per day or less.

However shocking such statistics might seem, the numbers have improved over the last 25 years. In 1990, 1.9 billion lived at or below the level of absolute poverty so the number counted as the poorest of the poor has declined by 43 percent. The second level, however, has shown less improvement but still some. Since 1981, 400 million have advanced beyond this second poverty level.

Most of those who have risen from absolute poverty live in China, where 753 million people between 1981 and 2011 had so advanced. But in many countries abject poverty is an ongoing fact of life. In Sub-Saharan Africa, for example, there are many countries designated LEDCs or *least economically developed countries*. The extreme poverty rate in 2011 in these countries was 47 percent.

Yet an even greater number of people in the world suffer from what is termed "energy poverty," which according to the World Bank means two things: "Poor people are the least likely to have access to [modern energy resources and technology]. And they are more likely to remain poor if they stay unconnected."[1] Statistics bear this out.

More than a billion people have no access to electricity and hundreds of millions more have access only intermittently. More than two and a half times as many (as Fig. 5.1 shows) have cooking facilities that are inefficient and can create lethal levels of indoor air pollution. An American consumes, on average, more than 40 times as much energy as an Eritrean; almost 20 times as much as a person from another of the very poor regions of the world, Nepal. A Nepalese citizen consumes on average one-third as much electricity per year as the best American *refrigerators*. Table 5.1 shows examples of countries classified in each of the World Bank income categories, from low income (LI), lower middle income (LMI), upper middle income (UMI), and two high income categories (HI), as to whether or not the country is a member of the 34-nation Organization for Economic Cooperation and Development (OECD, designated here by *), which works

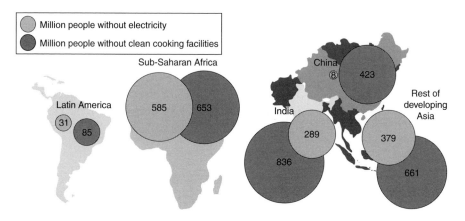

Figure 5.1 The reality of energy poverty.
Source: The World Bank, "A World Free of Energy Poverty," Alex McPhail, Hydropower Procurement Workshop presentation, June 2012.

Table 5.1		
Country	GDP 2013 (US$ per capita)	Energy consumption (MBTU per capita)
Bolivia (LMI)	2,550	30.9
Cambodia (LI)	1,001	14.7
China (UMI)	6,992	85
Congo, DR (LI)	440	12.3
Ecuador (UMI)	6,032	37
Eritrea (LI)	689	6.5
Finland (HI*)	49,310	244
India (LMI)	1,455	24.7
Japan (HI*)	46,679	140.6
Korea, South (HI*)	25,998	209
Mexico (UMI)	10,201	61.2
Nepal (LI)	691	14.5
Portugal (HI*)	21,507	80.7
Russia (HI)	14,487	209
Saudi Arabia (HI)	24,646	269
Singapore (HI)	55,980	187
Switzerland (HI*)	84,733	127
United States (HI*)	52,980	270

All data are from the World Bank: http://data.worldbank.org (2015)
Note: Energy values converted from kilograms-oil-equivalent (koe), 1koe = 39,652 BTU.

to promote economic growth and development worldwide. All those in the lowest income category have per capita GDP of $1035 or less; high income countries must have $12,616 or more. All GDP figures are calculated in the equivalent current (2013) US dollars; energy consumption per capita is in millions of BTUs (MBTU).

Not surprisingly, the disparity of energy consumption between HI and LI countries shown in Table 5.1 is great. But it is also important to recognize that often the distribution of income and wealth in many countries may be misleading as to the level of absolute poverty and energy poverty as well. Nigeria, for example, is a lower middle income country, but an estimated 40 percent of its population of 174 million live in poverty, many at the absolute level, and 44 percent have no access to electricity. It is also noteworthy that some upper income countries such as Russia consume nearly as much energy per capita as the United States with much lower incomes; a Russian consumes on average two-thirds as much energy as an American with an income one-third as great, and Saudi Arabians with less than half the per capita income of Americans nonetheless consume nearly the same per capita amount of energy. Of course, both Russia and Saudi Arabia are major producers of oil and gas and in fact the Saudis subsidize energy purchases, encouraging greater consumption of gasoline. Russia, like the United States (and Canada as well), is also a vast continent-spanning country and great distances mean a greater expense of fuel to ship goods and travel from place to place.

To some extent Table 5.1 does not tell the whole story of energy consumption since these statistics are for energy that is sold in the marketplace. In very poor countries especially (although also in some middle income nations) people use gathered wood and dried animal dung for fuel. In the African LEDC, of Mali, for example, over 90 percent of energy comes from such "fuelwood." In most of the other LI countries in both Africa and Asia, most energy resources consumed are nonmarket. But these resources do not solve the problem of energy poverty; in fact they serve to highlight the extent (and meaning) of energy poverty in the developing world.

For development, fuelwood does not provide sufficient energy to make cement. Dried dung does not serve to smelt metals or run electric generators. Conventional technologies for power plants or industrial processing are still overwhelmingly geared toward fossil fuels. Wood can be used in theory, but it would require whole new industries with enormous investments in new technology and new capital equipment – investments poor nations are not able to make. It can be argued that not every nation needs to have widespread automobile use or large-scale industrialization. But today technology for agricultural production, communications and data processing, or raw material extraction depends on commercial energy resources as much as manufacturing industries do. Without it, a country cannot compete in world markets for commodities and services with countries that employ modern technology. And without entry into world markets, economic growth in these poorer countries is likely to be slow or nonexistent. Although income and energy use are not absolutely linked, there is little doubt that income growth in the poor nations will require significant increases in energy consumption (Birol 2007).

The Future?

How much energy will we need to permit the kind of economic growth the developing world needs? There are various forecasts, but there is general agreement that energy demand worldwide will grow – especially in the developing world. As Figure 5.2 shows, the US Energy Information Administration predicts that by 2035 overall demand will rise by more than one-third, most of that as a result of increased demand in the developing world (especially India and China); by 2040, consumption is expected to be almost 50 percent higher than it is in 2016. The BP Energy Outlook for 2035 sees a rise of close to 50 percent. In that time they project a population increase of 1.5 billion people and a more than doubling of world GDP. This implies, however, that energy consumption growth will not be tied to economic growth. The forecast is for a substantial growth in productivity so that GDP per capita will rise by more than energy consumption per capita. Indeed, the OECD countries, as Figure 5.2 shows (and as the BP Energy Outlook notes) will see little increase in energy consumption, but these countries will account for 25 percent of the increase in world GDP.

How much can we count on these projections? It must be noted that predictions about energy production and consumption (as noted in previous chapters) have been wrong and often, spectacularly so. Still, it seems inevitable that, as population grows and as nations seek greater material wealth and well-being, energy consumption must rise. The question then becomes, not just "by how much?" but rather, "how?" That is, what resources and technologies are developing countries most likely to deploy in striving for development goals?

Global Energy Consumption
quadrillion Btu

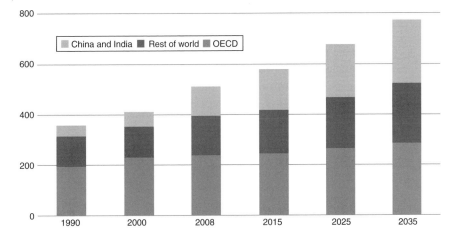

Figure 5.2 Global energy consumption.

World Energy Supply

If an LDC can meet its own energy needs with indigenous resources, is it better off than those that cannot? One would think so. A large local supply means that the ups and downs of the oil and natural gas markets are not a factor in a country's development. But the question is not so simply answered. Nigeria is an oil exporter and for the most part extremely poor. Indeed, as we show later in the chapter, resource-rich countries that have sufficient quantities to export as well as to fuel development at home become dependent on those resources and the wealth they bring. Instead of development, there is internal conflict over where the revenues are to go – conflict that actually retards growth. Moreover, market ups are followed by market downs – unanticipated market slumps that leave the country in debt that it can no longer repay. Some of the countries that are richest in oil and gas have not developed nearly to their potential; they are victims of what is sometimes termed the "resource curse." Of course, oil and gas are not the only viable energy resources. As we will discuss, other resources of the LDCs, such as water and wind, could contribute significantly to local energy consumption, without the problem of a "curse," though not without their own potential problems.

World Fossil-Fuel Markets and Supplies

World Oil

Oil is the dominant fuel in worldwide use today (Yergin 1991, 2011); about 40 percent of all commercial fuel is oil based. After World War II, increasing supplies and low prices helped oil gain ascendency in world energy markets. But it achieved, and has maintained, dominance for other reasons as well. It can be easily transported and stored, and it has been the fuel that is most technically appropriate for automotive transportation. As the automobile became more important worldwide, so did oil. By the start of the 1973–4 energy crisis, oil made up 48 percent of the US commercial fuel market. By then, oil and oil-based products had become basic staples of modern life. No country in the world existed without some demand for oil, and virtually the entire world was affected by the vicissitudes of the global oil market.

In 1973, world oil production was 58.3 MBD (million barrels of oil per day, which is equivalent to 21.3 BBL/year). Despite rising prices, world production jumped to over 65 MBD by the late 1970s, fell back in the early 1980s and then rose steadily through the 1990s and early 2000s, passing 80 MBD and then 90 MBD in the 2000s. By 2015 it was at over 90 MBD for oil alone, but 95 MBD when liquids associated with natural gas production are included, in total an equivalent of over 34 BBL/year.

As we might expect, a disproportionately large share – about 49 percent – of this was being consumed by the industrial nations of the OECD. China alone accounted for 12 percent of the remainder, while all of Africa consumed less than 5 percent.

Although demand is ubiquitous, supply is not (see Table 5.2). OPEC members make up the largest group of oil producing countries. Although the producers' group is in reality only a handful of nonindustrialized countries, OPEC produced over 30 MBD (11 BBL/year), or close to one-third of the world's daily production by about the year 2010. Previously, during the late 1990s, with demand in decline and production outside of OPEC growing, OPEC's ascendancy over the market ended and its percentage of daily output had fallen, while the oil production of the rest of the world slowly rose.

The OPEC countries are concentrated largely in the Middle East, as can be seen from the listings of oil production in Table 5.2, although they also include such non-Middle Eastern nations as Angola and Indonesia. The Middle East has the largest *reserves* and *estimated ultimately recoverable resources* (URR) of any of the regional groupings shown (see Table 5.3).

When OPEC was the center of world oil production in the 1970s, it effectively set the prices for oil and, of course, raised them several times over. The impact of OPEC's control of the market on industrial nations was notable, contributing heavily to inflation and recession in the industrial countries and stimulating conservation. But the effect of OPEC's domination of the oil market on *developing lands* – on the oil exporters as well as the importers – was even more dramatic.

First, there was initially a tremendous transfer of wealth to the exporting countries. Suddenly, some previously poor LDCs became extremely wealthy, accumulating vast sums of money termed *petrodollars*. In some countries, the dollars were immediately earmarked for development projects to make the transition from poor to rich permanent; in others, the money was transferred by elites to their supporters and often wound up in foreign bank accounts. The development process was further encouraged by banks in the industrial countries. These were happy to loan billions based on expectations of ever-rising prices of oil. In the meantime, the poor importing countries struggled to pay their energy costs, and tried to plan for the future with ever-rising energy prices a fact of life.

Table 5.2 World crude oil production, 1973, 1993, and 2015 (millions of barrels a day)

Countries	1973	1993	2015
Oil exporting countries (OPEC)			
Saudi Arabia	7.6	8.1	10.1
Angola	0.2	0.5	1.8
Iran, Islamic Republic of	5.9	3.4	2.8
Venezuela	3.4	2.3	2.4
Nigeria	2.0	1.9	1.8
United Arab Emirates	1.5	1.8	2.8
Iraq	2.0	0.4	3.8
Libya	2.2	1.5	0.4
Indonesia	1.3	1.3	0.7
Algeria	1.1	0.8	1.1
Ecuador	0.2	0.3	0.6
Kuwait	3.0	0.9	2.8
Qatar	0.6	0.4	0.6
Natural Gas Liquids	0.3	0.6	6.5
Total OPEC	31.1	23.5	38.2
OECD	14.5	19.7	24.8
including			
United States	11.0	7.2	13.5
Canada	2.1	1.6	4.5
Norway	—	2.3	1.9
UK	—	1.9	1.0
Australia	0.4	0.5	0.5
Other	0.3	0.4	—
Other producers	9.1	9.3	—
Former USSR (FSU)	8.6	8.9	13.5
Mexico	0.5	2.7	2.8
Total non-OPEC#			57.1
Total	58.3	57.5	94*

The total includes many countries that are not listed here. The list is merely representative and includes most major producers.
Sources: The World Bank (1985), Oil & Gas Journal Data (1994), *IEA (2015).

Table 5.3 World crude oil reserves and ultimately recoverable resources (URR) in BBL

Region	Reserves	URR
North America	233	250
Europe–Eurasia*	155	480
Africa	129	280
Middle East	811	1050
Asia/Pacific	43	120
World total[#]	1700	2615

Sources: Reserves, BP Statistical Report 2015; *includes the FSU; URR includes "conventional" resources only; unconventional URR adds 2455 BBL. Resources, McGlade and Elkins (2015).

But OPEC's loss of control of the market in the 1980s and economic policies in industrial countries that curbed inflation combined to hurt oil exporters far more than the importers. Of course those that had not borrowed substantially (such as Kuwait) were affected only marginally. But borrowers, including OPEC members Nigeria and Venezuela, suffered. Outstanding debts drained so many national financial resources that growth was hampered. Countries instituted austerity measures that lowered people's already low living standards and often were still insufficient to produce enough revenue to pay interest on the debt.[2]

Some countries could not pay their debts at all, much less under the terms they accepted when they had taken the loans. Consequently, a world financial crisis persisted throughout the 1980s. The crisis also affected the industrial powers. Even though they enjoyed general prosperity during the period, they held most of this debt and the debt problem continued to threaten the stability of the world's banking system. By the late 1980s, industrial governments were still concerned about finding solutions and were considering forgiving more than $100 billion in debt owed by the poorest African nations, many of whose debts to foreign banks and governments were significantly higher than their gross domestic product.

While the industrial powers clearly gained the upper hand over OPEC in the 1980s, the tables turned again in the 2000s. Moreover, OPEC remains the largest group of potential producers with the largest natural endowment of oil. Even though it does not control the market, it has remained (and will likely continue to be) the linchpin of the oil trade. Fracked wells in the United States have given America some market leverage that will likely cap increases in price, since at high prices shale oil production and enhanced oil recovery (see Appendix C) become attractive financially. Still, for the foreseeable future most attention related to the oil trade is likely to be focused on the Persian Gulf region, where supertankers pick up crude oil daily for delivery to markets mostly in the industrial world. As long as the world has a major need for oil, OPEC and especially Saudi Arabia will be in the position to assert some temporary control over supply and price.

Relatively low oil prices in the mid-2010s have helped the LDCs, but they are still largely at the mercy of the oil market. It should be noted that, at times in the past, OPEC members considered reducing the impact of price hikes on the poorest nations – either through aid grants or by selling oil through special contracts at discounted prices. But as many exporters suffer through another period of low prices (at the time of this writing in 2017) there are not enough petrodollars to meet most countries' national budgets, much less provide aid to others. A retu n to high prices could again hurt the importing nations of the underdeveloped world especially – unless they are able to utilize more of their indigenous resources.

World Coal

While oil is the most utilized fossil-fuel resource worldwide, coal is the largest. In Chapter 2, we noted total world coal resources in terms of billions of metric tons of coal. But we can better compare the size of coal and oil resources by looking at them in terms of the amount of energy – for example, number of BTUs – that is extractable from both. On that basis, we can consider a ton of coal as equivalent to a certain number of barrels of oil. In Table 5.4, world coal has been converted into *billions of barrels of oil equivalent* (BBOE), and we can plainly see that on this basis as well, coal is the largest fossil-fuel resource. It has an energy equivalent more than double that of oil as measured by reserves (R) and more than 4 (or possibly as much as 15) times larger in terms of the (estimated) *remaining undiscovered resources* ($Q_\infty - Q_d$). Yet coal, with an R/P ratio 2.4 times (in years) that of oil, is utilized far less.

We have elsewhere noted several reasons why coal consumption has been limited, even though it is the cheapest of the fossil fuels. Historically, coal lost out to oil, in large part because oil is so much easier to ship and to utilize in internal combustion engines. Natural gas, when it came into widespread use, also displaced coal in many industrial applications. In recent years any trend toward coal has also been slowed by concerns about greenhouse gas emissions and other pollutants (see Chapter 6). Coal is recognized as the greatest emitter of carbon dioxide (the principal greenhouse gas) per unit of heat delivered of all the fossil fuels, because it has the highest ratio of carbon in its chemical makeup. With the increased awareness of possible climate change resulting from global greenhouse warming, environmentalists and some governments have opened campaigns against carbon emissions with coal combustion as a major target. In fact, because of the environmental impacts, the United States, the European Union, and major lending organizations have been reluctant to finance construction of new coal-burning electric power plants in developing nations. Nonetheless, coal, because of its abundance and low cost, remains an important resource worldwide, and especially in the developing world.

As with oil, the distribution of world coal resources is as important a question as the total mass amounts. Only in this instance the largest supplies of the resource are

Resource measures[a]	World coal resources		World oil resources	World natural gas resources
	Bmt units	BBOE		
Production (P)	7.42 Bmt/year	35.4 BBOE/year	34 BBL/year	20.5 BBOE/year
Reserves (R)	887 Bmt	4250 BBOE	1700 BBL	1202 BBOE
R/P ratio	119 years	119 years	50 years	55 years
Remaining resources ($Q_\infty - Q_d$)	4610 Bmt*	22,000 BBOE	5060 BBL**	4109 BBOE[#]

Table 5.4 World coal resources compared with oil and natural gas

[a] See Chapter 2 for strict definitions.
* Estimates range from 4000 Bmt to 17,000 Bmt.
** Includes all unconventional resources including oil shale and natural gas liquids.
[#] Estimates, including unconventional gas resources, are quite varied and are as high as 9824 BBOE (Aguilera et al. 2014), which is still less than the low value for coal resources.
Sources: BP Statistical Review of World Energy 2015; USEIA 2015; McGlade and Ekins 2015.

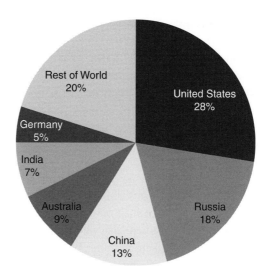

Figure 5.3   World distribution of recoverable coal resources.
Source: USEIA.

not in LDCs; they are in the United States and in Eastern Europe, particularly Russia (Fig. 5.3). South Africa, Australia, and Germany also have significant supplies, as do China and India, although the latter's reserves are only about half as much as China's and one-quarter of the reserves in the United States.

Except for India and China, the developing world does not possess reserves close to those of the nations noted above. Nevertheless, several countries have indigenous supplies that are significant enough to account for domestic consumption and in some cases substantial exports. Indonesia is perhaps the most notable example. Though reserves are only 22 Bmt, Indonesia has become one the world's largest coal exporters, selling 410 Mmt (million metric tons) in 2014, almost double the export sales of Russia and the United States combined. Only Australia has been exporting a comparable amount of coal (375 Mmt in 2014).

There are, it should be noted, significant reserves in several developing countries: Vietnam, Pakistan, and Mongolia in Asia; Colombia and Brazil in Latin America; and Bosnia in Europe – to name some of the most prominent. Several of these countries, Colombia in particular, already have a sizable export trade (it ranks fifth in exports among all countries), but many more developing nations are dependent on imports of coal.

Actually, the largest importers are in Asia; China, India, Japan, and South Korea together account for imports of around 850 Mmt/year. Indeed, China, which is the world's largest producer, over 3 Bmt/year, is also the world's biggest importer (292 Mmt in 2013) and of course the largest consumer of coal.

There is a substantial world coal trade. In all, the coal trade was over 1.3 Bmt in 2014, representing about a quarter of all coal consumption. Whether (and by how much) the coal trade will continue to grow will depend on a number of factors, including environmental concerns. China, for example, is reportedly looking to cut consumption to reduce the serious pollution problems in its cities. At the same time, China is still building new coal power plants and other developing countries are looking to low-cost coal to generate electricity where to date there has been little or none. There are, as of this writing, more than 2000 coal power plants under construction or in the planning stage (Plumer 2016). Should they all be built, coal exports will boom in the coming years, though fears of environmental damage will grow along with them.

World Natural Gas

Natural gas resources make up about one-third of the world's conventional fossil fuels. Even more than coal, it is best suited for use in the country or region where it is produced. While it can be transmitted at not prohibitively high costs through overland pipelines, overseas export requires liquefaction (creating liquefied natural gas or LNG). This process is not only expensive, it requires special ships and creates new safety concerns. LNG is so flammable and explosive that an accident can literally level a harbor area.

Yet natural gas is probably the most desirable fossil fuel. It is cleaner burning than oil or coal and has wide applications – from space heating to industrial heat to electric generation and is the lowest (fossil-fuel) emitter of carbon dioxide when burned. It also can be used in the petrochemical industries for the creation of organic compounds, including fertilizers.

The world's (recoverable) natural gas resources are much smaller in terms of equivalent thermal energy values than coal (see comparisons in Table 5.4). Still, at present rates of consumption, the remaining resources are likely to be relatively abundant at least through the twenty-first century, although, as we discussed in Chapter 2, it is unclear just how much natural gas is ultimately recoverable.

But some significant new reserves are being found, many of them in developing countries. At present, natural gas deposits exist in more than 50 developing countries, although the exact size of these reserves has yet to be determined. Nevertheless, these resources may be critical for economic development in several countries.

The world distribution of natural gas is depicted in Figure 5.4 and a breakdown for the developing world is shown in Table 5.5. As we can see, along with the most oil, the Middle East and the former Soviet Union (FSU) have the most natural gas. Much of this gas is *associated with* oil, that is, it exists in the same geological deposits as petroleum and escapes when the oil is extracted. Natural gas reserves that exist independently of oil are called *nonassociated* deposits. Nonassociated gas deposits have been discovered in such diverse locations as Bolivia and Bangladesh since extensive exploration began in the 1980s.

Newly discovered natural gas deposits hold out the hope of an energy supply path to economic development for a number of LDCs. Besides having a variety of uses, conventional sources of natural gas are also relatively easy and cost effective to extract and develop. The World Bank has noted that natural gas requires the lowest capital outlay of all commercial energy sources; natural gas can be most reliably utilized and distributed to domestic markets in LDCs that have the resource.

Region	Country	Recoverable reserves (TCF[a])	
		1984	2014
Asia/Pacific	Brunei	7.0	13.8
	India	14.8	47.8
	Indonesia	30.2	104.7
	Malaysia	48	83
	Pakistan	15.8	26.6
	China	30.3	155.4
	Thailand	8.5	9.0
	Bangladesh	7	9.3
	Regional estimate	188.6	540.4
Middle East	Abu-Dhabi	31.2	215
	Bahrain	7.4	3.3
	Iran	480	1193
	Iraq	29	111.5
	Kuwait	35.2	63.5
	Qatar	62	855.3
	Saudi Arabia	125.2	290.8
	Regional estimate	774	2818
Africa	Algeria	110.2	159.1
	Angola	1.5	9.7
	Egypt	10.2	77.2
	Ivory Coast	3	1
	Libya	21.4	54.7
	Nigeria	34.8	180.7
	Cameroon	4.2	4.7
	Morocco	0.2	0.05
	Tunisia	4.2	2.3
	Tanzania	0.2	0.2
	Congo (DRC)	0.04	0.04
	Regional estimate	193.7	606
Central and South America/Caribbean	Argentina	24.4	13.4
	Ecuador	3.5	0.2
	Mexico	75.3	17.0
	Venezuela	54.5	196.4
	Trinidad	13.1	13.1
	Bolivia	4.9	9.9
	Brazil	2.7	16.2
	Chile	2.4	3.5
	Regional estimate	186	294
Total (including developed world)		3208	6606

Table 5.5 Proven natural gas reserves in the developing world, 1984–2014

[a] One trillion cubic feet (TCF) of natural gas has the energy of 10^{15} BTU (1 quad) or 100,000 GW-hr (electric equivalent).

Sources: USEIA (2015); BP Statistical Review of World Energy 2015.

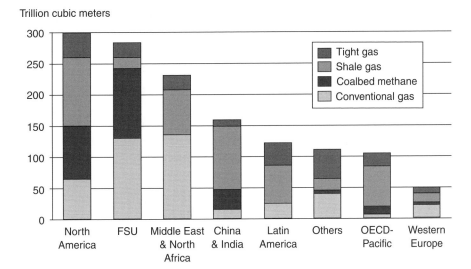

Figure 5.4 Worldwide natural gas resources.
Note: 1 cubic meter (m^3) equals 35.3 cubic feet.

Conventional and Nonconventional Fossil Fuels

Oil, coal, and natural gas resources exist in deposits scattered throughout the world. But there are more countries with little or no energy resource endowments than those with abundant resources. Some analysts hope that the have-not nations may be able to exploit nonconventional fossil fuels (Chapter 2). This category includes nonconventional natural gas, oil/tar sands, methane hydrates, and oil shale (described in Appendix C). Some, such as fracked natural gas and oil sands, can be profitably extracted if the price of oil and gas are high enough. At the time of this writing, some of that production was suspended because of low prices. In some cases, however, nonconventional resources are called that because the techniques of exploration and the technologies of extraction have not yet been developed sufficiently to allow large-scale production at a cost approaching that for conventional sources (see Chapter 2). As noted in Chapter 2, and in other parts of this chapter, if we could exploit these resources fully, the world's supply of available fossil fuels would increase immensely and the distribution would be far more widespread (see Table 5.6). While there are substantial nonconventional resources in the developing world, most of the nonconventional resources are found in the developed world.

That said, it should be noted that the extent of nonconventional resources, especially in developing countries, is mostly hypothesized and numbers change as more information becomes available. In Argentina, for example, there are extensive shale deposits and at the time the table above was completed these resources had not been estimated. More recently the EIA placed an estimate of 802 trillion cubic feet

Table 5.6 Ultimately recoverable fossil-fuel resources (estimates) in the developing world			
Region	Oil (BBL)	Gas (TCF)	Coal (Bmt)	
Africa	280	1589	50	
CSA*	360	1059	35	
Asia**	75	883	15	Conventional
China/India	90	353	1200	
Middle East	1050	3708	15	
Africa	70	1236	N/A	
CSA	450	1942	N/A	
Asia	5	520	N/A	Nonconventional
China/India	110	1413	N/A	
Middle East	353	706	N/A	

* Central and South America.
** Excludes China, India, and developed countries such as Japan and Taiwan.
Source: McGlade and Ekins (2015).

potentially in an Argentine shale formation, an amount that would increase the projection for the CSA substantially.

It is also important to recognize that although resources may exist, we must reemphasize that they are not always extractable in a cost-effective way and might not be for years, even if an immediate effort were made to develop new technologies (see Chapter 10). Indeed, if an LDC were to try to develop its nonconventional resources at present, the project would probably retard economic growth, not improve it. The financial resources would be better spent on building a manufacturing or agricultural capacity or improving a national infrastructure. Whether desirable or not, advanced technologies are more likely to come from industrial nations, which are in a better position to spend more on research and development. Thus, the direction of technological development probably will be affected by the way future need is perceived in the industrial world.

World Renewable Energy Resources

Lack of fossil fuels does not alone determine a country's indigenous supply of energy resources. There are also renewable energy resources, a diverse category that includes hydropower, wind and solar energy, and harvested biomass, particularly wood.

In the industrialized world, most interest in renewable energy arises out of a desire to displace fossil fuels for economic, environmental, or even aesthetic reasons. In much of the underdeveloped world, renewable resources play a fundamental role. Wood offers the best example. It is an extremely important fuel in many nations, providing the basic energy requirements of living: cooking and home heating. In many rural LDC villages, wood fuel is the primary, if not the only, available energy resource. The demand for wood was unaffected by the oil scarcity that hit the industrial world a few times during the past half century, although its availability is affected by different kinds of forces – such as population growth and migration. As we will see in considering several of the world's renewable resources, there may be both opportunities and serious problems in expanding the use of indigenous renewable resources in developing countries.

Hydropower

Hydropower is presently the largest and best developed renewable energy resource. A conventional technology for generating electricity (as we saw in Chapter 3), it provides economic service throughout the world. In the developed world, hydroelectricity utilization grew during the 1970s energy crises. In the developing world, hydropower projects have taken on an almost symbolic character. They are viewed as a symbol of progress, a representation of modern technological development. Huge dams with their human-made lakes have a special grandeur, although they also require significant amounts of money, which the LDCs often cannot raise.[3]

Because of the wealth of information on regional watersheds throughout the world, hydro-resources are the best defined renewable resource. The resources are defined in terms of *potential hydroelectric capacity* (see Chapter 3) measured in kilowatts, megawatts, or gigawatts (1 GW = 1000 MW) of electric output capacity, and actual output is measured in kilowatt/megawatt/gigawatt/terawatt hours (1 TWh = 1000 GWh). At present, world hydrocapacity is around 930 GW and annual output is approximately 3800 TWh. Note that a typical large thermal power plant has an output capacity of about 1000 MW, or 1 GW; therefore, the world's hydrocapacity is the equivalent of more than 900 large thermal-powered electric generating plants. Although some individual hydro facilities are massive and produce 1 GW or more, the overall category encompasses small hydroprojects as well; for example, there are 92,000 small hydrostations in China alone. Hydro has the advantages of being renewable and of producing no carbon dioxide or other greenhouse gases.

In Figure 5.5, as we can readily see, the world potential for additional hydroelectric generation is very large: more than four times the current level. Note that a great deal of the potential is in the developing world; in both Africa and Asia only a small fraction of the potential hydrogeneration is being utilized. Five developing countries in particular – Congo (DRC), India, Indonesia, Peru, and Tajikistan – possess major untapped hydro-resources.

However, as can be readily observed in the figure, hydro-endowment is anything but uniform, particularly with regard to need. Asia, taken as a whole, has half the world average of water runoff per capita. Though Africa in aggregate has substantial hydro potential, the vast arid strip known as the Sahel in the upper middle part of the continent is nearly devoid of any water resources. Rainfall in many regions is subject to wide variations; thus, any hydroelectric plants depending on rain runoff will suffer the same variations and will require thermal-generating equipment as a backup source to avoid the sporadic curtailment of electricity. Even with significant rainfall, there is no guarantee that a country will have significant hydroelectric resources. Hydro also needs a hilly or mountainous terrain, provided the landscape is not too rugged or remote. Such particular requirements exclude many regions of the world.

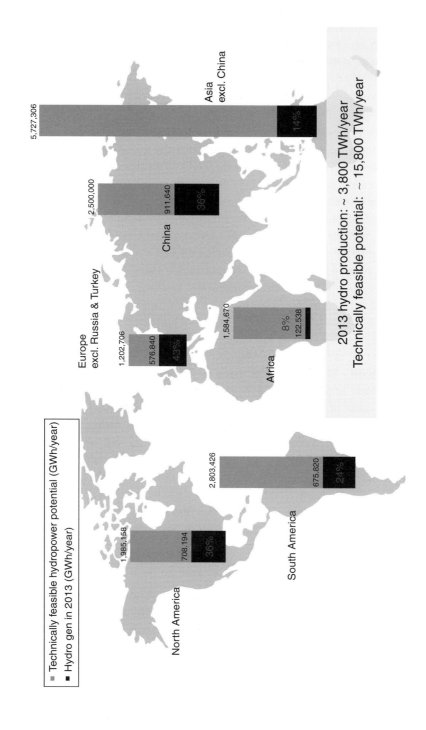

Figure 5.5 Hydropower: potential and actual.

In most potential sites, however, new hydrocapacity is cost competitive with thermal power plants. Large hydroprojects cost no more to build and often less to maintain than any thermal plant of comparable size, and also there is no fuel cost (see Chapter 9). But there can be a major expense that will reduce hydropower's cost competitiveness: long-distance transmission lines. Frequently, hydrosites are far from population centers, thus requiring the expense of high-voltage electric transmission lines. In some cases, it has paid to transmit hydrogenerated electricity long distances. In the DR Congo, for example, large hydro-resources (potentially 40 GW) exist on the Congo River 1700 km (over 1000 miles) from the country's rich copper resources.

At the same time, hydro creates environmental and social issues of its own. There are some spectacularly large projects, such as the Itaipa on the border of Brazil and Paraguay, which have led to the uprooting of communities as the reservoir behind the hydroelectric dam floods, at times, whole villages, taking away arable land or disrupting other means of livelihood such as river fishing (Rosenberg et al. 1995).

Still, there are many potential hydrosites in the developing world that can be employed without major disruption. As the need grows, some will be developed, although countries will have to find the financial resources to do so.

Geothermal Power

The interior heat of the Earth is an energy resource available to many countries. Called *geothermal energy*, this resource in the form of steam can generate electricity and provide heat for purely thermal applications, such as space heating and industrial processing (see Appendix C). Potential geothermal power sites are located in several parts of the world. The present world use of geothermal energy is almost 13 GW in generating plants and thermal applications across 24 countries. The potential is limited, but an increase in utilization of geothermal resources is possible. According to the Geothermal Energy Association, as of early 2015 an additional 12 GW was in the planning stage. Although geothermal energy is inevitably a limited resource worldwide (there has to be access to the interior heat of the Earth), there is potential for more development in Central and South America, parts of Southeast Asia, and in East Africa.

Wind and Solar Power

Although the technologies exist to convert wind or sunlight into electricity or to convert sunlight into useful thermal energy, wind and solar energy are only now beginning to be widely applicable. The principles of solar and wind energy conversion are reviewed in Chapter 11, but there are several points we should make here. First, given the current state of solar and wind technologies, the wind does not blow *strongly enough* or *long enough* nor is the sunlight *strong or regular enough* to make either a useful energy resource in many regions worldwide. Second, even where these resources are available and sufficiently strong, they are always *intermittent*; that is, they vary over time and are totally unavailable at certain times of the day and under certain weather conditions. For some applications, such as electric power, that require a continuous supply, an intermittent supply of energy is not acceptable as a *sole* source.

Therefore, these technologies must be used together with either *conventional energy sources* or *energy storage* to assure the continuity of supply. For other applications, however, an intermittent supply can indeed be tolerated in a working technology.

Windmills have been used since ancient times to drive pumps for irrigation and land reclamation (in the Netherlands especially). For these functions, the continuous availability of the wind energy is not as important as the average supply over time; it must be sufficient to pump the amount of water needed over the given time period. Modern irrigation pumps that are electrically driven from wind or solar (photovoltaic) generators are proving useful in the developing nations, especially in agricultural areas that are too remote from power lines. Solar and wind power are also useful to industries such as oil refining or water desalinization, where an intermittent supply can be tolerated as long as annual production is sufficient and where the product itself (refined oil or desalted water) can be stored if a continuous flow of output is needed.

For applications where interruptions in supply cannot be tolerated, technological developments will be needed to make energy storage inexpensive and efficient (see Appendix C). Currently, storage *is* expensive and wind and solar technologies thus have limited use, especially in poorer countries. To pay for all of the additional equipment required for sufficient storage or to provide backup sources run by conventional fuels, a poor country would place itself at a disadvantage by adopting wind or solar technologies. Large-scale solar- or wind-based energy would increase the costs of industrial production and take more from hard currency earnings than conventional systems. A heavy reliance on solar or wind energy using the present technology would slow a poorer nation's economic development.

Environmentalists who wish to limit the expanded use of fossil fuels have advocated the large-scale introduction of solar and wind technologies at dispersed sites in developing nations. These advocates point out that they are attractive energy sources that are environmentally benign, and completely renewable – and any country that adopted them could avoid polluting their environment. But if an LDC must rely solely on these energy sources, it will remain poor for the immediate future. Conventional technology is cheaper and will help an LDC develop. Although the environment is important, it may be hard to convince LDC planners to be concerned with large-scale ambient air pollution when its people are suffering from indoor air pollution from dung and wood fired stoves – pollution that could be greatly reduced by stoves run by electricity.

The potential for worldwide use of these two renewable energy resources will depend on geography, the cost of conventional energy sources, and the future prospects of the technology. Solar devices will always depend on significant and *consistent* sunlight and wind machines on strong, *regular* winds.[4] These realities limit the usefulness of these technologies. But as they advance, they will work more efficiently and cost less, increasing their economic viability as an energy source in the regions where feasible, especially if the cost of fossil fuels rises. There is some prospect for significant change in the next 20–30 years in terms of both relative costs and technological advances. But this time horizon, short as it may seem, is an important period for the developing world, a time in which nations may either grow or fall further and more hopelessly into poverty.

Forest and Biomass Energy

The final major category of world renewable resources is called *biomass*. Biomass resources are primarily various types of vegetation, either cultivated (e.g. sugar cane) or not (e.g. most wood). Wood – timber – is by far the largest resource. The world's total forest area is about 4×10^9 hectares (1 hectare = 10^4 square meters), out of a total land area of about 13×10^9 hectares. Table 5.8 shows that, as of 2000, the "sustainable" amount of biomass energy was potentially around 100 exajoules (which is very close to 100 quads). Of that, about 42 percent was "woody biomass." Of course, the potential is for the most part not being met, and world production was around 40 quads – not insignificant but well below potential. Note that in Asia production was 108 percent of the sustainable yield, meaning that forest lands and other forms of biomass were being over-harvested. As Figure 5.6 shows, areas can be deforested by excessive fuel gathering or animal husbandry.

Table 5.7 shows that energy crops – such those used for ethanol and biodiesel – were approaching a level of production that rivals that for wood. We will discuss the use of biomass especially for transportation fuel in Chapter 6 and Appendix C. But here it is useful to note the distribution of different forms of biomass. In Asia, for example, most biomass has been wood and straw, whereas in Africa, wood and biofuel crops provide most of the biomass energy. The latter has grown along with the rising demand for biofuels in the industrial world.

Wood – gathered from forest residues – is the prime fuel for many of the world's poorest countries; for example, 90 percent of the Sub-Saharan African population

Figure 5.6   A deforested area in Nepal caused by excessive fuel gathering and cattle grazing.
Source: Earl (1975)

Table 5.7 Biomass energy potentials and use in different regions (EJ/year) (EJ = 10^{18} J)								
Biomass potential	North America	Latin America	Asia	Africa	Europe	Middle East	Former USSR	World
Woody biomass	12.8	5.9	7.7	5.4	4.0	0.4	5.4	41.6
Energy crops	4.1	12.1	1.1	13.9	2.6	0.0	3.6	37.4
Straw	2.2	1.7	9.9	0.9	1.6	0.2	0.7	17.2
Other	0.8	1.8	2.9	1.2	0.7	0.1	0.3	7.6
Potential, sum	19.9	21.5	21.6	21.4	8.9	0.7	10.0	103.8
Use	3.1	2.6	23.2	8.3	2.0	0.0	0.5	39.7
Use/potential (%)	16	12	108	39	22	7	5	38

1 EJ (exajoule) = 0.95 quads
Source: Parikka 2004.

Figure 5.7 Wood kiln for charcoal production.
Source: npstockphoto, iStock/Getty Images Plus.

depend on wood as a fuel. Wood is either burned directly or is turned into a derivative fuel, such as charcoal. Charcoal is often used in industrial applications. But it is not adequate for the industrialization needs of the LDCs. (Charcoal-burning kilns, as depicted in Figure 5.7, have a low efficiency.) In many cases, the potential energy production as measured by annual forest-growth increments is not sufficient to supply the growing fuel needs of those countries; harvesting is already exceeding annual growth.

Hopes for continued and even expanded use of wood biomass have focused on both changing policies and technology. It has been proposed that people in poor countries be given (or have the opportunity to buy very inexpensively) efficient wood-fired cooking stoves (see Goldemberg *et al.* 1987). The use of these stoves, which would embody existing technology, could reduce indoor air pollution as well as deforestation, especially if an effort was made to manage wood supplies. (Increased rural electrification in the developing world would also reduce deforestation and could eliminate indoor air pollution.)

There is hope for expanded use of forest resources through the use of forestry management or forest plantations (Anderson & Fishwick 1985). Such operations, which may be called *energy-crop plantations*, balance harvesting with replanting and land fertility to achieve a sustainable production. In Asia, most notably, there were 115 million hectares of forest plantations as of the year 2000 (Parikka 2004).

Finally, it seems that more general policies and planning will be needed to deal not just with biomass energy management, but with related issues of population pressure, land use, and alternative fuel development. Without such planning and the financial investment to carry it out, energy poverty will continue and economic growth will be severely hindered in many countries lagging at present in development.

Nuclear Energy Worldwide

Nuclear energy (see Chapter 7) has also been introduced in parts of the developing world. Its status and prospects for the future depend not on fuel resources but on the status of nuclear technologies themselves. As we will discuss in Chapter 8, innovative designs are being developed in many places, mainly in the industrialized world, but also in India and China. The production of uranium-based fuel does not at present threaten to deplete significantly available reserves. Because newer nuclear technologies are not in commercial use even in the industrial world, their ultimate impact on resources and on energy development worldwide is uncertain.

Worldwide, there are, as of this writing, 437 nuclear power plants in operation and another 66 under construction. While construction of nuclear plants has only recently resumed in the United States, it has continued in China, India, Brazil, and Argentina. Concerns over the safety and costs of such plants, however, have dampened development, especially in Germany where nuclear power is to be phased out by 2022. But interest in nuclear energy has increased in the past decade as it does not emit carbon dioxide or any other greenhouse gas. Some prominent environmentalists who had previously opposed nuclear power are now strongly for its revival in industrial countries and its expansion generally throughout the world (Karecha and Hansen 2013).

There is one central point in considering the opportunity for development of nuclear energy outside the industrial countries: the technology of nuclear power at present is too complex for most countries to develop their own capability, and even those LDCs with uranium resources often can do no more with their natural wealth than sell it to the industrial world. Nuclear technology has been monopolized by the industrial nations and, since there is at least some convergence of the technology of nuclear power and nuclear weapons, the great international powers are not willing to transfer their nuclear capability without controls. To the extent that the developing countries can obtain nuclear energy, they generally must do so only under restrictive conditions designed to prevent the proliferation of nuclear weapons. Although questions have been raised about the effectiveness of those conditions, they nonetheless limit the availability of nuclear power. At the same time, given the extremely high costs of nuclear power plants, the need for expertly trained personnel to run the plants, and the issues connected with nuclear safety and waste disposal, nuclear power seems a reasonable option only in nations that have already made a great deal of progress toward industrialization.

Resources and Development: "Curse" or Good Fortune?

In the 1970s it was assumed that countries that were self-sufficient in energy resources had a great advantage over those that did not. In the United States there was much hand-wringing over America's dependence on world energy markets and a belief that this dependence made the country vulnerable to oil blackmail. That is, either we changed our policies on international issues to be in accord with the views of oil producers or they would no longer sell oil to us.

But the 1980s presented a different picture (even if many in the United States seemed not to notice that the oil market had changed). In fact, it was oil exporters who were in economic trouble, and the situation remained that way for the next two decades. Many had predicted in the 1970s that the oil exporters would soon dominate the world economy in every respect. The World Bank predicted that by 1985 OPEC nations would have a surplus of $1.2 trillion from oil sales as prices were expected to continually rise. Real economic growth in OPEC countries was expected to be at least 5.1 percent per year. Venezuela's president in 1979 claimed that soon people in North America would be driving cars made in Venezuela, from Venezuelan aluminum, powered by Venezuelan oil. "We will look like you," he said to an interviewer from the United States (Karl 1999).

But the forecasts were far off the mark. By the mid-1980s Venezuela was in turmoil. Low oil prices had left the country unable to pay the sizable debts it had taken on to make the president's dream a reality. Over the next 15 years, Venezuela faced a coup, budget shortfalls, and unemployment. Many other exporters were similarly affected. Some were even worse off. In Iraq, for example, per capita income fell to the level it had been in the 1940s. Nigeria, which had been ranked thirty-third in the world in per capita income by the end of the 1970s, fell to thirteenth from bottom of more than 200 countries 20 years later.

The problem was that oil-rich countries had come to rely so much on oil revenues that all other development activities were pushed into the background. When oil prices were low, there were insufficient funds for development; when they were high, much of the wealth was distributed to supporters or squandered on grandiose "mega-projects" that had nothing to do with the comparative advantages of the country.

This problem is known as the "resource curse" or the "paradox of plenty" (Mikesell 1997, Karl 1999, Mehlum 2006). Part of the problem is that, as exports grow, local currencies appreciate, making imports (particularly manufactured goods) cheaper and thus forcing out local producers so that only the key resource (oil in most cases) is left to trade internationally.

The main problem was the weakness of national institutions. As Karl (1999) noted, billions of dollars were circulating "in the context of weak administrative structures, insecure property rights, nonexistent judicial restraints, deep divisions and strong political ambitions. This is not a formula for economic efficiency."

Rule of law in some of these countries may exist on paper but not in fact. Instead, when times are good there is a scramble to acquire the riches pouring into the country and those well connected typically make off with them. Open political competition combined with the rule of law would likely divert most if not all of resource windfall to the general welfare of the nation. Indeed, Norway is an example of a country that became wealthy because of natural resources but maintained the rule of law and other basic institutions of fair play. Consequently, it has experienced exceptional economic growth in times of high resource prices and adequate growth when prices are low. The economy, already diversified when oil and gas were found to be abundant off Norway's coasts, has never had the kind of volatile ups and downs of nations such as Nigeria and Venezuela.

The key then is for resource-rich countries to change their institutions. But, as Nobel prize-winning economist Douglass C. North has pointed out, that is no easy matter. Those with access to power within the existing system have no incentive to change it (North 1990). Of course, during hard times there might be more general incentives for change, but in states as Karl (1999) described them, the systems themselves lack the flexibility to allow for the kind of institutional change that will make widespread development possible. Change when it does come is often disruptive, even violent, retarding development all the more.

It is also noteworthy that some of the fastest growing countries of the past half century are resource poor: South Korea, Taiwan, and Singapore are prime examples. But this does not turn the matter upside down: that is, it does not mean that resource poverty is a blessing and resource abundance, necessarily a curse. At the bottom of the ranking of per capita income there are resource-rich countries such as DR Congo and resource poor ones, Eritrea, for example. As in the case of Norway (and also to a lesser extent Botswana) resources can lead to general prosperity; the absence of resources need not be a barrier to growth.

For poor countries then, having some energy resources, such as uranium or natural gas, clearly does not assure economic development. In fact, in abundance they can be a negative factor. But since development in poor nations is critically dependent on the availability of energy, these nations somehow need to obtain supplies of energy resources that will not drain all of their financial resources or upend society. Instead, they must utilize energy to raise up their populations from desperate poverty. Unfortunately, there is no easy answer as to how to get there.

References

Aguilera, R.F., R.D. Ripple, and R. Aguilera. 2014. "Link between Endowments, Economics and Environment in Conventional and Unconventional Gas Reservoirs." *Fuel*, 126, 224–238.

Anderson, D. and R. Fishwick. 1985. *Fuelwood Consumption and Deforestation in African Countries: A Review.*The World Bank, Washington, DC.

Birol, F. 2007. "Energy Economics: A Place for Energy Poverty in the Agenda?" *The Energy Journal*, 28(3), 1–6.

Earl, D.E. 1975. *Forest Energy and Economic Development.* Clarendon Press, Oxford.

Goldemberg, J., T.B. Johansson, K.K.N. Reddy, and R.H. Williams. 1987. *Energy for a Sustainable World.* World Resources Institutes, Washington, DC.

Grossman, P.Z. 2013. *U.S. Energy Policy and the Pursuit of Failure.* Cambridge University Press, Cambridge, UK.

Karl, T.L. 1999. *The Paradox of Plenty: Oil Booms and Petro-States.* University of California Press, Berkeley, CA.

Kharencha, P.A. and J.E. Hansen. 2013. "Prevented Mortality and Greenhouse Gas Emissions from Historical and Projected Nuclear Power." *Environmental Science and Technology*, 47, 4889–4895.

McGlade, C. and P. Ekins. 2015. "The Geographical Distribution of Fossil Fuels Unused When Limiting Global Warming to 2°C." *Nature*, 517, 187–190.

Mehlum, H., K. Moene, and R. Torvik. 2006. "Institutions and the Resource Curse." *The Economic Journal*, 116(508), 1–20,

Mikesell, R.F.1997. "Explaining the Resource Curse, with Special Reference to Mineral-exporting Countries." *Resources Policy*, 23(4), 191–199.

North, D.C.1990. *Institutions, Institutional Change and Economic Performance.* Cambridge University Press, Cambridge, UK.

Parikka, M. 2004. "Global Biomass Fuel Resources." *Biomass & Energy*, 27, 613–620.

Plumer, B. 2016. "Hundreds of Coal Plants are Still Being Planned Worldwide: Enough to Cook the Planet," Vox article, at http://www.vox.com/2016/4/5/11361390/coal-plant-pipeline-china-india.

Rosenberg, D.M., F. Berkes, R.A. Bodaly, R.E. Hecky, C.A. Kelly, and J.W.M. Rudd. 1997. "Large-Scale Impacts of Hydroelectric Development." *Environmental Reviews*, 5, 27–54.

Yergin, D. 1991. *The Prize.* Free Press, New York.

Yergin, D. 2011. *The Quest.* Penguin Books, New York.

NOTES

1 Blog post at the World Bank website blog, *Voices*, by Managing Director Sri Mulyani Indrawati, 7 July 2015. At http://blogs.worldbank.org/voices/.

2 Countries often needed financial help from the International Monetary Fund (IMF), which has required strict adherence to austerity measures in order to receive additional loans.

3 They also can pose great hazards. The failure of a dam in China (the Banqiao Reservoir Dam) in 1975, caused the deaths of over 160,000 people.

4 It is important to note that a hot climate does not guarantee sufficient sunlight to generate a substantial amount of electricity. Also, a desert climate, which does typically have steady sunlight, often produces dust, which reduces the solar equipment's efficiency.

PART II

POWER GENERATION: THE TECHNOLOGY AND ITS EFFECTS

6 Fossil Fuels: Impacts and Technology

Introduction

Reliable electricity is a fundamental requirement of a modern industrial economy. Studies show it is more important to economic development and wealth creation than oil consumption or total energy consumption (see, for example, Ferguson et al. 2000.) In Chapter 3, we explained the basic technology of electric generation as well as the composition of a steam power plant and a hydroelectric facility. Over the next four chapters we will explore the resource demands, additional technological issues, economic costs and societal impacts of electricity production and use. But we start with fossil fuels. About two-thirds of the world's electric power is generated by fossil fuels, mainly coal and natural gas, although about 5 percent is generated by burning oil. Most of the rest is generated by hydro and nuclear power, about 16 and 12 percent respectively. We will look at nuclear power in the next two chapters, but here we will focus primarily on the two resources that provide most of the fuel for electric power.

In earlier editions this book, we focused exclusively on coal in this chapter because it has long been noted as an abundant resource. In fact, in the 1980s, when the first edition of this book was prepared, it seemed that coal might be the only fuel the United States could rely on for more than another decade or two. Oil production seemed to have peaked in the United States and natural gas supplies also appeared to be dwindling. As we have seen, coal is by far the largest fossil-fuel resource available. It has the resource capacity to supply the world's energy needs for at least a few centuries. While superabundant energy sources, such as nuclear fusion or large-scale solar technologies, may be developed, at this time coal is the most abundant (and lowest cost per unit of energy of any) conventional energy source.

But the second reason that we chose coal is that its environmental impacts are the most severe. In routine operation, from mining to combustion to disposal of wastes, there are hazards, creating potentially large spillover or externality costs – externalities, as noted in Chapter 4, being the unintended costs to society from use of a technology or resource. In recent years, the impression of coal's hazardousness has grown because during combustion it emits the most carbon dioxide (CO_2) of any fossil fuel – emissions that transcend borders and last for generations and are believed to be the main contributor to human-caused climate change. In the 2010s no environmental issue is of greater concern than the effect of human activities on the long-term state of the climate. In this chapter, we consider the issue, although in Chapters 8, 10, and 11 we will explore in greater depth the question of what to do about it.

While the effects of CO_2 from fossil fuels are global issues, coal can also have local pollution impacts that can be damaging in the short run to the health of those living nearby as well as to forests, lakes, and wildlife many miles away. In the developing world, as we have noted, coal is often the fuel of choice but it is in the cities of those countries that are burning coal that air pollution is most evident and dangerous. It is estimated that 38 percent of the population of China (over 500 million out of 1.4 billion people) breathe air that would be considered unhealthy in the United States. These types of coal-related externalities, and how to deal with them, will also be a major focus of this chapter.

For the United States, it should be noted, coal's importance is declining. As of our last edition (1998) coal provided over 50 percent of the fuel for electric generation and a good portion of the fuel for industrial process heat. In all it accounted for 21.7 quads of energy consumption, up 40 percent from 1980. It was approximately equal to the energy we were deriving from natural gas, although the latter was used also in home space and water heating. Usage of these two fuels stayed roughly comparable through 2008, when a notable shift away from coal and toward natural gas began. By 2013, coal consumption had fallen to around 18 quads while natural gas had risen to nearly 27.

There were two reasons for this: first, natural gas supplies, as we noted in Chapter 2, were not dwindling. Indeed, reserves were growing every year and the price of gas in the 2010s has been very low. At the same time, concern over global climate change led the Environmental Protection Agency to promulgate new rules that added additional costs to coal-fired electric power plants.[1] As Figure 6.1 shows, electric companies have begun to transition away from coal to natural gas. Gas, in addition to its abundance, burns far more cleanly than coal and the switch from coal to natural gas has offered a straightforward way for electric power companies to cut polluting emissions. Consequently, this chapter will look at the issues surrounding the use of coal, and will also consider the question of whether natural gas is at least one way to address those issues, or whether it just replaces one set of problems with another.

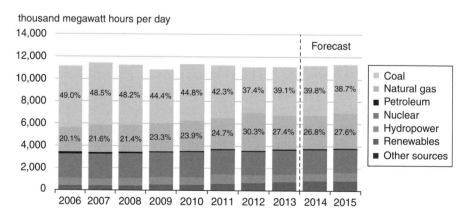

Figure 6.1   US electric generation by fuel type.
Source: USEIA.

Coal Use in the United States: Past, Present, and Future

Prior to World War II, coal was used extensively, not only in electric generation and industrial heating but also in railroad transportation. But following the war, coal use was displaced largely by oil for transportation. Oil was not only readily available, but proved easier and cheaper to transport and store, and in most cases was cleaner to burn than coal. Despite large annual increases in energy consumption in the United States, coal demand actually declined by about one-third from the 1940s through to the 1960s. In fact, the level of demand in 1943 (651 million tons) was not surpassed until 1975. During the 1970s energy crises, coal supplied less than one-fifth of the total US energy demand, including less than half of the fuel for electric generation and only about one-sixth of the fuel for industrial heating. These small shares of total energy supplies by coal were typical of other industrialized countries as well during that period.

With the price of oil and gas rising, and fears of depletion of those fuels growing, coal use increased. In fact, in 1978 the US Congress passed the Powerplant and Industrial Fuel Use Act, which essentially required any new fossil-fired power plant to use coal instead of oil or natural gas, and to convert oil-fired and gas-fired facilities to coal. Initially, conversion was somewhat slow because of the same disadvantages in transportation and handling that had contributed to a decline in coal use 50 years earlier. Nevertheless, by 1986 coal supplied about 56 percent of the energy for electric generation (over 15 quads, nationally). By 1995, US consumption exceeded 1 billion tons (short tons equal to 0.907 Bmt) per year, more than double its level in 1965. This trend continued into the mid-2000s, but then the trend began to reverse. From 2002 to 2013, the number of coal plants in operation in the United States fell by 18 percent and coal consumption declined overall by about 25 percent.

United States consumption of coal, however, fell more than production as America became one of the world's leading coal exporters. Nevertheless, the expectation is that both consumption and production will fall further. Figure 6.2 shows production from 2008 to 2013 with projected production to 2016. By 2013, according to the EIA, US coal production had fallen below 1 billion short tons to 984.8 million short

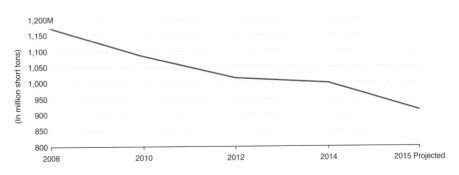

Figure 6.2 Decline of US coal production (1 short ton = 0.907 metric tons).
Source: USEIA.

tons (893.4 million metric tons) for the first time in two decades. As the figure indicates, production was projected to decline by an additional 100 million tons.

The Impacts of Coal

Coal usage affects water, air, land, and wildlife, as well as human health and safety, and therefore has been subjected to more and more environmental and land use restrictions. These curbs have been imposed on most aspects of its use, from *extraction*, *transportation*, *preparation*, and *combustion* to the final *disposal* of its by-products, as shown in Figure 6.3. At every step of the process, the impacts, if uncontrolled, can have serious consequences.

Before we discuss the specifics of the impacts of coal, we should make two points. First, not all coal has the same properties and the impacts vary according to the kind of coal in question. And second, its impacts are affected by where coal is found – in terms of its depth at a particular site and in terms of its geographic location for mining.

Coal resources may be divided into four categories: *lignite, sub-bituminous, bituminous,* and *anthracite*. Table 6.1 gives data on representative US samples of each category. It shows the most energy per pound in anthracite coal. But anthracite reserves are relatively small and it has been used in few electric generating plants in

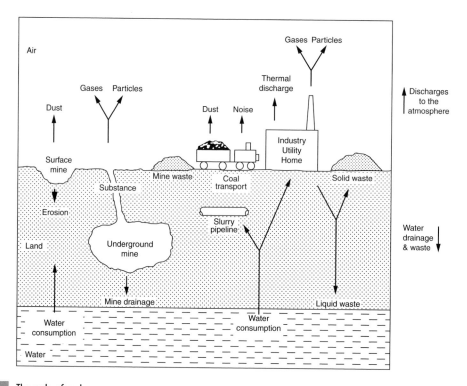

Figure 6.3　The cycle of coal use.

Source: Adapted from Wilson (1980) and Office of Technology Assessment, US Congress (1979).

Table 6.1 Representative types of coal (power plant data)				
Type of coal rank	Lignite	Sub-bituminous	Bituminous	Anthracite
Representative locations	McLean, North Dakota	Sheridan, Wyoming	Muhlenberg, Kentucky	Lackawanna, Pennsylvania
Physical composition (% by weight)				
Moisture	37	22	9	4
Volatiles	28	33	36	5
Fixed carbon	30	40	44	81
Ash	6	4	11	10
Total	101	99	100	100
Chemical composition (%)				
Sulfur	0.9	0.5	2.8	0.8
Carbon	41	54	65	80
Other	58	45	32	19
Total	100	100	100	100
Energy (BTU/lb)	7,000	9,610	11,680	12,880

Table 6.2 US coal: sulfur content and reserves (by coal type)				
	Lignite	Sub-bituminous	Bituminous	Anthracite
Sulfur content (%)	0.7	0.5	2.2	0.6
Estimated reserves (Bmt)	38.8	161.5	233	6.8

Sources: US Department of Energy, Coal Data; Cost and Quality of Fuels: 1979; and USEIA US Coal Reserves: 2012.

recent years. Clearly, bituminous coal, which is second highest in energy content, has a tremendous advantage in size of reserves (Table 6.2). Indeed, not only are the reserves enormous but they are widespread, as Figure 6.4 indicates. But bituminous coal also has by far the highest sulfur content and that increases some of the negative impacts, as we will see.

Lignite and sub-bituminous coal present other problems. Lignite has the lowest heat content and crumbles easily, which makes it difficult to transport and store. Sub-bituminous coal, which is low in sulfur and has higher energy content than lignite, would seem ideal. But here is where the issue of location enters; more than 90 percent of the sub-bituminous reserves are in Montana and Wyoming (see Fig. 6.4). To distribute this around the country requires transporting it as much as 2000 miles. As with any commodity, the further it must be shipped the greater are the costs for its use. It also matters where in the ground that coal is found. Is it near enough to the surface so that top soil can be removed and the coal simply scooped up (see the next section)? Or must shafts and tunnels be dug and the coal extracted by underground mining techniques? It should be noted that both surface and underground mining have environmental impacts, but the mining techniques used will affect both the costs of extraction and the extent of the spillovers.

Coal Extraction, Preparation, and Transportation

Though coal production and consumption in the United States have been falling, America still produces several hundred million tons of coal per year and will probably continue to do so for the foreseeable future. Yet the extraction of coal is a dirty and potentially dangerous process. Underground mining, principally in the Appalachian regions of the Eastern United States (Fig. 6.4), has harmed local water supplies and caused other local disruptions. As water seeps into coal seams, particularly in abandoned mines – where there is no attempt to deal with drainage – it reacts with compounds in the coal. Acids (such as sulfuric acid) form and then leach into underground aquifers and drain into local rivers and streams, polluting them. In Appalachia, grassroots organizations and government agencies have been fighting water pollution for decades – a fight that has not ended.

Underground extraction can also threaten the stability of the ground. Years of digging can result in land subsidence at the surface – an actual caving in of the ground. Buildings have been damaged or destroyed by land collapses, often many years after a mine had been exploited and closed. According to the US Geological Survey (USGS), as of 2000, more than 17,000 square miles in 45 states, an area roughly the size of New Hampshire and Vermont combined, had been directly affected by subsidence. Cumulative damage to property by land subsidence is in the billions of dollars.

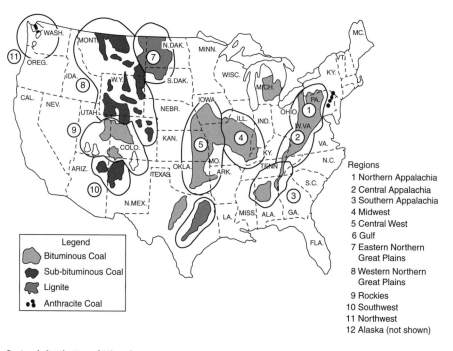

Regions
1 Northern Appalachia
2 Central Appalachia
3 Southern Appalachia
4 Midwest
5 Central West
6 Gulf
7 Eastern Northern
    Great Plains
8 Western Northern
    Great Plains
9 Rockies
10 Southwest
11 Northwest
12 Alaska (not shown)

Legend
Bituminous Coal
Sub-bituminous Coal
Lignite
Anthracite Coal

Figure 6.4   Regional distribution of US coal resources.
Source: Energy Information Administration, US Department of Energy (1981).

However damaging underground mining has been to the environment and to the public at large, these dangers have been slight compared to those historically faced by underground coal miners. Underground mining is one of the most dangerous jobs in the world. Miners have been killed and injured by mine cave-ins and explosions, which still occur around the world, though infrequently in the United States. Much more damaging to US miners is *black lung disease* (pneumoconiosis), a condition that results from prolonged inhalation of coal dust. As of 2014, the rate of black lung disease had risen back to the level of 1974, notwithstanding congressional efforts (such as passing the Coal Mine Health and Safety Act of 1969) to protect miners' health.

Surface mining of coal has turned out to be less hostile to workers, but it has produced more environmental damage (National Academy of Science 1974). Prior to congressional passage of the Surface Mining Control and Reclamation Act of 1977, surface coal mining operations had scarred thousands of acres in Appalachia and the Midwest – marring them to the point that they could not be reclaimed for other uses. This was especially true where the technique of *highwall mining* was used. In this method, which was frequently employed in Appalachia, the coal was dug out of the side of a mountain, beginning where the coal seam reached the surface. Typically, the top soil and rock and vegetation covering the seam – called the overburden – were simply scooped up and dumped down the side of the hill, leaving a steep slope. At best, this method left unsightly mounds called *spoil piles*. But, in fact, rocks from the overburden sometimes tumbled down hills into fields and even backyards of homes. Meanwhile, the digging continued in toward the center of the mountain until there would be a sheer wall of rocky rubble, as much as 100 feet high and miles in length, where soil and vegetation had been – a wall that would be left, sheer and barren, once the seam had been fully removed.

Some have pointed to the environmental and human costs of coal mining as the main reason to slow the expansion of coal use. In fact, many of these problems can be alleviated, if not solved, by technology. But it is important to note that coal extraction may also present social and economic issues that have no easy solutions.

Perhaps the most important social question connected to the expanded extraction of coal is *land use*. In the United States, some farmers and Native Americans who own land under which lie vast fields of coal have opposed mining these fields regardless of the economic incentives offered by mining interests. If personal values are given a higher priority, many farmers would prefer to keep their land undisturbed, if for no other reason than because the land has been farmed by their families for generations. A number of these farmers have refused outright to sell or lease their land for surface mining, because to them the traditional tie to the land takes precedence over profit.

Coal also must be transported from the mine to the point of use. Actually about 10 percent of all coal is burned at the mine mouth to generate electric power. But the rest must be shipped. As we have noted, the cleanest North American coal is in the western states, far from present points of major use. In the United States, coal is transported in river barges or by truck or railroad. More than half is shipped by rail. Western coal is commonly placed in what are called *unit trains* – trains of 100 coal cars or more – each train carrying a total of 10,000 tons on each trip from the mine.

These trains travel on established rail lines that typically pass through towns and cities. Between 2010 and 2014 there were more than 50 accidents in the United States involving coal trains, including five over a 3-week period in the summer of 2012.

Alternatives to the rails are *coal slurries*, where coal is pulverized, mixed with water, and sent through a pipeline. This has advantages, not least of which is that coal-fired boilers often use pulverized coal. After the water is removed and the coal is dried, the slurried coal is ready for use. Slurry pipelines are a cheaper means of transport than rails. One slurry pipeline can take the place of a tremendous amount of rail traffic. The disused Black Mesa coal slurry pipeline is 273 miles (439 km) long and transported 5 million tons of coal annually from the coal deposits at Black Mesa in Arizona to a power plant in Nevada. This would have required 150 rail cars per day if it were done by railroad.

But the Black Mesa line, which was shut down at the end of 2005, was unusual, and attempts to build more like it have met resistance – particularly over issues of land use and water rights, especially the latter (Begaye 2005). Coal slurry lines require massive amounts of water; the Black Mesa pipeline used 2700 gallons per minute. Water rights have been an especially critical issue in the Western coal lands because water in the West is relatively scarce. Ranchers, farmers, and others who rely on local sources have opposed use of water for slurries. In 1978, Western congressmen led the fight against a bill that would have permitted construction of slurry pipelines, and the Black Mesa line was closed mainly because Native American tribal councils prohibited the coal company from access to the Navajo Aquifer.

As for the preparation process of coal for use, it consists of crushing and cleaning coal for improved and cleaner combustion. A potential for environmental damage is present because preparation facilities produce large volumes of waste and often consume considerable quantities of water. On the whole, however, in normal operation, the environmental impact of the preparation process itself is relatively low.

Coal Combustion

Combustion converts coal into useful heat energy, but it is also the part of the process that engenders the greatest environmental and health concerns. Coal combustion releases a variety of air pollutants, including suspended particulates, sulfur dioxide (SO_2), nitrogen oxides (NO_x), carbon monoxide (CO), and fine particles. Of course, carbon dioxide (CO_2) is also produced in combustion, and in fact in greater amounts per unit of energy produced than any fossil fuel.

The United States has been successful in reducing or even eliminating some coal pollutants that are still found elsewhere in the world. In fact, SO_2 remains *the* major air pollutant from coal combustion today. Electric generation and industrial heat account for about 90 percent of all SO_2 emissions, almost 70 percent of which comes from power plants. Sulfur dioxide causes widespread material damage to building exteriors (the treasured Taj Mahal in India, for example) and can destroy works of sculpture made of limestone or marble. It also aggravates metal corrosion. A power plant without controls emits about 14,000 tons of SO_2 per year.

Sulfur dioxide is a particular concern to public health because it produces respiratory irritation in the surrounding population. In high concentrations, it can increase acute respiratory ailments and, where local concentrations are very high, it can lead to a significant number of deaths. It should be noted that the evidence for most of these fatalities is *epidemiological*, that is, derived from public health statistics. As a consequence, the victims usually cannot be identified as having died directly from pollution. However, it has been observed that the death rate, especially among vulnerable groups such as the elderly and the chronically ill, suddenly rises during periods of high levels of SO_2 locally in the atmosphere. Indeed, during the past century, air pollution levels in a few locales have been so high that hundreds of deaths have resulted. In London in 1952, for example, 3900 more deaths than were expected occurred during four days of very high levels of pollution; SO_2 levels were especially high (see Wilson et al. 1980). In China's cities in the 2010s, where levels of SO_2 can be exceptionally high, it is estimated that upwards of 600,000 people die from the effects of air pollution each year.[2]

Present air quality standards as established by the US government limit pollutant densities within specific localities to what are called *threshold values*. At these levels, according to the concept, health effects begin to become significant for each pollutant (Table 6.3). Other industrialized countries have similar air quality standards. Differences of opinion persist, however, as to the adequacy of these standards for public health protection.

Other pollutants whose effects are less well understood include the nitrogen oxides (NO_x, where: x is an integer denoting the particular oxide compound) and the related photochemical oxidants, particularly ozone (O_3). (The photochemical oxidants result from the reaction of nitrogen oxides with hydrocarbons.) About half of these come from stationary sources, such as coal-burning furnaces and boilers, whereas the other half are emitted by mobile sources, such as cars and trucks. Heavy concentrations of the two in a locality cause *smog*, as has been experienced in places around the world from Mexico City to Moscow. Smog produces respiratory distress and eye irritations as well as aesthetic degradation and visibility limitations.

So-called *fine particles* and carbon monoxide are the remaining air pollutants of coal. Particles in the stack gases of combustion vary in size from 0.01 μm (μm = 10^{-6} meters) to 10 μm in diameter. The smallest particles, in the range of 0.01–1 μm, are the most dangerous to health. They are thought to be deposited in the respiratory system and add to pollution-related respiratory distress. Carbon monoxide (CO) is a natural product of combustion of the carbon in coal as well as petroleum. It impairs the oxygen-carrying capacity of the blood, thus straining the cardiovascular system and can, in high concentrations, contribute to heart disease.

Although air pollution is the most apparent local effect of coal combustion at the point of use, coal combustion also affects the environment in remote regions. Two major pollutants of fossil-fuel combustion, and coal burning in particular, are SO_2 and NO_x, which are the primary sources of *acid rain*. The acid in acid rain is about one-half to two-thirds sulfuric acid, and most of the rest is nitric acid (see Fig. 6.5a). According to most theories, the acids form high in the atmosphere where the oxides are transformed into sulfates and nitrates that then react with moisture to form acids. The acids then fall with the rain.

Table 6.3 Health effects, threshold concentrations, air quality standards, and margins of safety for fossil-fuel pollutants

Pollutant	Lowest best judgment estimate for an effects threshold	Adverse effect	Standard	Percentage safety margin[a] — Lowest best judgment estimate for an effects threshold	Percentage safety margin[a] — For all best judgments estimates
Sulfur dioxide	300–400 µg/m^3 (short-term)	Mortality harvest	365 µg/m^3	None	None to 37
	91 µg/m^3 (long-term)	Increased frequency of acute respiratory disease	80 µg/m^3	14	14 to 212
Suspended sulfates	8 µg/m^3 (short-term)	Increased infections in asthmatics	None	None	None
	8 µg/m^3 (long-term)		None	None	None
Total suspended particulates	70–250 µg/m^3 (short-term)	Aggravation of respiratory diseases	260 µg/m^3	None	None to 15
	100 µg/m^3 (long-term)	Increased prevalence of chronic bronchitis	75 µg/m^3	33	33–133
Fine particles (<10 µm dia.)	(short-term)	Deep penetration into lung tissues, aggravates asthma and bronchitis	150 µg/m^3		
	(long-term)		50 µg/m^3		
Nitrogen dioxide	141 µg/m^3 (long-term)	Increased severity of acute respiratory illness	100 µg/m^3	41	41–370
Carbon monoxide	23 mg/m^3 (8 hour)	Diminished exercise tolerance in heart patients	10 mg/m^3	130[b]	130–610[b]
	73 mg/m^3 (1 hour)		40 mg/m^3	82[b]	82 to 788[b]
Photochemical oxidants	200 µg/m^3 (short-term)	Increased susceptibility to infection	160 µg/m^3	25	None to 363

[a]Safety margin = effects threshold minus standard divided by standard × 100.
[b]Safety margins based upon carboxyhemoglobin levels would be 100 percent, for an 8-hour period; with a range of 100–40 judgment estimates. µg/m^3: micrograms per cubic meter; mg/m^3: milligrams per cubic meter.
Source: NAS, NRC (1977), EPA (1994).

    The heaviest concentrations of acid rain in North America have been recorded in the Northeastern United States and Eastern Canada. But atmospheric scientists feel that most of the constituents of acid rain are produced in the US Midwest and that prevailing wind patterns carry the chemicals several hundred miles eastward. In any case, regions of the Northeast have noted unusually acidic rain. Normally, rain has an

(a)

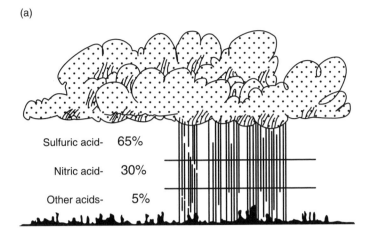

Sulfuric acid-   65%

Nitric acid-   30%

Other acids-   5%

Typical northeastern US acid rain components

(b)

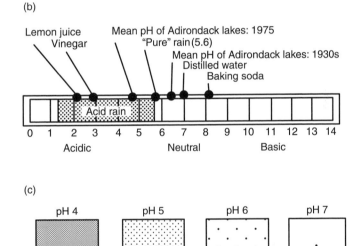

(c)

| pH 4 | pH 5 | pH 6 | pH 7 |

Acid concentration

Figure 6.5 Acid rain.
Source: US Environmental Protection Agency (1980).
(a) Acid rain components. Pure rain has a pH of between 5.6 and 5.7. These pH values take into consideration the amount of acidity created by the reaction of rainwater with normal levels of atmospheric carbon dioxide. Acid precipitation (rain, snow, sleet, or hail) has a pH 5.6 or below.
(b, c) The pH scale. The pH (potential hydrogen) scale is a measure of hydrogen ion concentration. Hydrogen ions have a positive electrical charge and are called cations; ions with a negative electrical charge are known as anions. A substance containing equal concentrations of cations and anions, so that the electric charges balance is neutral, has a pH of 7. However, a substance with more hydrogen ions than anions is acidic and has a pH less than 7; substances with more anions than cations are alkaline and have pH measures above 7. Thus, as the concentration of hydrogen ions increases, the pH decreases with increasing acidity.

acid balance, with a value of 5.6 measured on a scale called the *potential hydrogen* (or pH). But rain with pH values ranging down to 5.0 or lower (increasingly acidic; see Fig. 6.5b, c) is regularly recorded in some places, with serious consequences for lakes, rivers, and forests. As we discuss below, the Clean Air Act of 1970 and particularly the amendments to it in 1990 have considerably reduced the pollutants responsible for acidic precipitation, and recovery of some of the damaged areas has occurred. But, as the 2011 National Acid Rain Precipitation Assessment prepared for Congress noted, "Despite the environmental improvements reported here, research over the past few years indicates that recovery from the effects of acidification is not likely . . . without additional decreases in acid deposition [SO_2 and NO_x]." Analysis of the annual precipitation pH in 2008 showed that throughout the Northeastern United States, the pH, though improved since 1994, was still lower than 5.0.

Coal Disposal

Every coal-burning plant produces air-born solid waste products (such as fly ash from the flue and bottom ash from the furnace), all of which must be disposed of. Even though such products represent only a small percentage (by volume or weight) of the coal input, the absolute amounts are huge because of the massive quantities of coal used. For example, a typical 1000 MW power plant operating for a day will burn close to 10,000 tons of coal. If solid wastes weigh only 10 percent, then 1000 tons of waste must be disposed of for each day of operation.

The problem of coal waste disposal revolves around the issue of land use. Large landfill areas must be provided, either adjacent to each generating site or at sites available for low-cost transportation (rail or water). These can be difficult to find, especially when a plant is located in a densely populated region where landfill sites are becoming over-utilized. Care must be taken in selection of these waste sites for both aesthetic and environmental reasons. The particular environmental concern, here and with much of coal technology, is the risk of seepage of harmful chemicals into groundwater.

Control Technologies and Their Costs

It is possible, albeit costly, to reduce substantially most of the environmental impacts from the use of coal, although with coal consumption decreasing in the United States it is not clear how much more should be invested in coal spillover abatement.

There have been efforts to control the problems of extraction. To prevent subsidence, tunnels may be back-filled with mining spoils or with refuse from the preparation process, or deliberately collapsed in a controlled fashion.

Miner health and safety improvements have been required since the 1969 legislation, and the number of mining fatalities has fallen dramatically. Better ventilation has reduced mine explosions, and because it lowers the level of suspended coal dust it may also help to reduce instances of black lung in miners. The use of sprays of water at the coal seam reduces coal dust levels as well. In fact, one study suggested that further improvements in dust control could lower pneumoconiosis rates by more than 90 percent.

The process of land reclamation depends in part on the kind of land that must be restored. To reclaim farmland, for example, the topsoil needs to be saved during the mining process. In Germany, the subsoil too is stored and returned. This eases the process of reclamation, but adds so much to the cost that it is only cost effective where coal seams are especially thick. Mine pits can be refilled not only with soil but with debris. The coal industry has experimented with using coal ash (a useful technique for *mine-mouth* power plants), as well as municipal sewage, as fill to reclaim lands. Sewage actually may increase the land's fertility and ease revegetation. The overall cost of land restoration from surface mining can vary tremendously. To restore mined lands to agricultural pursuits such as pasturage is more costly, generally, than for forest. In the West, the cost has been estimated to be as little as $2000 per acre ($2014), while farmland in Ohio can cost up to eight times as much per acre to restore.

Combustion Controls

The most important, and costly, problem of coal technology is the control of combustion emissions. The basic question is this: How much air pollution control and abatement is desirable, given the cost to achieve it? One might think that 100 percent removal of pollutants is the most desirable, but it might well be prohibitively expensive (and possibly technically infeasible). Of course, significant pollution control is possible; indeed, it is already required by law. But the cost of using the technology could rise considerably, depending on the level of control that a community or society demands.[3]

There are a number of different ways to reduce pollution from coal burning. Many utilities have relied on low-sulfur coal (Table 6.1), which over the last decade has helped reduce – although not eliminate – the need for new pollution control technology. Because of transportation costs, low-sulfur coal may add to the expense, although when we consider the cost of pollution control, the savings on capital equipment may more than offset it.

But local, state, and national air quality standards have made investment in pollution control technology mandatory, with respect primarily to three pollutants: SO_2, NO_x, and suspended particulates. Current, national clean air standards require that SO_2 emissions be reduced over 90 percent. The best available technology for eliminating SO_2 is the *scrubber*, or flue-gas desulfurizer. It was the *specified technology* for desulfurization in the amended Clean Air Act of 1977.[4] As Figure 6.6 shows, the effect of pollution control technology is quite dramatic.

With flue-gas desulfurization (FGD), coal combustion gases pass through a wet alkaline solution, usually of calcium oxide (lime) or calcium carbonate (for instance, limestone). The solution is sprayed through the scrubber, creating an alkaline slurry which reacts with the sulfur and removes it from the emitted gas (Figure 6.7). While this is effective in lowering SO_2 emissions, it requires large

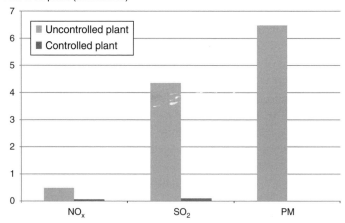

New coal-fired plant (Ibs/MMBtu)

Figure 6.6   Pollutants from coal-fired plants.
Source: National Energy Technology Laboratory.

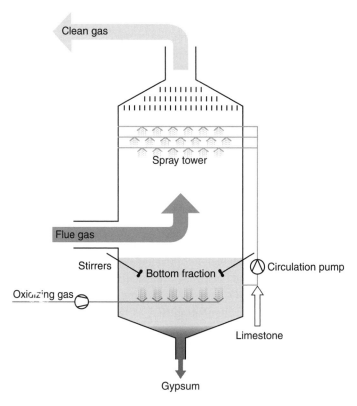

Figure 6.7   Schematic design of the absorber of a flue gas desulfurization unit.

quantities of water (1000 gallons per minute for a 1000 MW power plant) and leaves a sludge that presents a second disposal problem. It is also expensive, depending on the size of the plant as much as US$300/kW, although larger plants lower the cost closer to US$200/kW. The process is principally designed to remove SO_2 from emissions, but it also removes particulates and nitrogen oxides to some extent. On the whole, though, significant removal of nitrogen oxides requires a different treatment, which can occur either after or during combustion.

Suspended particulates from coal-fired power plants has largely been eliminated in the United States since the mid-twentieth century, using filtering and *electrostatic precipitators*. Such devices have long been used to remove suspended particulates from coal smoke. The technology works by the simple principle of electrostatic attraction. The particles in the smoke become ionized by a strong electric field between two electrodes of an electrostatic precipitator. Once ionized, the particles are drawn to and captured on the electrodes. Periodically, the particles are removed from the electrodes and disposed of. Fabric filters, sometimes called *baghouses*, do a similar job. Essentially, the filters are like vacuum cleaner bags through which the flue gases must pass. The bags then filter out the particulates (see Wilson et al. 1980). Through the use of these two devices, coal-burning utilities have captured the larger, more visible particles, and removed the elements that used to cause a smoky appearance in most cities. This manifestation of pollution can still be observed in many developing countries where particulates are uncontrolled.

Selective catalytic reduction (SCR) systems are the most effective technology for reduction of NO_x. This is a process where a reducing agent, usually ammonia, is injected into the stream of gases that emerge from the burner into the stack. The mixture of gas with ammonia (or urea) is absorbed onto a catalyst (usually metal, either vanadium or titanium) enabling the chemical reduction of NO_x to its basic constituents: nitrogen and water. The systems add about $100/kW to the cost of a new coal-fired power plant. According to one study in 2007, the technologies to control SO_2, NO_x and particulates increased the cost of a conventional 550 MW coal-fired power plant by $324/kW, or 21 percent.[5]

Of course, fossil fuels all emit carbon dioxide when they are burned because carbon is the major part of their chemical composition. Biomass fuels, such as wood, also have carbon in their composition, but create less CO_2 in combustion than fossil fuels because of the carbon fixation that has taken place in the growth of the wood. We discuss the CO_2 issue later in the chapter.

Cap-and-Trade for Sulfur Dioxide

Pollution control technologies are expensive, but an innovative way to limit costs of pollution abatement lies not in technology but in the way that pollution control rules are administered. The Clean Air Act of 1970 prescribed initial rules for pollution control. The newly formed Environmental Protection Agency (EPA) was to devise ambient air quality standards and performance standards for "new sources" such as newly built coal-fired power plants. It also called for states to develop plans to reduce pollution and for control over polluting emissions from mobile sources (i.e. cars and trucks). Though standards were established, it was not always clear how they would be achieved. With respect to automobile emissions standards, not only did cars need to be fitted with a pollution control device (called a catalytic converter) but gasoline had to be reformulated so that the device would not be degraded by a particular fuel additive, lead.

Considerable progress was made in reducing pollution over the next 20 years. Sulfur dioxide and NO_x levels fell per BTU of coal consumed; in the case of SO_2 emissions fell by more than 20 percent. But this was not deemed sufficient as the acid rain problem was still growing and experts believed that a solution required far greater reductions in the pollutants most responsible for it. In any case, by 1990 the level of emissions had stopped falling.

In 1990, the Clean Air Act was amended again. Title IV of the amended act created the Acid Rain Program, which focused specifically on the reduction of SO_2 and NO_x emissions from power plants. But the act contained an important innovation for implementing this program. Under the rules, the EPA set a maximum limit for emissions – a "cap" – and gave allowances (essentially a form of property rights or, at least, legal privileges) to each of the sources covered by the program to emit a certain amount of the pollutants. But the cost of emissions reduction varied among these sources. New power plants, for example, often had more efficient burners that consumed more of the sulfur and nitrogen oxides through combustion; they could achieve the required emissions reductions and then some with little or no additional cost. Others, with older equipment, might have had to spend hundreds of millions of dollars to upgrade their facilities.

But the new rules allowed the efficient plants to sell – that is, "trade" – their allowances to the inefficient ones. This meant each of the former, now with fewer allowances, took upon itself the task of reducing emissions by more than its initial allowances permitted, while at the same time allowing inefficient plants to exceed their allowances at a fraction of the cost of new equipment.

From 1995 to 2004, the trading program was clearly successful. Sulfur dioxide emissions from electric power plants fell 36 percent – from 15.9 million to 10.2 million tons – even though electric generation from coal-fired power plants had increased by 25 percent over the same period. By 2007, emissions had fallen further still (to 8.95 million tons), and by 2010 emissions had fallen to 5.1 million tons (Schmalensee & Stavins 2012). The results of the cap-and-trade program can be seen in Figure 6.8. Note that there were different phases. Initially, in Phase I, the program only included power plants of 100 MW or more. In Phase II, all electric generating facilities that burned fossil fuels were included – 3572 units in all. Nitrogen oxides were included in the plan but not in the trading regime. But one can readily see that the cap-and-trade system worked remarkably well, with SO_2 falling altogether by 75 percent from its 1990 level by 2012.

By the mid-2000s, with emissions now so much lower, the SO_2 program had outlived its purpose. Regulatory changes led to the cessation of the cap-and-trade market.[6] But the success of the program for SO_2 has had the effect of making cap-and-trade the default choice for pollution control of many policymakers. Any pollution issue, from mercury to carbon dioxide, has elicited calls for a cap-and-trade regime. But it is important to note two characteristics of the SO_2 program that made it

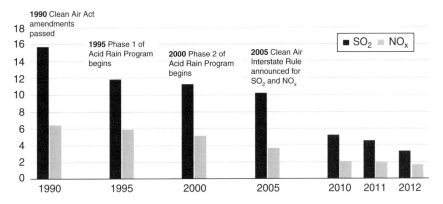

Figure 6.8   Reductions in SO$_2$ and NO$_x$ post-Clean Air Amendments of 1990.
Source: USEIA.

work so well – characteristics that are not always present with respect to other pollutants. First, the pollutants are emitted high in the atmosphere and are carried along with the jet stream patterns, mixing with other emissions along the way and then falling with rainfall often dozens or even hundreds of miles from the source. This means the reduction in *total* emissions is useful. If SO$_2$ pollution were local only, there would be local pollution "hot spots" around the inefficient plants and a cap-and-trade system would not be helpful; this was observed with respect to mercury emissions, which did not lend themselves to a cap-and-trade program.

Second, the sources were fixed and their emissions could be monitored in real time. In other words, it was possible for the EPA or other regulators to know exactly how much SO$_2$ was reaching the atmosphere every moment of the day. This is information that is often not possible to obtain with pollutants from other sources. In those cases, technical standards – one can easily tell if a pollution control device such as an automobile catalytic converter, for example, is installed – may be more efficient (Cole & Grossman 1999). Monitoring is a problem for the utilization of cap-and-trade for CO$_2$, where so many sources, stationary and mobile, are involved. Nevertheless, cap-and-trade was a very effective administrative innovation in reducing both pollution and cost with respect to sulfur dioxide.

Natural Gas: "Bridge" Fuel?

There is a relatively low-cost technology that substantially reduces the externality problems of coal. It is the replacement of coal in electric power generation with a high-efficiency combined-cycle generating system fueled by natural gas (see Chapter 3). Combined systems produce virtually no particulates and miniscule amounts of sulfur dioxide. They do produce some nitrous oxides, but according to a study the NO$_x$ emissions are 40 percent lower from a combined-cycle gas

(CCG) facility than from a coal plant with full abatement equipment (de Gouw et al. 2014). In other words, natural gas electric power stations provide a solution to many of the combustion externalities of coal.

Of course, the United States benefits from an abundance of natural gas – an abundance not found throughout the world. In many places, for example Japan, natural gas must be brought in by ship in liquefied form. Typically, liquefied natural gas (LNG) costs far more than the gas itself. For example, in early 2016, LNG unloaded in Japan was selling for US$7.75/per million BTU (1000 cubic feet equals 1 million BTU, MBTU), while in the United States the gas was selling for around $2/MBTU. Still, the benefits of natural gas are evident in the United States, where air quality has been substantially improved by a switch to natural gas.

There are some environmental problems with natural gas. Burning it does not create particulates, but drilling for it can. When it is produced through fracking, there have been, as noted in Chapter 2, mostly unsubstantiated claims of groundwater contamination as well as erosion and land disturbance in the vicinity of production.

Still, on the whole, natural gas is far less of a burden on the environment than coal and, as Figure 6.1 showed, it is becoming the fossil fuel of choice in the United States for power generation. It has been argued that natural gas-fired electricity should be included in "clean" energy development plans. Others see natural gas as a "bridge fuel,"[7] a cleaner replacement for coal to be utilized until the world can transition to a completely nonfossil-based electric system (see Chapter 11). It should be noted that some analysts are concerned that because natural gas is plentiful and cheap as well as relatively clean burning, that the "bridge" would instead be a barrier slowing the development of non-CO_2 emitting technologies. But in the meantime, the switch to natural gas lowered annual energy-related US CO_2 emissions by about 600 million metric tons from 2005 to 2014.

Carbon Dioxide's Potential Global Impact

Combined-cycle generating technology as well as control systems can lessen most of the impacts of burning fossil fuels. The issue of global carbon dioxide (CO_2) concentrations, however, is different.

Carbon dioxide is a constituent of the atmosphere of the Earth and a universal product in the cycle of life. It is absorbed by plants and generated and released by natural decay processes (see Fig. 6.9). The level of CO_2 in the world depends on the balance of these natural processes. Over time, world levels of carbon dioxide change very slowly.

Carbon dioxide, however, is also a product of the combustion of hydrocarbons coal, oil, and gas. It cannot be eliminated from fossil-fuel combustion and CO_2 emissions can be removed or reduced to very low levels only at great expense and trouble.[8] Essentially, CO_2 emissions will take place as long as we use these fuels. But, it is argued that the worldwide combustion of coal and other fossil fuels is causing global atmospheric CO_2 concentrations to rise. Careful scientific measurements conducted over decades have confirmed a slow-trend increase in the global concentrations

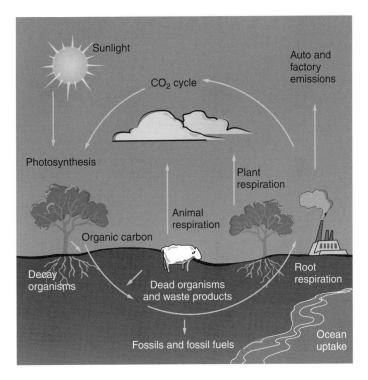

Figure 6.9 The carbon cycle.
Source: Adapted from Smith (1966).

of CO_2. In absolute terms, fossil-fuel burning is adding about 36 Bmt/year of CO_2 to the world's atmosphere. Although that may seem like a great deal, much more comes from other sources: an estimated 150 Bmt is emitted annually from the oceans and about 65 Bmt from terrestrial plant decay (which can be accelerated by human-caused deforestation). The increase in CO_2 concentration has actually been measured at around 2.1 parts per million per year (see Fig. 6.10). Yet the basic hypothesis is that this incremental addition is enough of a burden on the atmosphere to cause widespread harm by the year 2050 in the absence of worldwide controls. We are on a path to double the amount of CO_2 in the atmosphere from pre-industrial levels.[9]

If doubling occurs, then the Earth's average temperature will likely increase because of what is called the *greenhouse effect*. As the Sun's energy warms the Earth each day much of the heat escapes back into space through the atmosphere; but if the atmosphere contains a high level of CO_2, less heat (in the form of infrared radiation) is radiated out. Like the glass of a greenhouse, the atmosphere lets in the solar energy (of all wavelengths), but does not let all of the heat out. The natural greenhouse effect makes our planet inhabitable by sustaining the temperature at which life has evolved. It is caused by CO_2 and other natural gases in the atmosphere, such as methane. If CO_2 concentrations in the atmosphere continue to increase, however, it is feared that additional heat will be trapped and higher temperatures will result globally.

There is widespread concern about human-caused global warming – *anthropogenic global warming* (AGW) or the broader concept of anthropogenic *climate change*. This

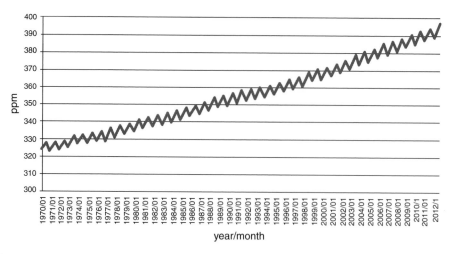

Figure 6.10   Atmospheric carbon dioxide. Data from Mauna Loa Observatory, Hawaii.
Source: http://cosmoscon.com/2012/06/09/global-temperature-and-co2-update-june-2012/. Data compiled by Climate Research Unit and the University of Alabama Huntsville.

concern became a very public issue in the 1980s. In 1988, the environmental program at the United Nations joined together with the World Meteorological Organization to form the Intergovernmental Panel on Climate Change (IPCC), which was charged with assessing through

> ... a comprehensive, objective, open and transparent basis the scientific, technical and socio-economic information relevant to understanding the scientific basis of risk of human-induced climate change, its potential impacts and options for adaptation and mitigation. IPCC reports should be neutral with respect to policy, although they may need to deal objectively with scientific, technical, and socio-economic factors relevant to the application of particular policies.

The first IPCC report (AR1) was released in 1990, and led to creation of the United Nations Framework Convention on Climate Change (UNFCCC), an international treaty to deal with the international character of climate change, its causes, and possible consequences, as well as actions to forestall and/or adapt to those changes.

Over the next 23 years, four more IPCC reports were issued, with AR5 appearing in 2014. The reports have been the basis for ongoing meetings among nations termed Conference of the Parties (COP), the first of which was held in Berlin in 1995 and laid the groundwork for the 1997 international treaty for the reduction of greenhouse gas emissions, primarily CO_2, the Kyoto Protocol. Though seemingly a major step toward a reduction in CO_2 emissions, the Protocol was at best of limited value. Developing countries were exempt even though such countries as China and India were accelerating their development of coal-fired power plants as well as energy intensive industries. The United States refused to be a party to the Protocol as well. Emissions (as Fig. 6.10 reveals) did not decline in the ensuing years.

But what happened to temperatures? The first IPCC report as well as subsequent ones gave reason for general alarm: the world was warming. According to the IPCC,

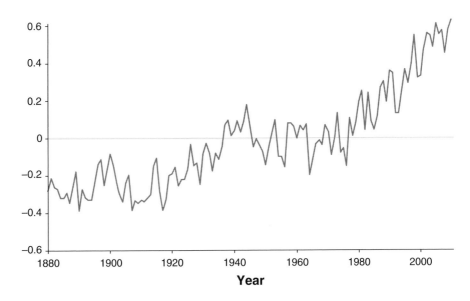

Figure 6.11   Global temperature anomalies: deviation from the mean. GISS temperature graph.
Source: NASA GISS at www.giss.nasa.gov.

a doubling of CO_2 concentrations in the atmosphere would lead to an increase of somewhere between 2 °C (degrees Celsius) and 5 °C (3.6–9 °F (Fahrenheit)). The only question was just how "sensitive" Earth's climate would be to a doubling of CO_2. Computer models consistently projected continually rising temperatures along with CO_2. The consequences of this were widely seen as potentially catastrophic, especially at the high end of projections.

But did actual temperatures validate projections? As Figure 6.11 shows, higher global temperatures were evident, especially in the period from the 1970s through the mid-1990s. But then, note that the temperature appeared to flatten out even as CO_2 was consistently rising (Fig. 6.10). Some have argued that global warming paused; others using different data sets that it continued to rise albeit more slowly. In fact, it is difficult to get one reading of global temperatures to which everyone would agree on its accuracy. But it also leaves unclear the question of climate sensitivity.

Still, the prospect of a global temperature change is disturbing. According to most climate scientists, if the Earth warms by more than 2 °C, the consequences could be severe. Higher temperatures (more than 3 °C) could lead particularly to severe regional effects, such as prolonged droughts, crop failures, and increased desertification. Melting ice caps in Antarctica and Greenland could lead to rapid increases in sea levels. Cities such as New York might have to construct seawalls to prevent flooding, and poor coastal countries such as Bangladesh could be inundated. Some scientists have argued that it is at least possible that climate change could lead to large regions of the world being uninhabitable, where human and plant adaptation is not possible because of the extremes of temperatures and/or weather-related catastrophes. Indeed, some have suggested that Earth itself may no longer be habitable because of human-induced change.[10]

There has been a general consensus among scientists worldwide that human activities have had (and will continue to have) an influence on the climate. As the IPCC has noted there are various factors that can contribute to climate change. These include: manufacturing processes that emit chlorofluorocarbons (CFCs); agriculture, which produces methane; deforestation, which reduces carbon-fixing vegetation; but most especially CO_2 emissions from fuels for energy. The emissions from energy production and consumption are thought to make the largest contributions to climate change.

Mitigation of CO_2 is, however, problematic both because of the cost and because of the disruption of lifestyles in a world so dependent on fossil fuels. Moreover, many issues remain unclear: for example, how severe will the negative effects be if there is no mitigation; how general (that is over the whole of the planet or just regions) will the impacts become; or when will serious problems begin? Indeed, it is argued by many experts that at least for the first few decades the effects of warmer temperatures might well be positive. Higher temperatures will mean longer growing seasons, and increases in CO_2 will enhance plant growth. Later, most scientists and economists hypothesize that impacts will have costs. But these, too, are speculative. A comprehensive review of evidence in the UK led by Nicholas Stern (2007) concluded that by the year 2100 average world GDP per capita would be 20 percent lower than if the world did not warm by as much as the consensus forecasts expect. But as Richard S.J. Tol, an economist who has contributed to the IPCC reports, notes, the losses projected in the *Stern Review* are not so severe: "Put in context, a century of climate change is about as bad as losing a decade of economic growth – or as bad as lowering the economic growth rate from 2.0% to 1.8% per year."[11] In the meantime, building the infrastructure for a vast new system of noncarbon electric power generation would cost trillions of dollars in the United States alone.[12]

But there remain important uncertainties with respect to climate change. Climate forecasts of global temperatures, moisture, and other atmospheric factors are based on computer models. These models, while very sophisticated and elaborate still require development. In general they have failed to predict temperature movements accurately, to date typically over-estimating warming. Studies have reevaluated climate sensitivity, with some arguing it is much lower than previously thought and that a doubling of CO_2 might in fact produce less than a 2 °C rise. The IPCC in its 2014 report (AR5) reduced its low-end projection to less than 2 °C for a doubling of CO_2, and for the first time did not hazard a "best estimate" for the temperature impact of doubling.

There are an enormous number of variables that might affect the climate. More research is needed on the various dynamic mechanisms in the atmosphere that can affect the passage of thermal radiation (heat) from the Earth's surface back out to space, such as radiation feedback from clouds, upper-atmosphere water vapor, and the radiation-scattering properties of non-natural aerosols.

One controversial issue is the effect of the increased amounts of water vapor that would occur if global increases in temperature did take place. The majority view in the scientific community is that an increase in atmospheric water vapor will lead to an increase in (heat) radiation containment, although some dissenters have said not. This is an important question since, if the majority are correct, increased water vapor becomes a "positive feedback" effect for the AGW phenomenon, meaning that the more that world temperatures increase, the more that water vapor reinforces (or amplifies) greenhouse warming. If the critics are correct, then the extent of the warming would be much less. Since computer models typically include feedback mechanisms, understanding the true phenomenon is critical to the validity of these projections.

Scientists have sought to gain a greater understanding of this problem through evidence from the Earth's geological past. Indications of temperature change locked in rock and ice provide evidence of past cycles of global warming and cooling, but so far investigators have not found clear indications of how the climate is likely to change over the next century. Further, computer climate models run backward through time are not generally able to replicate the historical record, calling predictions of change into some doubt. In addition, there is still much to be learned about the adaptation mechanisms of biological organisms to changes in CO_2 concentrations and higher temperatures. Just how marine and terrestrial components respond to increased CO_2 input from fossil fuels cannot be determined with certainty.

A considerable number of scientists maintain that we cannot afford to wait and see, where doing nothing could result in devastating impacts. According to this view, we should take preventive action soon *because* the processes are of unknown magnitude and the potential damage is uncertain. In late 2015 leaders from 195 nations met in Paris for COP-21 and agreed to a nonbinding pact to reduce greenhouse gas emissions in the near future, toward a goal of massive reductions in the years ahead – with a general goal of keeping temperatures from rising less than 2 °C.

Adaptation to a warmer world was considered important enough to be on the agenda, but it was not deemed an alternative to mitigation of CO_2 now. Of course, one solution would involve greatly reducing the use of fossil fuels. As noted, however, that would entail massive economic dislocation. Still, the carbon dioxide threat has been widely seen to warrant action. Some steps, albeit limited ones, have been taken. General policies in support of slowing and eventually stabilizing the emissions of greenhouse gases have been adopted by most industrialized countries, especially in Europe. The most common actions have focused on the promotion of energy efficiency of the sort described in Chapter 4 and increase in the utilization of renewable energy technologies (Chapter 11). With respect to the developing world especially, the objective is to reduce fossil-fuel use (or at least limit its growth) without hindering economic growth. At the Paris conference, the wealthy nations pledged to provide $100 billion in aid to poor countries for this effort. However, it should be noted that the agreement was nonbinding and the aid, due in 2020, is at best uncertain.

Combined-cycle natural gas plants do lower AGW emissions. Gas emits only 40–50 percent as much CO_2 per unit of delivered power as coal. Indeed, one of the reasons for the switch to natural gas from coal in the United States is that the conversion is relatively straightforward. Gas-fired facilities are much cheaper than nonemitting power generation from nuclear, wind, or solar. For that reason also, as we have discussed, some argue that natural gas should be seen as a bridge fuel to reduce greenhouse gas emissions now while nonemitting alternatives are developed further (see Chapters 8, 10, 11). However, critics of this approach note that methane is also a greenhouse gas and that methane often leaks from natural gas wells (chemically, natural gas is mostly methane). A few studies claim that leakage essentially undermines the idea of expanding use of natural gas (Howarth et al. 2011). But as Hausfather (2015) argues, this may just be a matter of stricter rules on well development and operation. If so-called "fugitive" emissions can be contained there would be definite gains with respect to climate change by a switch from coal to natural gas.

In many countries there are incentives (subsidies) for renewable energy, especially solar and wind. Some countries (as well as some states and localities in the United States and Canada) have also attacked the CO_2 question by employing disincentives to lower carbon emissions. These have included a "carbon tax" whereby fuels are taxed according to their carbon content.[13] Economists have generally supported a carbon tax as the most direct way to mitigate CO_2 emissions. A cap-and-trade system for CO_2 emissions was also put into place in Europe with limited success – although a reform of the market is ongoing, with a new set of benchmarks set for the 2020s. One problem is that CO_2 allowance prices have diverged widely from a few dollars per ton of CO_2 to more than $1000. Still, a cap-and-trade system as well as a carbon tax should make high carbon fuels more expensive. As we saw in Chapter 4, higher prices discourage consumption. In this case, the result should be lower consumption of coal since it has the highest proportion of carbon in its chemical structure. A switch from coal to another fuel, such as natural gas, would reduce CO_2 emissions as would upgrading coal plants to the supercritical technology discussed in Chapter 3. Often fuel or equipment switching can reduce emissions sufficiently to avoid, or at least greatly reduce, any taxes.

Of course, incentives would have to be much greater and more encompassing for the developed world to stabilize emissions at a quantity that is significantly below current levels. It would have to include a large-scale switch from fossil-fuel use altogether to renewable sources such as solar and wind and/or the expansion of the use of existing noncarbon technologies such as nuclear energy and hydropower. This kind of a transition would be enormously expensive, but the rationale is often presented from a legal idea that stems from a philosophical perspective – as the next section explains.

The "Precautionary Principle"

What actions, if any, should be taken with regard to preventing potentially dangerous climate change? Should we try now to reduce drastically our emissions of CO_2 and make a rapid transition to renewable energy technologies, as some experts, pundits, and politicians urge? Should we reduce emissions gradually, preparing the way for a transition but in the meantime working on adaptation to a warmer world? Given that there are many uncertainties, should we not worry about emissions for the time being, seeking instead to continue and broaden economic development, while maintaining research into climate science?

Some scientists and social scientists would argue for the first of these actions – drastic reduction and rapid transition. Others might prefer the second. But the argument for concerted action now employs a concept used frequently in legal and policy discussions of major environmental issues, called the "precautionary principle" (PP). This principle has been a part of the environmental policy discussion since the early 1970s, but its roots lie in moral philosophy.

The most common definition of PP is: "When an activity raises threats of harm to human health or the environment, precautionary measures should be taken even if some cause and effect relationships are not fully established scientifically."[14] The basic idea was present in the climate change debate from the beginning. At the first climate summit of the UNFCCC in Rio de Janeiro in 1992, the participants signed an agreement (The Rio Declaration) that said, in part:

> In order to protect the environment, the precautionary approach shall be widely applied by States according to their capabilities. Where there are threats of serious or irreversible damage, lack of full scientific certainty shall not be used as a reason for postponing cost-effective measures to prevent environmental degradation.[15]

Basically, the precautionary approach requires three elements: Threat of "serious" harm to people and/or the environment; uncertainty with respect to impacts and/or cause; and perceived need, and ability, to protect those in danger from harm (Gardiner 2006). Thus, with temperatures rising to potentially dangerous levels, from this perspective, humans are morally obligated to act.

It is not hard to justify a precautionary approach with respect to environmental issues, especially climate change. Nevertheless, there are many questions that arise with respect to its application. Who is to determine what constitutes "serious" harm – especially in conjunction with vast expenditures required to mitigate it? And what is the basis for that decision? Often, decisions about environmental actions depend on a cost-benefit calculation, but as we discuss in Appendix B, cost-benefit analysis (CBA) has many problems, especially where the harm is intergenerational.

More directly problematic is the question of trade-offs. The point here is that resources are limited and doing something in one area means foregoing doing it in another (i.e. there is an *opportunity cost*). It has been argued that there are more immediate and urgent problems in the world than climate change – the greatest harms of which are expected to be decades away. So, for example, in developing countries people have limited access to health care, sanitation is often poor, children do not receive an education, and food at times is scarce. It is the case that people in the developing world do not rate climate change as a major concern.[16]

Much of the climate change literature suggests that there need not be trade-offs. "Helping the poor does not foreclose the option of mitigating climate change," writes one scholar (Gardiner 2010, p. 61). The Rio Declaration also seems to indicate a belief that one form of action does not preclude another. Principle 3 says, "The right to development must be fulfilled so as to equitably meet developmental and environmental needs of present and future generations." Of course, in principle both present and future needs could be attended to, but in reality, if the cost of mitigation is in itself very high (as many experts claim it will be[17]), governments in rich nations will find it politically difficult to foot both sets of bills.

Much depends on how one views the current situation. Many climate scholars refer to "catastrophic climate change," implying that however bad conditions are for the poor today they will be much worse a decade or more hence when AGW makes life for them intolerable (and bad for those in rich countries as well). Consumption of fossil fuels, it has been argued, is fundamentally "unjust," given possible harms that are not only massive but perhaps irreversible (Caney 2010). Or as Norman et al. (2015) argue, "We have only one planet. This fact radically constrains the kinds of risks that are appropriate to take at a large scale. Even a risk with a very low probability becomes unacceptable when it affects all of us – there is no reversing mistakes of that magnitude." In that case, the moral imperative would be to treat climate change above all else. Indeed, some scholars have framed the PP as a brake against the possibility of catastrophe (see the discussion in Manson 2002). At the same time, some of the catastrophic scenarios behind the advocacy of the PP go well beyond anything deemed plausible by the IPCC. And many scientists would argue that damage will not be planet-wide but will vary among locations around the globe even if temperatures increase by the high end of forecasts.

Gardiner (2006, 2010) has sought to reframe the idea of the PP from the perspective of the moral philosophy of John Rawls (1971). A Rawlsian Precautionary Principle would apply the "maximin" principle, which means actions should maximize the minimum, that is, provide the greatest benefits to those with the least. In this context, when actions have uncertain outcomes (both good and bad), choose the course of action that leads to "the least bad worst outcome" (Gardiner 2006). This is a strategy that can be applied with respect to a problem such as climate change only with the following qualifications: there must be great uncertainty; there is a lack of motivation for personal aggrandizement by decision-makers; and there is a situation involving grave, catastrophic risks (Gardiner 2006). This seems to fit the dilemma of climate change action and would suggest the need for a robust response by wealthy countries.

But once again, there are the questions in evaluating the seriousness of the harm, the efficacy (as well as the cost) of proposed solutions, and the opportunity cost of undertaking them. This will be debated, but like many environmental problems, doing nothing (or little) is in itself a decision, and must be seen as such, with potential costs that with a positive probability might be catastrophic. But even granting that there is a moral imperative to address it, a positive probability does not settle the questions: How, when, or even, where we should act?

Precaution, Technology Transfer, and Aid to Poor Nations

What does the precautionary principle mean in terms of the divide between rich nations and poor? What can rich, industrialized nations do to help countries and peoples that are poor – desperately in many cases – develop modern energy, especially a modern reliable electric system, without worsening the potential of climate change? What *should* rich countries do from a moral as well as a self-interested perspective? Aid to poor nations may be regarded as imperative; so might preventing catastrophic climate change. Of course climate change would affect all people, rich as well as poor – though the latter probably would be harmed more heavily. As noted, the wealthy countries of the world have pledged to dispense

$100 billion per year in aid to developing countries in order for them to mitigate carbon dioxide emissions and adapt to the impacts of climate change.[18] The rich need the poor to help fight climate change, but more than a billion people need electricity which, as we have noted, is a precondition for modernization and the poor need the help of the rich to acquire it.

If rich nations are seeking to induce least developed countries to reduce emissions, such as the poorest nations in Africa, the people of those countries may fairly ask in return: "What emissions?" Except for relatively rich (and definitely resource-rich) countries such as Libya and South Africa, emissions across the continent are for obvious reasons exceptionally low. How can they be expected to reduce emissions when they produce so little?

The aid many advocate would be in accord with the Clean Development Mechanism of the (now expired) Kyoto Protocol, thus favoring solar and wind primarily to "insure low carbon [economic] growth" (Lema & Lema 2013). Yet there is a question about whether "clean energy" is what the poor need (as well as can utilize) most immediately. The first imperative is perhaps to provide the basic technology of electric power generation and aid in the development of the infrastructure to provide universal access. Right now, that infrastructure as well as the technologies of electric power production are lacking entirely in much of the developing world.

Economic development generally implies rising emissions – something one can see as the result of industrialization in countries such as South Korea, which were once very poor like most of Africa. But the aim of a least developed country is likely to be utilization of forms of energy that are lowest cost and require the least technological sophistication to operate and maintain. Some in the developed world believe that the least developed can "leapfrog" over fossil-based energy technology and go straight to wind, solar, and biofuels (for example, Goldemberg 1998, Chu & Majumdar 2012).[19] But in reality wind and solar, as intermittent sources, require substantial backup and computerized control networks to provide the level of continuous supply needed for real modernization of an economy. For many countries, therefore, the form of energy that is preferred is the cheapest and most basic: coal. As noted in the previous chapter, this is true especially in Asia. Bangladesh, Vietnam, Pakistan, and Indonesia, to give a few examples, have built and have planned many new coal-fired electric power plants. They are not necessarily ignoring renewables (biomass accounts for a great deal of developing country energy resources), but coal is inexpensive and the technology to utilize it is well understood and easily managed (Bryce 2014).[20] The key to a nation becoming environmentally concerned, it has been observed, is general prosperity and thus emissions reductions can be expected only after the least developed countries can actually begin the process of development (Grossman & Krueger 1995).

There is also a question with regard to aid of how well a nation's institutions can absorb an inflow of money and technology. This has been a concern with the UN's Millennium Development Goals, which were passed by the UN General Assembly in 2001, intended as a general program to lift up the world's poorest. The issue is usually framed as one of good governance. Governance is defined as "purposeful and authoritative steering of social processes . . . [including] activities of government and non-government actors" (Biermann et al. 2015). Put a bit differently, countries with weak institutions, such as uncertain property rights, are often unable to use aid successfully for the purpose intended (Easterly 2002, Moyo 2009). As a United Nations Development report in 2014 noted, "the quality of governance plays a defining role in supporting" the Millennium Development Goals and the successor program, the Sustainable Development Goals.

Good governance then is an essential precondition for aid generally, all the more so it would seem where there is a wish to leapfrog to a new energy paradigm that the rich countries themselves are just beginning to master. But this raises an obvious question: How does good governance come about? It is often easy to identify problems but hugely difficult to change them. Certainly it does not come about simply by a promise of aid. It may be effective in some instances to make continued aid conditional on change, but every existing political/social system, however disorganized or corrupt, has winners who will aim to maintain the same institutions that have benefited them (North 1990), thus delaying good governance. At the same time, does that mean aid should be withheld until governance improves? To do so means that extremely poor people remain extremely poor. It might be argued that the institutional problems of developing countries suggest that transfer should consist of the most basic technology with tight oversight on financial arrangements. That would perhaps be an argument in favor of the development of fossil-fuel-based electricity for the poorest of the poor – at least for a generation. But that is a solution that could have consequences for the world's future.

References

Begaye, E. 2005. "The Black Mesa Controversy." *Cultural Survival Quarterly*, 29 (4), retrieved from https://www.culturalsurvival.org/publications/cultural-survival-quarterly/united-states/black-mesa-controversy.

Biermann, F. et al. 2015. "Integrating Governance into Sustainable Development Goals," *Policy Brief #3, United Nations University–Institute for Advanced Study of Sustainability*, Tokyo, Japan.

Bryce, R. 2014. *Not Beyond Coal: How the Thirst for Low-Cost Electricity Continues Driving Coal Demand*. Energy Policy & the Environment Report, No. 14, Center for Energy Policy and the Environment, Manhattan Institute, New York.

Caney, S. 2010. "Human Rights and Global Climate Change." In R. Pierik and W. Wouter (eds.), *Cosmopolitanism in Context: Perspectives from International Law and Political Theory*. Cambridge University Press, Cambridge, UK.

Chu, S. and A. Majumdar. 2012. "Opportunities and Challenges for a Sustainable Energy Future." *Nature*, 488, 294–303.

Cole, D.H. and P.Z. Grossman. 1999. "When is Command-and-Control Efficient? Institutions, Technology and the Comparative Efficiency of Alternative Regulatory Regimes for Environmental Protection." *Wisconsin Law Review*, 1999, 887–938.

de Gouw, J.A., D.D. Parrish, G.J. Frost, and M. Trainer. 2014. "Reduced Emissions of CO_2, NO_x, and SO_2 from U.S. Power Plants Owing to Switch from Coal to

Natural Gas with Combined Cycle Technology." *Earth's Future*, http://onlinelibrary.wiley.com/doi/10.1002/2013EF000196/full.

Easterly, W. 2002. *The Elusive Quest for Growth*. The MIT Press, Cambridge, MA.

Ferguson, R., W. Wilkinson, and R. Hill. 2000. "Electricity Use and Economic Development." *Energy Policy*, 28 (13), 923–934.

Gardiner, S.M. 2006. "A Core Precautionary Principle." *The Journal of Political Philosophy*, 14 (1), 33–60.

Gardiner, S.M. 2010. "Ethics and Climate Change: An Introduction," *Wiley Interdisciplinary Reviews: Climate Change*, 1 (1), 54–66.

Goldemberg, J. 1998. "Leapfrog Energy Technologies." *Energy Policy*, 26 (10), 729–741.

Grossman, G.M. and A.H. Krueger. 1995. "Economic Growth and the Environment." *The Quarterly Journal of Economics*, 110 (2), 353–377.

Hausfather, Z. 2015. "Bounding the Climate Variability of Natural Gas as a Bridge Fuel to Replace Coal." *Energy Policy*, 86, 286–294.

Howarth, R.W., R. Santoro, and A. Ingraffea. 2011. "Methane and the Greenhouse-gas Footprint of Natural Gas from Shale Formations." *Climate Change*, 106, 679–690.

Intergovernmental Panel on Climate Change (IPCC), World Meteorological Organization/United Nations Environment Program. 1990. *Climate Change: The IPCC Scientific Assessment*. J.T. Houghton, G.J. Jenkins, and J.J. Ephraums (eds.), Cambridge University Press, Cambridge, UK.

Intergovernmental Panel on Climate Change (IPCC), World Meteorological Organization/United Nations Environment Program. 2014. *Climate Change 2014: Impacts, Adaptation and Vulnerability. Part A: Global and Sectoral Aspects. Contribution of Working Group II to the Fifth Assessment Report of the IPCC*. C.B. Field et al. (eds.), Cambridge University Press, Cambridge, UK.

Lema, A. and R. Lema. 2013. "Technology Transfer in the Clean Development Mechanism: Insights from Wind Power." *Global Environmental Change*, 23, 301–313.

Lomborg, B. 2001. *The Skeptical Environmentalist: Measuring the Real State of the World*. Cambridge University Press, Cambridge, UK.

Manson, M.A. 2002. "Formulating the Precautionary Principle." *Environmental Ethics*, 24 (3), 263–274.

Moyo, D. 2009. "Why Foreign Aid is Hurting Africa." *Wall Street Journal*, Retrieved from http://www.wsj.com/articles/SB123758895999200083.

National Academy of Science. 1974, *The Rehabilitation of Western Coal Lands*, Ballinger, Cambridge, MA.

Norman, J., R. Read, Y. Bar-Yam, and N.N. Taleb. 2015. "Climate Models and Precautionary Measures." *Issues in Science and Technology* (Summer), 16–17.

North, D.C. 1990. *Institutions, Institutional Change and Economic Performance*. Cambridge University Press, Cambridge, UK.

Perkins, R. 2003. "Environmental Leapfrogging in Developing Countries." *Nature Resources Forum*, 27, 177–188.

Rawls, J. 1971. *A Theory of Justice*. Harvard University Press, Cambridge, MA.

Schmalensee, R. and R.N. Stavins. 2012. "The SO_2 Allowance Trading System: The Ironic History of a Grand Policy Experiment." *HKS Faculty Research Working Paper Series RWP12- 030*, John F. Kennedy School of Government, Harvard University.

Stern, N. 2007. *The Economics of Climate Change: The Stern Review*. Cambridge University Press, Cambridge, UK.

Wilson, R.S. et al. 1980. *Health Effects of Fossil Fuel Burning: Assessment of Mitigation*. Ballinger, Cambridge, MA.

NOTES

1 As of this writing, a change in administrations from that of President Obama to that of President Trump left uncertain the fate of some of the EPA regulations. The regulations were created under the heading of the Clean Power Plan, a policy that was never explicitly authorized by Congress. It remains to be seen how (or if) these regulations are handled by the Trump administration.

2 In 2014, the estimated toll was 670,000; Geoffrey Smith, "The cost of China's dependence on coal – 670,000 deaths a year." *Fortune*, November 14, 2014. At http://fortune.com/2014/11/05/the-cost-of-chinas-dependence-on-coal-670000-deaths-a-year/.

3 From economics, the desirable amount of pollution control is at the point where the last dollar spent on control leads to exactly one dollar reduction in the damage costs of pollution.

4 Also see *Acid Rain Program*, EPA, April 14, 2009, at www.epa.gov/airmarkets/progsregs/arp/index.html.

5 "The Facts About Air Quality and Coal-Fired Power Plants." Institute for Energy Research, 2007. At http://instituteforenergyresearch.org/media/pdf/the-facts-about-air-quality-and-coal-fired-power-plants-final.pdf.

6 Utilities had bought $3 billion in emissions credits that were no longer of any value.

7 See, for example, John D. Podesta and Timothy E. Wirth, 2009, "Natural Gas – A Bridge Fuel for the 21st Century." Center for American Progress, at http://72.167.124.53/files/webfmuploads/CAP%20EFC%20NG%20Memo%208-08-09.pdf.

8 One possibility is carbon capture and sequestration (CCS), a technology that has reached the demonstration stage. But it is costly and not obviously successful in removing CO_2 from combustion and storing ("sequestering") it. For a look at CCS technology and progress, see https://www3.epa.gov/climatechange/ccs/index.html.

9 The pre-industrial level of CO_2 in the atmosphere is said to be about 275 parts per million (ppm). Currently, the level of CO_2 is about 400 ppm.

10 David Auerbach, 2015. "A child born today may live to see humanity's end, unless. . ." Reuters.com blog, at http://blogs.reuters.com/great-debate/2015/06/18/a-child-born-today-may-live-to-see-humanitys-end-unless/.

11 Richard S.J. Tol blog post, December 2, 2015. At http://richardtol.blogspot.com/2015/12/the-climate-spectator-is-no-more.html.

12 In a 2011 interview, Vaclav Smil put the cost for the United States alone for infrastructure and replacement capacity at $2.5 trillion. At www.pwc.com/gx/en/capital-projects-infra structure/pdf/gridlinesrev_0815pgs.pdf.

13 Ireland, Sweden, and Finland have some form of carbon tax. Chile, British Columbia and Quebec (Canada) have also instituted some form of tax on carbon.

14 Wingspread Statement (1998).

15 Rio Declaration (1992), Principle 15, available at www.unep.org/documents.multilingual/default.asp?documentid=78&articleid=1163.

16 See, for example, the commentaries by Bjorn Lomborg (2001) who notes that people in poor countries do not consider climate change as important. An example can be found at http://www.wsj.com/articles/this-child-doesnt-need-a-solar-panel-1445466967.

17 According to one report from Imperial College, London, UK, to cut CO_2 emissions by 50% would cost $2 trillion per year by 2050.

18 Developed countries at the Paris climate conference COP21 in 2015 made this pledge although the agreement is nonbinding and the first $100 billion is not due until 2020.

19 Other scholars have questioned leapfrogging, for example, Perkins 2003.

20 This has not been well received by governments and aid organizations in the developed world. The United States, for example, has refused to help finance new coal plants anywhere in the world. Japan has taken a different view and is financing the construction of coal plants that use the latest supercritical technology (Chapter 3).

Nuclear Fission Technology

Perspective

The Promise

In 1953 the future of nuclear energy seemed bright, with the launch of the Atoms for Peace program. These promises were to contrast with the dire consequences of the atomic bomb developed at the end of World War II. Commercial nuclear energy would be a *clean*, *abundant*, and *inexpensive* means to generate electricity, according to its proponents. One early advocate went so far as to claim that nuclear power would "be too cheap to meter." Nuclear power development accelerated during the late 1950s and into the 1960s, and by the 1980s there were 104 operating nuclear plants in the United States.

However, the hopeful vision did not materialize. Beginning in the 1960s, nuclear power came under a cloud of controversy that carried into the decades following. Moreover, far from being an inexpensive source of energy, nuclear power has often proven too costly to undertake. By the mid-1980s, construction of nuclear plants had nearly ceased in the United States. Many orders for new plants had been cancelled throughout the world. Accidents at the Three Mile Island plant in the United States (1979) and the Chernobyl facility in what was then the Soviet Union (1986), caused grave doubts about the future of nuclear energy, most notably in Germany and several other European countries as well as the United States.

Following an almost 30-year period in which no reactors were built in the United States, in the early 2000s, six new nuclear power plants were approved and are expected to be operational by 2020. Indeed, there has been a renewed interest in nuclear power because it emits no carbon dioxide. As the next chapter will explain, several new reactor designs, if they function as hoped, will operate with essentially no chance of a dangerous accident, little chance for diversion of nuclear materials for the fabrication of nuclear weapons, along with an absence of carbon dioxide or any other type of greenhouse gas emissions that could contribute to climate change. In this chapter, however, we look at the technology and the kind of nuclear reactors that operate in the United States and provide about 20 percent of the nation's electric power. It should be noted that the conventional light water reactor (LWR) in use for the past 50 years, while not without its own problems, also emits no carbon dioxide or other greenhouse gases.

A Short History[1]

Following World War II and the successful development of the atomic bomb by the United States, US government officials sought to find peaceful uses for nuclear energy. The Atomic Energy Act of 1946 (AEA) sought to address the promise for peaceful nuclear energy expressed by leading scientists and the press at the close of the war. An agency of the US government, the Atomic Energy Commission (AEC) was established to direct nuclear activities. But there was a fundamental problem in launching a program for the peaceful uses of nuclear energy: all of the information was highly classified so that scientists and engineers could not utilize much of the existing information unless they did so while working for the government. Nevertheless, although AEC programs of the early 1950s produced no reactor prototypes specifically designed for *civilian* electric power generation, it did oversee the development of a reactor for *military* applications. Admiral Hyman Rickover, one of the leading proponents of nuclear power development, is credited with the successful construction of a small boiling water reactor (BWR), which was used to propel a submarine. He led a design group, in a cooperative AEC–Westinghouse Electric Corporation effort, which completed the reactor for use by the US Navy in 1954.

By this time, interest in developing nuclear power generation had grown to the point where it was acknowledged by President Dwight D. Eisenhower in his Atoms for Peace proposals at the United Nations in 1953 and in a second Atomic Energy Act passed by Congress in 1954. In 1957, a commercial reactor, based generally on Rickover's naval design, was developed. This reactor was a pressurized water reactor (PWR),[2] and it was installed at Shippingport, Pennsylvania, for the Duquesne Electric Light and Power Company, a privately owned electric utility. This first reactor generated only 60 MW (though later it was modified to produce 150 MW). The first truly commercial design of the Westinghouse PWR was the Yankee plant at Rowe, Massachusetts, in 1958 (Fig. 7.1), and over 20 more such PWRs were installed in the 1960s. A BWR commercial design was introduced by the General Electric Company in 1969, following AEC-supported prototype development at the Argonne National Laboratory; soon after, several more BWRs were in commercial operation.

It should be noted, however, that these early commercial reactors were designed for the modest power range of about 100 MW. Later reactors were several times more powerful and those designed to come online in the late 1970s and 1980s were large-scale base-load (see Chapter 3) plants in the range of 1000 MW. The size of an individual nuclear generating plant is significant not only for the amount of electric load demand it supplies – 1000 MW provides enough generating capacity for about a million homes – but also, as we will see, for the implications of accidents.

As of this writing (2017), there are 99 commercial nuclear power reactors in 61 power plants operating in the United States; worldwide 444 nuclear power facilities are in operation with a total world generating capacity of a little over 386,000 MW. Another 66 are under construction, and many more are in the planning stage. As noted, nuclear generation accounts for almost 20 percent of electrical energy generated in the United States and about 11 percent worldwide (as of 2012), and these proportions have not changed significantly over the past few years, although they could increase as the 66 new plants come online.[3]

Figure 7.1   The Yankee atomic power plant at Rowe, MA. (Ceased operation in 1992.)
Source: Courtesy of energy.gov/US Department of Energy.

The Basic Physics

Nuclear fission concerns the reactions of nuclei from the heavy end of the periodic table. In the fission process, these nuclei are split into lighter nuclear parts and they release energy in the process. Today's commercial nuclear reactors commonly use the fissionable isotope ^{235}U, an element found naturally in low concentrations (less than 1 percent) in virgin uranium ore.

Another fuel element in use is plutonium (^{239}Pu), which is manmade. It results from neutron bombardment in the reactor of ^{238}U, the major constituent of natural uranium ore. The resulting (nonfission) reactions lead to the *breeding* or *conversion* of ^{238}U into ^{239}Pu. Thorium, which exists in nature much like uranium and in similar quantities, may also be a fertile element for these breeding reactions.

Nuclear Fission

Fissionable elements (for example ^{235}U) in the fuel rods of a commercial nuclear reactor are split when a neutron collides with one of the nuclei and is absorbed, as depicted in the artist's conception of Figure 7.2. The figure describes the fission process step by step, ending with the splitting of the uranium nucleus into lighter nuclei (called fission products)) and neutrons, all accompanied by the release of energy. The energy released through fission is, of course, a manifestation of the famous mass–energy equation of nuclear physics. The sum of the masses of the fission products is less than the aggregate mass of the original ^{235}U nucleus plus the incident neutron; the discrepancy in nuclear mass is manifested as kinetic energy. This kinetic energy is converted into useful heat in the reactor.

Chain Reactions

The energy released from a *single* fission reaction would be of no practical importance, because the amount is minuscule, less than 10^{-15} kilowatt hours (kWh, see Appendix A) equivalent electrical energy. But the consequences of nuclear fission are significant because of the possibility of a *chain reaction*. In a chain reaction, the neutrons liberated by the fission of one nucleus move on to create other fission reactions. Because more than one neutron is usually released in each fission reaction, the fission process may become multiplied (depicted schematically in Fig. 7.3). The successive collisions and fission reactions create a self-sustaining (or ever-growing) chain reaction, if the density and total mass of the fissionable fuel are great enough. A volume of fissionable material that can sustain a chain reaction is called a *critical mass*.

The chain reaction is at the heart of both a nuclear bomb and a nuclear fission reactor. But there are several distinguishing characteristics that determine whether a body of fissionable material is a *bomb* or merely a *reactor*. The most important is that in a reactor the rate of fission reactions can be controlled. The flow of neutrons is limited by *control rods* of materials, such as cadmium, that absorb neutrons. These control rods are placed between the fuel rods (see Fig. 7.4) and absorb the moving neutrons, thereby preventing the neutrons from inducing more fission reactions. These control rods can be inserted or retracted from the reactor core and can thus control the rate of the chain reaction or even shut it off. A fission bomb, by contrast, assembles a critical mass, which then releases its energy (uncontrolled); no further external control of its rate of reaction is then possible.

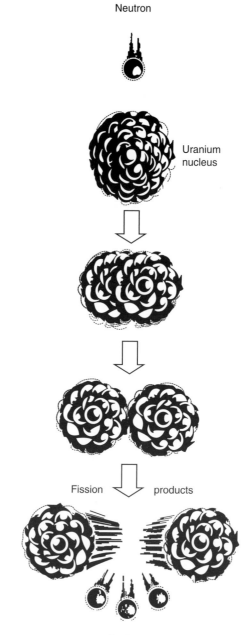

Neutron

Uranium nucleus

Fission ⬇ products

Neutrons

Figure 7.2 Diagrammatic view of the fission process. Here we see a four-part sequence, in which the neutron is first shown to be moving toward a collision with the uranium nucleus. Next, the neutron has collided and has been absorbed into the uranium nucleus. The nucleus is depicted at this stage as bulging out to the sides to suggest that the nucleus is now unstable and is ready to split apart. In the third image, we see the start of the splitting – or fissioning – of the nucleus into fragments called *fission products*. These products include the nuclei of lighter elements, such as ^{141}Ba and ^{92}Kr, and neutrons. Finally, the fission products fly apart as a result of the energy released by the nuclear fission process.
Source: Collier & Hewitt (1987). © 1987 Hemisphere Publishing Corporation

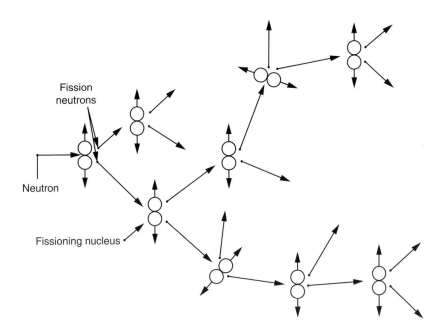

Figure 7.3   Schematic representation of fission chain reactions.
Source: J. R. Lamarsh. *Introduction to Nuclear Engineering* © 1983, Addison-Wesley Publishing Co., Inc.,
Reading, Massachusetts. Reprinted with permission.

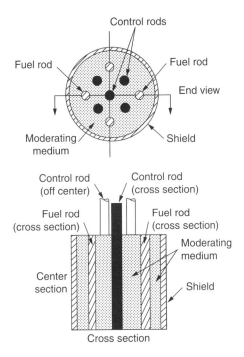

Figure 7.4   Idealized nuclear reactor.

An important feature of all nuclear chain reactions is the density of the fissionable element, because this determines the likelihood of a moving neutron colliding with a fissionable nucleus. For present-day US nuclear power technology, the fuel element must be *enriched* to a concentration of about 3 percent ^{235}U for the density of fissionable atoms to be sufficient to sustain the chain reaction.

Moderating Medium

The reactor vessel of currently operating US reactors contains the entire reactor core. It is important to understand that the chain reaction in any reactor takes place throughout the entire core from one fuel rod to another, and therefore the entire set of fuel rods determines a critical mass and not any individual rod. Between the rods, in most present-day reactors, there is a *moderating medium*, which for US reactors is simply water. The moderating medium in a typical reactor fills most of the space between the fuel rods (see idealized reactor shown in Fig. 7.4).

The purpose of the moderating medium is to *slow* the speeds of the fission-liberated neutrons. It is necessary to slow the neutrons in a moderated reactor to enhance their capture (collision and absorption) by the fissionable nuclei. Without a moderator a chain reaction could not be sustained because of the low density of ^{235}U even after enrichment. But the early nuclear physicists found that the neutrons at the slow end of the range of thermal speeds had a much better chance of being captured than those at higher speeds and so could better sustain a chain reaction.

In a moderating medium such as water, these captures are accomplished through multiple collisions (about 35 on average before capture) of the neutrons with the light nuclei, such as hydrogen (in the water molecule). In this process, a typical neutron starts out at high velocities after liberation from a nuclear fission reaction, but after several collisions has been slowed to the speeds of *thermal motion* in the medium (speeds at a factor of 10^{-6} of the initial speed from fission). These slowed neutrons are therefore called *thermal neutrons* and the reactors using moderators are often called *thermal reactors*.

A good moderating medium contains atoms with light nuclei, so that each collision of a neutron with an atom causes only a partial loss of the neutron's velocity. In addition, the medium should not contain nuclei that are likely to absorb the neutrons upon collision, causing their loss to the chain reaction. Besides ordinary water (H_2O), another good moderator is *heavy water* (D_2O, or *deuterated* water [4]), which absorbs fewer neutrons than ordinary water and is used in Canadian reactors (Canadian Deuterium-Uranium Reactors, called the CANDU reactors). Graphite also has been used as a moderator and was in fact the moderator of the earliest atomic reactors (called then atomic "piles"); it is still the medium used in US converter reactors for weapons material production. Abroad, notably in Russia and the UK, a few power-production reactors are graphite moderated. Graphite is actually the best moderator, but it has disadvantages for use in commercial reactors. Its most crucial disadvantage is its flammability; the possible consequences of this disadvantage, as we will see, were vividly in evidence at the graphite-moderated Chernobyl reactor in the former USSR.

Heat Removal

In our description of the fission process, we noted that not only do the fission fragments fly apart, but that there is also a release of energy. About 85 percent of this liberated energy goes into the nuclear fragments, which rapidly collide with other atoms within the fuel rod before escaping into the surrounding moderator. (Neutrons, on the other hand, mostly escape directly into the moderator.) When these nuclear fragments collide within the fuel rod, they give up their kinetic energy in the form of heat, thereby heating the rod.

This heat of fission must be removed from around the rod, to be put to use, *and* to cool the rod. In nuclear reactors, this process of heat transfer is accomplished by circulating a coolant past the fuel rods. In *light water reactors* (LWRs) – including both BWRs and PWRs – used in the United States, this coolant is the same water that serves as a moderator. In Canadian reactors, the heavy-water moderator is also the coolant, whereas abroad, for example in the UK, some other reactors use circulating gaseous coolants such as carbon dioxide.

A simplified depiction of the coolant flow within a reactor vessel is given in Figure 7.5, showing the input and output ports for coolant flow. We should visualize the flow as a vertical movement parallel to the fuel rods.

We find in US and Canadian reactors that heat is conducted to the water coolant and heats the water as in a (fossil-fired) conventional boiler. Indeed, it has been said that a nuclear reactor is merely an *exotic way to boil water*. There are two alternative ways in which the boiled water from the reactor can be used in a power plant. The coolant can itself circulate as steam through the turbine as shown in the BWR in Figure 7.6. Alternatively, the coolant can circulate through a separate heat exchanger in which the steam used to circulate through the turbine is generated. The advantages of this scheme (Fig. 7.7) are (1) the turbine is isolated from the radioactive contamination of the reactor and (2) the reactor can be operated under higher pressures and temperatures (see Chapter 3 and Appendix A). A PWR (Fig. 7.7) has a higher thermal efficiency than the BWR because the temperature of the working fluid is initially higher (see Appendix A).

The Technology

The Reactor Core

The core of an operating reactor is actually more complicated than the configuration shown in Figure 7.4 suggests. It has not just one set of rods, but a number of fuel assemblies, each one with a cluster of fuel and control rods. The cross section of a

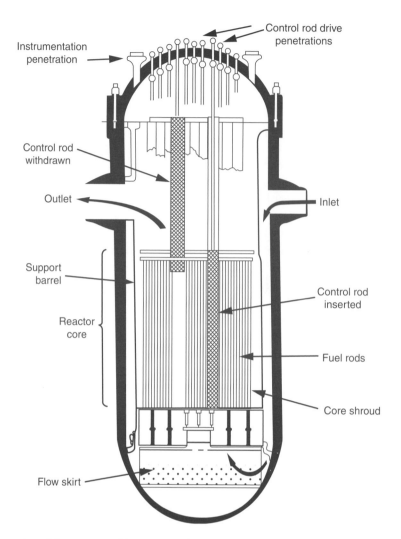

Figure 7.5 Cutaway view of the reactor vessel in a pressurized water reactor. The vessel is over 40 feet high, with an inside diameter of over 14 feet and wall thickness in excess of 8 inches; fuel rods in the vessel contain more than 100 tons of uranium dioxide.
Source: Reproduced with permission from *The Nuclear Fuel Cycle,* The Union of Concerned Scientists, The MIT Press, 1975.

typical PWR core (Fig. 7.8) shows the spaces provided for each assembly, of which there are usually hundreds. The insert in part (b) of the figure shows expanded cross sections of four fuel assemblies, with clusters of control rods dispersed through each assembly.

Figure 7.9 is an expanded cutaway side view of an individual fuel assembly containing a control rod assembly. Each fuel assembly can be loaded or withdrawn individually from the reactor core, as required by refueling operations. The control rods can be inserted or withdrawn during the control operations in a cluster for each fuel assembly. The entire cluster for an assembly is attached to a control-rod

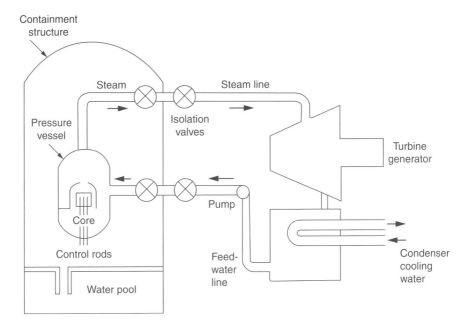

Figure 7.6   Boiling water reactor (BWR): schematic diagram.
Source: Energy Technologies and the Environment, June 1981, US Department of Energy, Report No. DOE/EP0026. Available from National Technical Information Service, Springfield, VA 22161.

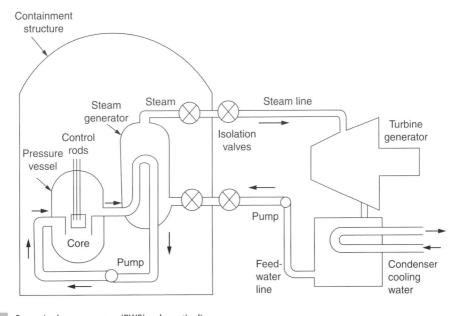

Figure 7.7   Pressurized water reactor (PWR): schematic diagram.
Source: Energy Technologies and the Environment, June 1981, US Department of Energy, Report No. DOE/EP0026. Available from National Technical Information Service, Springfield, VA 22161.

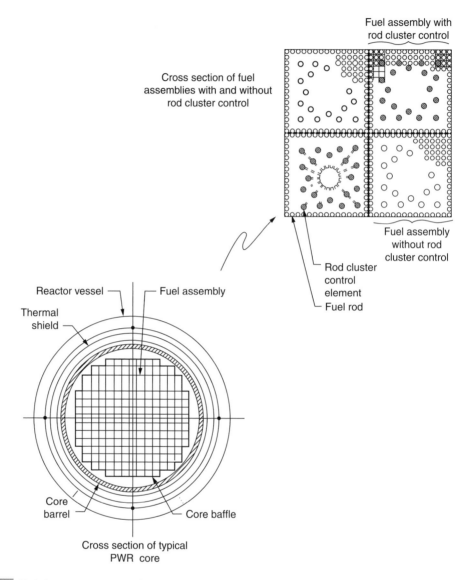

Figure 7.8   Typical reactor core cross sections.
Source: Courtesy of Westinghouse Electric Corp.

assembly structure (Fig. 7.9) and the rods slide in unison, each in a separate tubular guide. As we have noted, water flows in the spaces between and within the assemblies of fuel rods and control rods.

With this layout of the core in mind, we can now better visualize the processes that take place there:

1. Fission reactions in a fuel rod result in energetic neutrons and fission fragments (nuclei) being propelled away from the site of the reaction. Most of the high-energy neutrons escape from the fuel rod and penetrate the moderating

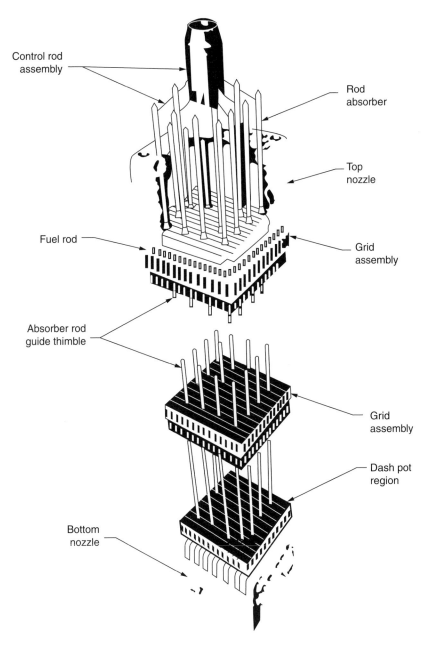

Control rod
assembly

Rod
absorber

Top
nozzle

Fuel rod

Grid
assembly

Absorber rod
guide thimble

Grid
assembly

Dash pot
region

Bottom
nozzle

Figure 7.9   PWR fuel element: cutaway view.
Source: Courtesy of Westinghouse Electric Co.

medium. The fission fragments, however, collide quickly with other nuclei in
the same fuel rod and give up their kinetic energy as heat.

2. Fission-liberated *neutrons* have penetrated the moderator, each enduring multiple collisions with hydrogen nuclei; each time bouncing elastically and
transferring a fraction of its kinetic energy. The moderating action of the water

slows the neutrons so that their capture by a fissionable (^{235}U) nucleus in another fuel rod becomes more likely.

3. Some liberated neutrons, in their random paths in the water between the rods, may instead encounter a *control rod* and be absorbed. As a result, these do not create another fission reaction.

4. Finally, the water flowing along the spaces between the rods acts as the *coolant* to carry away the heat generated within the fuel rods by the fission fragments.

Nuclear Reactor Kinetics

These are the basic processes within a nuclear power reactor. But to appreciate fully how a reactor works, we must consider one other aspect: the time-dynamic behavior of fission reactions. This behavior, along with the characteristics of neutron production and travel, is referred to as the *kinetics* of the nuclear reactor.

To understand reactor kinetics, we have to appreciate the balance of neutrons over time in a chain reaction. The number of fission reactions in a given time interval depends directly on the number of neutrons moving through each of the fuel rods – neutrons that are thus available for capture by *fissile* nuclei. These neutrons, of course, have already been liberated by fission reactions, usually from other fuel rods. The possibility of a self-sustained chain reaction exists because an average of 2.5 neutrons are liberated by each fission reaction. The chain reaction may either grow or remain at a constant rate of fissions per time interval, and the issue for nuclear engineers in building a nuclear power plant is to be sure that reactions remain self-sustaining or *critical* without growing out of control in the *supercritical* range. And the key to that question lies in the balance of neutrons available for fission in a given time interval. If the number equals that of the previous interval, the reactor will remain critical; if it is greater, it will go into the supercritical range.

In any time interval, the neutron population accounts for neutrons produced by fissions in the preceding time intervals, neutrons lost from nonfission processes (for example, those captured by ^{238}U nuclei), neutrons that are absorbed into control rods, and neutrons lost at the edges of the reactor. Finally, there is another group of neutrons that is crucial to the chain reaction (as well as to an appreciation of nuclear processes) called *delayed* neutrons. These are emitted not by the fission of ^{235}U nuclei, but by the decay of unstable nuclei from fission products, such as radioactive bromine and iodine, which are themselves created by the prior fissioning of ^{235}U nuclei. These delayed neutrons are distinct from the *prompt* neutrons, which are created immediately in each fission process, and their time delays range from about 1 second to about 1 minute.

In the ordinary LWR, fission results principally from capture of moderated neutrons that have been slowed by the medium between the fuel rods. The average lifetime of a neutron in the moderator is about 10^{-4} seconds, which serves as a slight delay in the generation of the next set of fission reactions in the chain reaction. However, since this delay is a small fraction of a second, the number of generations per second could be quite large (thousands, in fact). If this were the case, the chain reaction would have a rapid growth rate and the reactor would be difficult to monitor and control.

But delayed neutrons act as *pacemakers* to keep the reactor from going supercritical. Engineers design a delicate balance of prompt and delayed neutrons into the system by maintaining a flux of prompt neutrons that is just below the density required to take it into the critical condition. The delayed neutrons, however, allow the reactor to *go critical*. Thus, the delayed neutrons for the most part determine the reaction rate of growth when the control rods are withdrawn or the rate of decay when the rods are inserted. The long time delays (an average of 14 seconds for ^{235}U fission) determine the reactor kinetics.

The consequence of these long average delay times is a much slower growth rate of fission reactions than those that would result from prompt neutrons alone. An LWR requires several minutes to go from low to high levels of the chain reaction (and power). With such dynamics, the LWR operator can monitor the buildup of the reactions, override automatic controls (if necessary), and manually cut back on the criticality (the supercritical growth). It is fair to say, as Enrico Fermi, a pioneer of modern nuclear physics, once observed, that without delayed neutrons, there would be no nuclear power (Segre 1965).

There is the possibility, however, that under some conditions the delicate balance of delayed versus prompt neutrons in the reactor may be upset. This is the reason often given for the violent accident at the Chernobyl reactor. Some have argued that the reactor went into a *prompt criticality*, whereby supercritical growth was determined solely by the prompt neutrons. The reactor was reported to have soared from a thermal power output a fraction of its rated value to 100 times its rated value in *4 seconds*.

It should be noted that it is possible to build a working power reactor designed to operate from *unmoderated neutrons*, as we will discuss in the next chapter. Its fuel rods are packed more densely than those in a LWR, thus increasing the chance of neutron capture shortly after fission, while the neutrons are still at high speeds. However, the fast reactor, too, is paced by a minority of delayed neutrons and, again, the long delay times of these neutrons are controlled by slowing the changes of the chain reaction. In the event of a *prompt* criticality, however, the fast reactor's doubling time is much shorter than that of the moderated reactor, because the average lifetime of prompt *fast* neutrons is 1000 times shorter than delayed moderated neutrons.

Fissionable Fuel Resources

Fissionable resources are found in naturally occurring deposits of *uranium* and *thorium*. Resources for uranium and thorium may be estimated in much the same way as fossil-fuel resources are. Reasonably assured resources (RAR) of uranium for the world are about 7.5 million metric tons; about 200,000 tons are found in the United States. Global usage has been about 200 metric tons per year (mt/year) for each 1000 MW of nuclear power plant capacity, and world production is about 66,000 mt/year, since installed world nuclear generating capacity is about 386,000 MW.[3] These estimates result in an *R/P* ratio (see Chapter 2) of a little over 90 years for uranium.

A so-called "breeder" reactor (see Chapter 8) can increase the yield from uranium ore. The breeder enhances the creation of plutonium ^{239}U from ^{238}U through an intense neutron bombardment. The ^{239}Pu, of course, is fissionable and can be used as a fuel. The breeding process can create almost 100 times the amount of fissionable fuel from a given amount of natural uranium. Breeding, as a result, can theoretically provide energy resources comparable or larger in energy value than those of coal or natural gas. A similar breeding reaction can transmute natural thorium (^{232}Th) into ^{233}U (which also is fissionable) and thereby also provide energy resources comparable to coal.

Nuclear Plants: Operation

Normal Operation

Let us imagine the operation of a PWR under normal conditions. Technicians start the reactor by retracting the control rods (refer back to Figs 7.4 and 7.5). As the rods slide up toward the top of the reactor vessel, the flux of neutrons grows; soon, the chain reaction commences.

The more the control rods are retracted, the more are fission neutrons able to collide with other fissile nuclei in the chain reaction. The reactor as a whole passes from the *subcritical* stage (below the number of neutrons needed for a sustained chain reaction), through the *critical* stage, and temporarily into the *supercritical* stage to sustain growth. As the number of nuclear reactions increases, the thermal power rate grows as well. This growth continues until the full power level is attained, at which time controls should be adjusted to keep the reaction rate steady in time – called the *steady state*.

Maintaining a steady reaction rate is a delicate control problem and typifies the precise demands of nuclear technology. Not only must the balance of reactions be kept at a constant rate, but also temperatures must be controlled by the rate of circulation of the *coolant* (water for the BWR and the PWR). Maintaining a sufficient cooling rate is a critical requirement and makes an absolute demand for an *emergency cooling* system.

Under normal operations, the reactor is operated in the critical range and generates thermal energy. This thermal energy, in turn, is carried away in the PWR (Fig. 7.7) by the circulation of the coolant water. The water travels in a loop – the *primary loop* – from the reactor vessel to a steam generator and back to the reactor. The steam generator is a *heat exchanger* through which the reactor heat is transferred into the working fluid (steam) in a secondary loop, which is physically isolated from the coolant of the primary loop. The steam, in turn, flows to the blades of the turbine and the turbine acts as the prime mover for the electric generator. In the BWR (Fig. 7.6), the steam from the reactor directly turns the turbine blades.

The process, as designed, is therefore self-contained (see Fig. 7.10, containment and shielding) and discharges virtually no effluent to the atmosphere other than steam exhaust. There are, furthermore, no unsightly piles of fuel or wastes

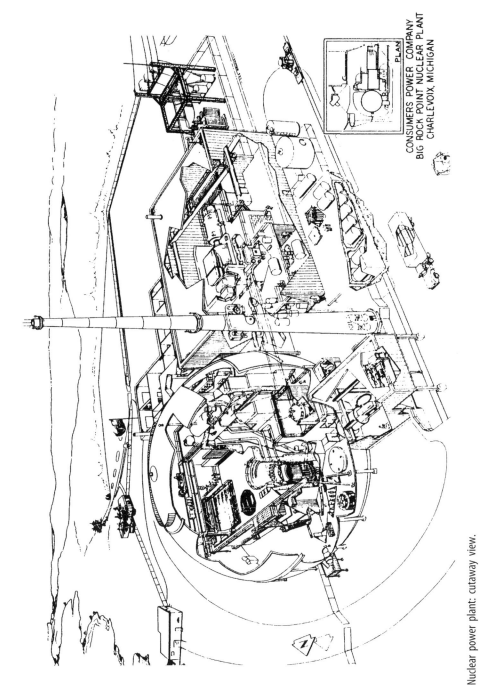

Figure 7.10 Nuclear power plant: cutaway view.
Source: Courtesy of Consumers Power Co., Charlevoix, MI.

CONSUMERS POWER COMPANY
BIG ROCK POINT NUCLEAR PLANT
CHARLEVOIX, MICHIGAN
PLAN

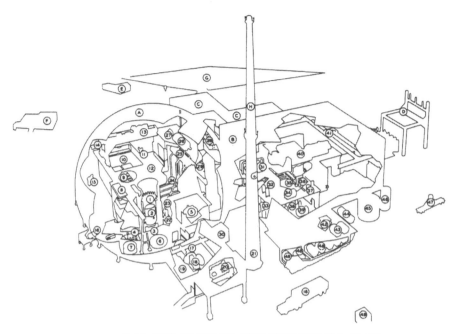

CUTAWAY DRAWING OF THE BIG ROCK POINT NUCLEAR PLANT

1.	Reactor Vessel	31.	Reactor Feedwater Pumps
2.	Reactor Core	32.	Sand Filters
3.	Control Rod Drives	33.	Condensate Demineralizer
4.	Control Rod Drives - Hydraulic System	34.	Neutralizer Tank
5.	Reactor Vessel Head	35.	Reactor Feedwater Heaters
6.	Reactor Vessel Shield Plug	36.	Main Turbine Condenser
7.	Reactor Vessel Head Thermal Insulation	37.	Make-Up Water Demineralizer Tank
8.	New Fuel Storage Area	38.	Condensate Pumps
9.	Refueling Platform	39.	Caustic Storage Tank (NaOH)
10.	Fuel Storage Pool	40.	Turbine-Generator Unit (75 Mwe)
11.	Transfer Cask	41.	Turbine Building Crane (75 Ton)
12.	Reactor Area Floor	42.	Liquid Radioactive Waste Strainer
13.	Overhead Semi-Gantry Crane (75 Ton)	43.	Liquid Radioactive Waste Hold Tanks
14.	Monorail Crane	44.	Acid Storage Tank (H_2SO_4)
15.	Heating and Ventilation Unit	45.	Condensate Storage Tank
16.	Escape Lock	46.	Demineralized Water Storage Tank
17.	Equipment Lock	47.	Spent Fuel Shipping Cask
18.	New Fuel Shipping Containers	48.	Liquid Radioactive Waste Tanks
19.	Unloading Dock	49.	Calibration Facility
20.	Unloading Crane		
21.	Ventilation Fans		
22.	Reactor Recirculating Pump		
23.	Downcomer	A.	Containment Sphere
24.	Steam Water Riser	B.	Turbine Building
25.	Offset Steam Drum	C.	Office Building
26.	Emergency Condenser	D.	Substation
27.	Liquid Poison Tank	E.	Information Center
28.	Control Room (Rear View)	F.	Screen Well and Pump House
29.	Personnel Lock	G.	Parking Area
30.	Containment Sphere Ventilation Room	H.	Ventilation Stack

Figure 7.10 *(cont.)*

outside the plant, nor any large lineups of railroad cars or barges around the plant. As such, the image of the fission plant is considerably different from that of the coal-fired plant, and these features are among the advantages enumerated by proponents of the technology. (The popular image of a nuclear power plant is typically of its cooling towers, which we learned in Chapter 3 are not peculiar to nuclear plants alone.)

During normal operations, virtually all radioactive materials are contained within the reactor system, and small amounts only of short-lived radioactive elements are allowed to vent into the atmosphere. Shielding of the reactor and associated components is required by law to protect the health and safety of workers inside the plant. This shielding, along with the limitations on radioactive discharges, is also designed to meet public health standards for the population outside the plant. The United States Nuclear Regulatory Commission (USNRC) is charged, under the Atomic Energy Act, with enforcing all standards of health and safety for nuclear power plants and related activities. The NRC has set a standard for the maximum radiation exposure level of the general public from a nuclear plant at less than 25 millirems per year (mrem/year), whole-body dose. Radiation dosage may be measured in *rems* (roentgen equivalent in man), which can be specified for the whole body or for specific organs (e.g. the thyroid) of the body. The ambient natural dosage at sea level of cosmic-ray radiation is about 50 mrem/year.

Radiation standards for *normal* operations concern *low-level* radiation. This term generally denotes dosage at or below ambient levels. At these levels, the impacts on occupational or public health do not have a clear-cut cause and effect relationship. Nevertheless, these standards have provoked controversy regarding thresholds of hazard. The concept of thresholds has already been encountered in our discussion of air quality standards (see Chapter 6). The question in both cases is: to what extent should environmental controls be required to limit levels of contamination to be below that which has a clear-cut cause and effect relationship on human health? We will discuss the implications of this policy issue in Appendix B.

Impacts of Normal Nuclear Plant Operation

Leaving aside the question of low-level radiation standards, let us consider the potential impacts of the process of nuclear-powered generation.

Extraction

Uranium mining is evidently not as accident prone an occupation as coal mining, even though much of it is underground. There is evidence, however, that radiation hazards have affected the overall mortality rates of miners, even though the mortality rate is much lower (as little as 10%) than of that for coal mining. However, uranium mining may have *public health* consequences in areas surrounding the mines; uranium ore *tailings* (residues) have often been left in open piles outside the mines and may present a health hazard.

Transportation (of Processed Uranium)

No measured impacts have been reported throughout the more than 50 years of nuclear plant operations as a result of the transport of the uranium ore from mine to processing plants. This step might be considered comparable to the shipment of coal to its site of preparation for combustion. In the case of nuclear fuel, however, there is another step: the shipment of processed fuel rods to the nuclear generating plant. Here, also, no measured environmental impacts have been reported. Still, concerns have been expressed by some municipal officials and others over possible traffic accidents that could cause spills of radioactive material.

Processing

Nuclear fuel processing consists of uranium *enrichment* and *fuel-rod fabrication*. Enrichment and fabrication can be compared logically to the *preparation* step in the coal cycle, although nuclear fuel processing is a vastly more sophisticated technological operation than coal processing. Occupational hazards for nuclear fuel industry workers appear to be lower than for coal miners, however, although controversies will likely continue over standards, especially as they relate to the question of *acceptable* levels of risk (see Appendix B).

The Reactor

As a step in the nuclear power process, the *fission* process in the reactor corresponds to the *combustion* step for coal. Here, we must be careful to specify that we are considering *normal* operating conditions, in contrast to abnormal or accident conditions. Under normal operating conditions, the environmental or public health impacts of nuclear plants are less than those of an equivalent coal-fired plant. Hazards for operating personnel, however, may be greater at nuclear plants than at coal-fired plants. Historical data on nuclear plant personnel dating back to the early days of the industry indicate radiation dosages averaging slightly less than 1 rem/year (1000 mrem/year), but the range of dosages runs up to several hundred rems/year. During these periods a statistically significant number of excess cancers were recorded for nuclear plant workers. Since 1969, however, the average annual dosage to nuclear workers has been reduced by over a factor of two and at the present time federal regulations permit no more than 5 rem/year (5000 mrem/year) for any individual nuclear worker – still well above the ambient level of radiation exposure.

Disposal

In this chapter, we will consider only the issue of *short-term storage* (on-site) of nuclear waste (particularly *spent* fuel rods), rather than *long-term storage* (away from the site), or reprocessing or *permanent disposal* – unresolved issues that we will discuss in Chapter 8. At the reactor site, the 25 mrem/year dosage standard is for the most part maintained, and we can conclude therefore that on-site storage of spent nuclear fuel appears to have a lower impact (for normal operating conditions) on the local environment than the disposal of coal wastes.

Another environmental concern is the thermal discharge from nuclear plants as well as coal-fired plants. As noted in Chapter 3, discharged heat has caused some ecological concerns because it can raise temperatures in natural bodies of water significantly, killing plant and animal life. The problem is controllable by the use of cooling towers (see Chapter 3), which help limit discharge temperatures to within regulated limits. There is little difference in the thermal effects caused by nuclear and coal-fired plants, although the LWRs do discharge more heat and thus require more cooling of the discharge to meet standards.

Unresolved Issues: Nuclear Plant Safety

Proponents of nuclear technology emphasize normal operating characteristics of nuclear power plants. By using that state, they are able to demonstrate that nuclear plants have lower local environmental impact and a better local visual impression (Fig. 7.1) than fossil-fired plants. Proponents have also claimed that only a massive shift away from the combustion of fossil fuels, primarily to nuclear power, can provide a technical response to the long-term dilemma of the carbon dioxide–greenhouse effect (see Chapter 6). Furthermore, nuclear proponents note that while risks exist with respect to nuclear, far more people have suffered health problems from coal (at every stage) than from nuclear power. Weighed against these arguments, however, are the controversies surrounding nuclear plant safety under *abnormal* conditions. (There are also concerns over *long-term waste disposal*, and *weapons proliferation*; these issues, however, will be treated in Chapter 8.)

When we talk about nuclear plant safety, we mean simply the prevention of *accidental* release of radioactive materials outside the plant, which would expose the public at large to the hazards of radiation sickness and cancer. Although plant safety is a complicated issue, we want to focus on the most important type of nuclear accident, called the *loss of coolant accident* (LOCA). As we have learned in our discussion of reactor operation, the coolant must circulate through the reactor to remove the heat generated by the fission reactions. If coolant circulation is interrupted for any reason, the accumulation of heat will increase the temperature, which, if allowed to continue, can lead to the *meltdown* of the reactor core. If not *totally confined* within the reactor containment building (Figs. 7.1 and 7.10), a *core meltdown* can release a devastating inventory of radioactive materials into the atmosphere. The radioactive inventory measured in curies (Ci, see Chapter 8) of a large commercial reactor is thousands of times more than the amount released by a Hiroshima-sized fission bomb, and the potential for catastrophe with most of the nuclear power plants in operation today is always present.

A LOCA can have any one of a number of initiating causes, many of which are mundane matters of plumbing – quite surprising given the "high-tech" image of the operation. For example, malfunctioning valves were reported an average of seven times per year for a group of 17 plants in 1972, early in the era of large nuclear plants. A malfunctioning valve also figured significantly in the accident in 1979 at Three Mile Island. Since that time, quality control on valves has reduced the number of such incidents, but the probability of this type of malfunction still exists and can never be reduced to zero.

There are other mechanical failures that can initiate an accident sequence: e.g. high-pressure pipes could rupture or pumps could fail. Such malfunctions have all occurred at many nuclear plants. Electrical equipment also fails: failures of circuit breakers, generators, batteries, transformers, and electric instruments or indicators have been recorded at plants, and could initiate or contribute to plant accidents. More importantly, failure of the electric supply for the cooling system leads to excessive heating of the reactor core and a LOCA situation (as at the Fukushima (Japan) power plant in 2011; discussed below).

The important aspect to recognize at the outset, in assessing such technological malfunctions, is that their occurrence can never be reduced to zero. The basic question, therefore, is not how to prevent individual malfunctions absolutely, but rather how to deal with them within the entire system in an acceptable way when they do occur. The approach then is how to engineer the system so that as an entity it functions in an acceptably safe manner. Such approaches have been used in aircraft safety and in mission planning for space flights.

Safety of operation became a major concern of the US nuclear program during the early 1970s. It was soon recognized that equipment failures could (and *would*) occur and that even the safety equipment designed to deal with such failures could fail. Consequently, studies of plant safety were initiated, although there was controversy over what constituted an adequate safety study. The best known reactor safety study was conducted by Professor Norman Rasmussen of the Massachusetts Institute of Technology for the reactor-safety section of the Atomic Energy Commission (AEC, which was reorganized into the NRC). The study, a probabilistic analysis of the complex workings of the plant and its safety systems, was reported in 1975 as the AEC document number WASH-1400. This report figured heavily in the safety debate over the following decade. It was known variously as *WASH 1400* or the *Rasmussen report*. (Our discussion of reactor safety will refer frequently to this report.)

The Engineered Safety System

For nuclear plants, the approach to safety generally has taken the form of the engineered safety system (ESS). The ESS executes five basic actions, as indicated schematically in Figure 7.11.

Briefly, these actions are the following, as shown in the figure:

1. Reactor trip (RT)
2. Emergency core cooling (ECC)
3. Post-accident radiation removal (PARR)
4. Post-accident heat removal (PAHR)
5. Containment integrity (CI).

In considering whether this entire ESS functions in an acceptably safe manner, analysts have reduced the system to a probabilistic assessment of the failure of each of its parts. The possible failures of the various parts must be analyzed in a sequence of distinct steps, since the consequences of any single failure are conditional upon what prior failures have occurred.

One possible outcome, for example, would mean that all five functions (see Fig. 7.11) operated successfully: the chain reaction was shut down (RT), the core received emergency cooling (ECC), the post-accident radiation was safely removed

(PARR), the post-accident decay heat was removed (PAHR), and the containment building allowed no radiation leakage to the atmosphere (CI). Another outcome, on the other hand, would represent partial failure of the ESS. Following successful RT shut-down, the emergency core cooling (ECC), radiation removal (PARR), and heat removal (PAHR) all fail, leaving the containment (CI) as the last defense against catastrophic discharge of radioactive materials. The worst, of course, would be the failure of *all* systems, including the reactor trip and the containment building, with the almost certain massive radioactive discharge into the atmosphere. No extensive operating experience existed and no testing had been performed by the time the (first) reactor safety study (WASH 1400) was issued. Prototype testing of a PWR, called a LOFT (loss of fluid test), was planned by the AEC/NRC in the 1970s, but had not been carried out by the time of the nuclear accident at Three Mile Island in 1979.

Some newer reactor designs, such as the Westinghouse AP1000, utilize "passive" safety components. Such systems depend on physical phenomena such as pressure differentials, convection, and/or gravity and do not rely on immediate operator response or redundant but still fallible technical equipment. In the event of a LOCA or other accident, the AP1000 and other new models are designed to need no human intervention for days (see Chapter 8).

The Chances of an Accident

Let us next examine how the probabilities were assembled in the Rasmussen safety study for the prediction of a major accident and its consequences.

1. The probability of an *initiating event* – The estimate is supposed to include all possible events that could lead to a LOCA. An estimate should include causes internal to the plant (for example, valve failures) as well as external causes (such as earthquakes). The Rasmussen probability estimate taken theoretically over all conceivable causes was 1 in 100 reactor years (expressed as 10^{-2} per reactor year).

2. The probability of the failure of the ESS, *exclusive of CI* – Here all major steps (RT, ECC, PARR, and PAHR; Fig. 7.11) have failed and major radiation release is prevented only by the integrity of the containment building. The Rasmussen estimate was 1 in 100 reactor years (expressed as 10^{-2} per reactor year).

3. The probability of CI (Fig. 7.11) failure – In this case, the integrity of the building has been broken due to forces developed in the accident sequence itself (for example, extremely high steam pressures). The estimate was, again, 1 in 100 reactor years.

4. The probability of *worst weather* – This refers to that combination of wind direction, precipitation, and so forth, that would promptly deposit the released radioactive material on the largest population in the vicinity of the plant. This was estimated as a chance of 1 in 10 (10^{-1}) at any time of occurrence of the accident.

5. The probability of an accident occurring at a site of the highest *population density* – The Rasmussen study averaged all nuclear plant sites in the United States to estimate the chances of such an accident occurring at a site with the highest population density. The estimate was 1 in 100 (10^{-2}) sites. (Subsequent to

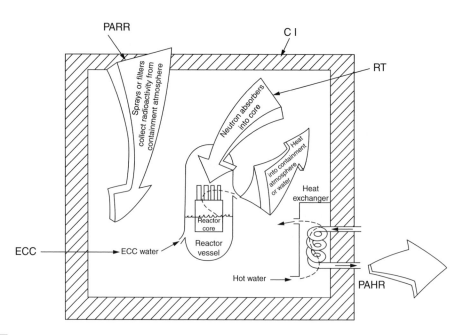

Figure 7.11 Schematic diagram of engineered safety system (ESS) for a water-cooled reactor (see text for abbreviations). *Source:* US Atomic Energy Commission (1975). WASH-1400 Washington, DC.

the Rasmussen study, accident probabilities and consequences have been estimated not hypothetically, but for specific nuclear plant locations.)

These probabilities of failure may logically be combined into composite failures simply by multiplying the individual probabilities, provided the individual *probabilities are independent* of one another. The Rasmussen report proceeded on this assumption and came up with an overall probability of an accident that the study group was convinced would appear acceptably low to the public.[5] They concluded, in summary:

1. The chances of a *reference accident* were considered, where such an accident was originally defined as a major accident, but without catastrophic release of radioactive materials. Such a case was later defined to be a *meltdown*:

$$
\begin{aligned}
\text{Meltdown probability} &= \text{Initiating event probability} \times \text{ESS failure probability} \\
10^{-4} &= 10^{-2} \times 10^{-2}
\end{aligned}
$$

   That is, the chances were 1 in 10,000 reactor years or 1 in 100 years for 100 reactors in operation.

2. The chances of a worst possible accident were defined as:

$$
\begin{aligned}
\text{Worst accident probability} &= \text{Meltdown probability} \times \text{CI failure probability} \times \text{Worst weather probability} \times \text{Population high probability} \\
10^{-9} &= 10^{-4} \times 10^{-2} \times 10^{-1} \times 10^{-2}
\end{aligned}
$$

That is, 1 in 1,000,000,000 reactor years or 1 in 10,000,000 years for 100 reactors (as noted, there were 99 nuclear reactors licensed operating in the United States as of 2016).

The executive summary of the Rasmussen report then claimed that the "chances of being killed by a nuclear accident are as remote as those of being hit by a meteor." Such claims, proponents of the technology contended, would lead a "reasonable" person to conclude that the risks of a nuclear plant accident are so low that they could be ignored. However logical these claims may have seemed at the time, they were *not* universally accepted by the public, for good cause as subsequent events would later prove. Notably, in 2011, the nuclear complex of six reactors at Fukushima, Japan, experienced catastrophic meltdowns following damage by a tidal wave (tsunami). This accident occurred when the remotely expected, but unaccounted for, magnitude wave came ashore at the site. The complex, containing six nuclear reactors, was located at an elevation above sea level *not* sufficiently above the very highest for tsunamis ever recorded along that part of the Japanese coast. The result was a disabling of reactor cooling systems and multiple loss-of-coolant accidents.

Consequences of an Accident

Included in the Rasmussen safety study results were estimates of the casualties in the various possible accidents, ranging from the reference accident to the worst possible accident. These casualties would occur, of course, solely from the effects of exposure to radioactivity (since there are no long-distance shock, blast, or heat effects from a runaway reactor of the moderated type, as there would be from a fission bomb). Radiation exposure at high levels causes death and illness due to prompt somatic effects and can induce cancer over the longer term. Somatic effects include *prompt fatalities* from massive doses of radiation.

Radiation doses above the level of 30 rem have a definite causal impact on human health. Doses above 100 rem cause radiation sickness and 200 rem is the threshold for *promptly fatal* doses. Doses in the range of 1000–3000 rem cause death within two weeks and exposure above 3000 rem results in death within hours. Besides the real danger from radiation exposure, psychiatrist Robert J. Lifton (1976) has noted that the mere idea of such exposure can have a profound emotional impact on people. Lifton described "primal fears about the integrity of the human body, as threatened by the invisible poison of radiation" (also Weart 1988). Some engineers and scientists believe that the psychological impact of the use of nuclear energy is irrational, and they have argued that it should be dismissed in the debate over nuclear energy. But even if it is irrational, should that alone determine its relevance? If people fear something for whatever unscientific reason, should they be forced by scientists or others in society to endure that fear? Should the rationality of the expert be the *final* word in societal debates? These questions are more philosophical than technological and, as we will see in Appendix B, they have no pat answers.

The consequences of a major nuclear plant accident to the surrounding area can be illustrated as in Figure 7.12. The graph, which was originally projected by the Rasmussen study in 1975, shows the spread of expected radiation dosages according to the radial distance from the accident. The thyroid dose curves are especially important because of the susceptibility of the thyroid gland to cancer-causing isotopes. Therefore, they are commonly used as indicators for emergency measures needed for the population. Although these curves were estimated before the actual experience at Chernobyl, they proved a fairly good picture of the dosages from the damaged reactor. The *actual data* from that accident (Fig. 7.13) show thyroid doses of 1 rem and more extending to a range of about 1000 miles. Other more detailed Chernobyl data suggest that an unprotected person living about 50 miles northwest of the accident site would have received a cumulative 30 rem thyroid dose over the 6 days following the accident.

The potential for the rapid spread of radiation from a nuclear accident site has spurred much debate over the issue of public evacuation. Following the accident at Three Mile Island, the NRC adopted a rule requiring public protection, including possible evacuation, within a 10-mile radius. But the range has been subject to dispute. Critics of the industry want the radius extended to 20 miles or more, while industry proponents have argued that 10 miles is unnecessarily great.

Even with an evacuation, the Rasmussen study group estimated heavy casualties for the worst possible accident. They projected 2300 fatalities, which critics called too low. A subsequent calculation (published in the *Washington Post,* November 1, 1982), on behalf of the NRC, supported the critics. It suggested that fatalities could reach tens of thousands.

By this measure we have never seen a "worst possible" accident. The worst to date was the one at Chernobyl in what was still the Soviet Union. According to earlier Soviet reports on the Chernobyl accident, there were 28 prompt fatalities due to radiation exposure at the time of the accident. In the ensuing years, there have been conflicting reports on the number of excess cancers and deaths from the accident. Estimates of deaths range from around 50 to upwards of 1 million. But the most authoritative analysis in 2005 (conducted nearly 20 years after the accident) by the World Health Organization[6] counted fewer than 50 deaths. The WHO estimated that about 4000 people have died, or will die, from exposure to the radiation from the damaged nuclear plant, but that remains speculative even today. It is the case that about the same number (4000) of excess thyroid cancers (mostly in people who were children or adolescents at the time of the accident) have been identified. But more than 99 percent were cured of the disease. Still, 30 years after, uncertainties remain as to the health effects of the Chernobyl disaster, but it does not seem that the worst fears will be realized.

Early in the nuclear program it was recognized that a major nuclear accident could cause huge legal claims for indemnity for loss of life, illness, and property damage against a plant operator. It was also clear that few, if any, privately owned electric utilities would opt for a nuclear plant with such a possibility, because the potential size of the claims could lead to bankruptcy. During the period of the initial promotion of nuclear power in the 1950s, there was thus a strong motive to remove this barrier, and this resulted in the Price–Anderson Act, passed by the US Congress in 1957.

The Price–Anderson Act has been controversial since it was enacted and the controversy has been renewed every time Congress considers its extension. The act limited the liability of a nuclear plant owner to a fraction of the possible claims for a

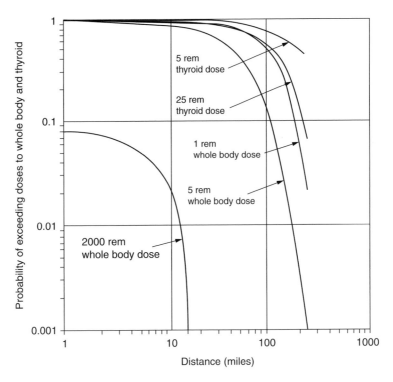

Figure 7.12   Expected radiation dosage versus distance from a nuclear accident.
Source: © 1980 by IEEE.

major accident and therefore appeared to leave the public at risk of great uncompensated losses. The measure initially limited liability to $60 million for the operator, with the US government responsible for additional amounts up to a maximum of $500 million; possible claims even then were anticipated to be as much as 30 times that figure. The Act has been extended three times, first in 1965, then in 1975, and was subsequently extended again (in 2003) to cover nuclear accidents through 2026. The liability limit for an individual operator is $375 million (which is not far from the $60 million limit if adjusted for inflation), with the nuclear power industry carrying additional liability up to a limit of $7 billion. Even though this is over ten times the original limits, critics point out that damages from a major accident could far exceed this figure.

Public Dispute: The Experts Disagree

Beginning in the 1970s, prior to the WASH 1400 study, many members of the technical community became concerned with nuclear reactor safety. There was a growing fear that standards of safety for nuclear reactors were being sacrificed to promote the fast expansion of nuclear power (Ford 1982).

Upon the publication of the draft of WASH 1400, several critical studies were made of the results. The American Physical Society (APS) sponsored the most

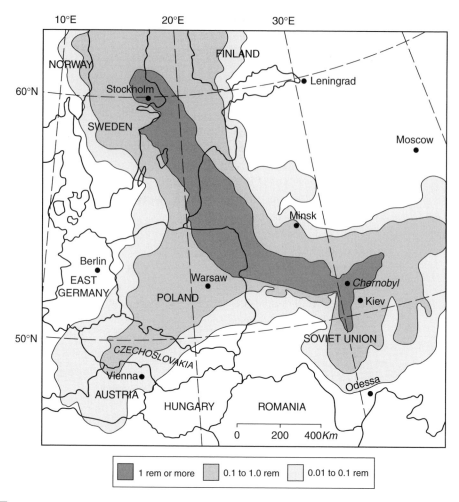

Figure 7.13 Estimated thyroid inhalation doses, 4 days after the accident at Chernobyl (countries as of 1986).
Source: Reprinted by permission of *the Bulletin of the Atomic Scientists*, a magazine of science and world affairs. © 1986 by the Educational Foundation for Nuclear Science, 6042 South Kimbark Avenue, Chicago, Illinois 60637.

prominent: the APS Study Group on Light-Water Reactor Safety. This committee, made up of professional physicists, criticized the method, omissions, and results of WASH 1400; for example, the Rasmussen study was said to have overlooked the high potential for *common-mode* failures. (Common-mode failures are those that do not occur independently of other failures.)

The nuclear industry, along with the AEC/NRC and several distinguished scientists, attacked the critics. But the debate, which pitted eminent scientists against one another, appeared to confuse the public and many elected officials rather than clarify the issues. After all, they asked, how could there be disagreement on a matter of scientific fact? But there was no agreement and no *fact* available to resolve issues of estimates and probabilities. As is discussed in Appendix B, experts have their limitations – a point made abundantly clear to the public by the actual failures of nuclear technology at Three Mile Island and Chernobyl and the meltdown of nuclear power facilities in the aftermath of a natural disaster at Fukushima.[7]

Three Mile Island

A major event in the history of nuclear power took place on March 28, 1979, when a high-pressure relief valve opened in reactor number 2 at the Three Mile Island (TMI) nuclear plant near Harrisburg, Pennsylvania. Although the valve opening was the *initiating event* of the TMI accident sequence, it did not actually indicate the fundamental problem. As an official investigation revealed, the accident sequence started because of human error. A major feedwater valve had been left closed erroneously at the end of routine maintenance the day before. Critics had warned years before of the inevitable potential for human error.

At TMI, the operators made drastic errors, and before they recognized and corrected them, there was a near meltdown of the reactor core. Even though there were no prompt casualties and it remains unclear whether there were any adverse health effects of the accident, the consequences of the TMI accident were catastrophic for the nuclear industry. The reactor core was severely damaged, and the entire area within the containment building had become prohibitively radioactive, hampering the means to repair it. The cost of the cleanup nearly bankrupted the operating utility – General Public Utilities (GPU). It has been estimated that the cleanup and repair of the station cost in the range of $1 billion ($1980). While this went on, GPU had to purchase electric power from neighboring utilities at an added cost over generating its own power.

A review of the safety systems at TMI showed that *human failures* were at least as important as mechanical failure. Actually, the human errors committed by the operators during the accident have been attributed to inadequate training, which could be called an *institutional* failing. However, the critics of the technology contend that human error can be held only to an irreducible minimum level, and that this minimum will determine the lower limit of accident probability at a nuclear plant.

It should be reemphasized that this accident was of the type that can occur even if the reactor is *scrammed* successfully (that is, the control rods are inserted and the reactor is made subcritical). Even with the chain reaction shut down, heat continues to be generated and a loss of coolant will cause core temperatures to rise. This type of accident, a simple LOCA (loss of coolant accident), is distinctly different from the accident that occurred subsequently at Chernobyl, described below.

Chernobyl

The Chernobyl accident has been called a case of a *runaway reactor*. The implication of this characterization is not only that the reactor's chain reaction was not shut down, but also that a supercritical condition took place in its core, causing rapid growth of the chain reaction.

Not only was the Chernobyl accident different from the TMI event in terms of the critical state of the reactor, but also as to the initiating cause. Whereas the TMI accident was initiated in the course of routine operation, the Chernobyl event occurred during a *special test* of the plant. The causes of the accident have been attributed largely to the nonroutine and ill-advised reactor conditions created for the test. Nonetheless, as we will see, the operator errors and misjudgments that occurred could well have happened elsewhere.

A great deal has been made of the design of the reactor at Chernobyl, which is typical of a number of operating nuclear plants in the former Soviet Union designated as the RBMK design (see Fig. 7.14). Indeed, this design has a problem-plagued history. According to reports, at times engineers have had difficulties in controlling dangerous concentrations of reactivity in RBMKs.

The poor ability for control arises in part from the *design combination* of graphite as a moderating medium and water as a coolant. This combination leads to a *positive feedback* effect on reactivity when a burst of heat causes *steam voids* in the water coolant. This comes about since the coolant water in the RBMK acts more as an absorber of neutrons than as a moderator. The nuclear fuel mixture in this design has a lower enrichment (about 2 percent ^{235}U) than a Western-designed LWR (about 3 percent ^{235}U) and depends on the superior moderating properties of graphite to achieve the chain reaction. Thus, when a void occurs in the water, more neutrons are allowed to penetrate the graphite moderator and, ultimately, reach another fuel rod for fission. The overall effect is to have bursts of chain reaction and heat production, leading to further increases in the chain reaction. A reactor in which this can occur is said to have *a positive void coefficient.*

The RBMK was also known to be susceptible to concentrations of high reactivity occurring at a point where one region of the core is in the supercritical state while the rest remain in the subcritical condition. Some control rods in fact were included in the design to stabilize these localized excursions of reactivity, especially along the long length of fuel rods in the 7-meter-high core.

Graphite itself has several undesirable features for use in high-temperature applications. First, it burns at temperatures above 700 °C when exposed to air. Normally, the operating temperature of the graphite is about 600 °C and the entire core is filled with inert gases, but the potential for combustion remains. Furthermore, it is possible that the combination of steam and carbon will generate hydrogen and carbon monoxide (both combustible). The potential for hydrogen generation is enhanced by an oxidation reaction of steam with the zirconium tubes in the fuel-rod channels (see Fig. 7.14 insert) if the temperatures rise above 1000 °C.

Ironically, the Chernobyl disaster resulted from a special test to improve safety responses to emergencies. In the test that initiated the disaster, the plant operators were obviously very intent upon achieving the test objectives, so much so that they violated safety rules of operation, several of which were strict prohibitions. Their purpose, generally stated, was to test the ability of the turbogenerator to supply electric power to the station for a brief interval following the cutoff of steam from the reactor. The supply of electricity was important in such an event to trigger the emergency core-cooling system and other engineered safety operations for a loss of coolant accident. There lies the irony of the outcome.

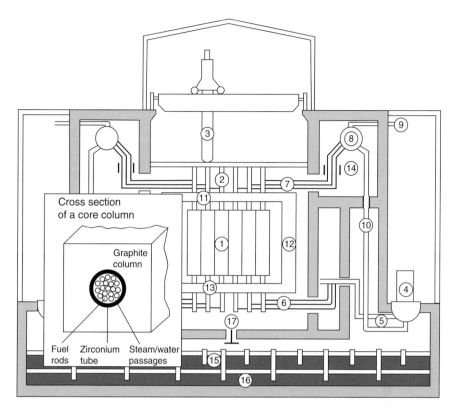

Figure 7.14 The Chernobyl reactor: a schematic view. Note in the cross section of the figure that the water coolant flows along the fuel rods within a tube contained inside a graphite moderator block. (1) Reactor core, composed of graphite columns. (2) Fuel and control rod channels. (3) Refueling machine. (4) Main circulation pumps. (5) Steam pipes. (6) Coolant inlet pipes for water. (7) Coolant outlet pipes for steam and water. (8) Drums for steam separation. (9) Steam pipes to turbines. (10) Water return lines. (11) Upper biological shield. (12) Lateral biological shield. (13) Lower biological shield. (14) Cladding failure detecting system. (15) Upper bubbler pool. (16) Lower bubbler pool. (17) Pressure relief valve.

Source: Reprinted by permission of the *Bulletin of the Atomic Scientists,* a magazine of science and world affairs. © 1986 by the Educational Foundation for Nuclear Science, 6042 South Kimbark Avenue, Chicago, Illinois 60637.

In conducting the test, the operators took the RBMK reactor into a very low-power operating condition, where the positive void effect is even more pronounced than at near full-power operation. In addition, when they encountered a momentary drop in reactivity, they retracted more control rods than recommended to reach the prescribed conditions of the test. Finally, the operators disabled two automatic reactor trip (RT) functions of the safety system and shut off the emergency core-cooling system (ECCS), so that the reactor was not only operating in a highly unstable condition, but also had most of its automatic safety features turned off.

Later analysis by a group of reactor experts, convened by the International Atomic Energy Agency (IAEA) in Vienna, concluded that a *prompt critical excursion* (where the chain reaction grows and is controlled by prompt neutrons) occurred, causing a steam explosion that ripped open the reactor core. The roof of the containment building collapsed and massive amounts of radioactive debris spewed into the air.

The amount of radioactive material discharged into the atmosphere was, in fact, unprecedented and much larger than most nuclear safety experts had predicted, approaching 100 megacuries (MCi). This included the *volatile radioactive isotopes* of iodine, tellurium, and cesium, which were carried great distances from the accident. Also, larger amounts than had been expected of heavier isotopes, such as strontium and plutonium, were detected in the fallout. The quantity and composition of the fallout from this accident was, of course, attributed to its violent nature. The accident at TMI, in contrast, resulted in the discharge of an estimated 15 Ci, because it was a meltdown type accident and was contained. Even if containment had failed at TMI, however, the discharge of radio nuclides would most probably have been a fraction of that at Chernobyl.

The radiation dosage to the surrounding population is depicted in Figure 7.12. About 135,000 people were evacuated from the immediate area (30 km radius) and about 25,000 were estimated to have received doses of at least 35 rem. Vegetables and dairy products were tested throughout Europe, and large quantities were confiscated. Contamination by cesium (^{137}Ce), with a half-life of 30 years, may make the immediate area around Chernobyl uninhabitable for decades.

Soon after the accident, experts in the United States and Europe began arguing about the implications of the accident for the future of nuclear energy. Proponents of nuclear power contended that such an accident could not happen in US reactors and that the accident was unique to the RBMK design. Even though these contentions were true for commercial power reactors in the United States, it was soon pointed out that some graphite-moderated reactors are used for military production of plutonium (see Chapter 8). Moreover, nuclear critics did not accept the arguments concerning dissimilarities between US and Soviet reactors, noting that certain aspects of reactor operation are generic to the technology.

For example, all presently operating power reactors worldwide have the same:

1. Reactor operation, using chain reactions, with *volatile kinetics*, which make control an exacting task and there is always the possibility of high-power excursions, including *prompt criticalities*, which are remote possibilities but cannot be entirely eliminated.
2. Absolute requirement for *heat removal* from a reactor in operation or immediately after shut-down, therefore exacerbating the threat of a LOCA.
3. Possibility of *human error*, carrying with it an irreducible probability of occurrence with a technology that is unforgiving of error (Tenner 1996).

Chernobyl had an impact on nuclear policies in Europe, especially in Germany. A poll taken before the accident showed about 60 percent of the German public in favor of nuclear power, about 20 percent against; in the aftermath of Chernobyl, the numbers reversed (Krohn and Weingart 1987). As a result, a "Green" political party grew and in 1998 received enough votes to enter a national coalition government. The Greens won approval of a phase-out of all of Germany's nuclear power facilities. This phase-out was partly reversed by a subsequent government, but it was reinstated in 2011, after the next major nuclear power accident at Fukushima, Japan. Germany is due to shutter its remaining nuclear reactors by 2022.

As we will see in the next chapter, some of the safety characteristics of nuclear technology are being addressed by the industry in new designs. But even if such new projects work as predicted and are adopted for future use, the issue of reactor safety will undoubtedly remain, because, unless other countries follow Germany's lead, TMI and Chernobyl reactor types will still be in operation for many years to come.

Fukushima

If previous major nuclear accidents (Three Mile Island and Chernobyl) were attributable to the design or operation of the reactors themselves, the meltdown accident at Fukushima Daiichi resulted as a principal consequence of its chosen location and the accounting for the possible consequences of that location. The six-reactor complex (see Fig. 7.15) is located at a site that is susceptible to inundation by *tsunami* waves, arising from its position on the Japanese coast. The complex is located on the east coast of Honshu Island, with a sea wall intended to protect the plants from tsunami surges. It is located on a bluff whose height was 6 m (19 feet) above sea level. The tsunami's peak surge was later determined to have been 15 m (49 feet) high, vastly over spilling the wall and inundating the plant. It would appear that the Japanese authorities had ignored the known history of earthquakes/tsunamis dating back to the year AD 860, when a tsunami wave had evidently swept 4 kilometers inland. More

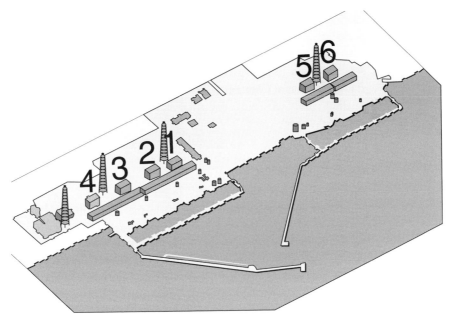

Figure 7.15   Fukushima 1 nuclear power plant site close-up.
Source: Wikipedia, 2011.

recently there were the *great tsunamis* of 1896 and 1933, which came ashore on Honshu just north of Fukushima.

In March 2011 a magnitude-9 earthquake produced a tsunami that swept ashore in the Tohaku region about 200 kilometers north of Tokyo (see Fig. 7.16). Emergency generators were knocked out, leaving no way to feed the cooling systems for the five reactors of the complex except battery-powered pumps, which functioned only until the batteries were discharged a few hours later. The sixth reactor's cooling system continued working and saved that reactor as well as neighboring reactor 5. The others were not so fortunate.

After battery power was exhausted, the supply of cooling water stopped and the reactors soon became overheated, even the three that were not in operation. A nuclear fission reactor still requires cooling after shut-down, because of the remaining *decay heat* of the spontaneous fission reactions, even after they have no more useful heat output. Without the circulating cooling water to take this decay heat away, the temperatures in the reactors built up. Left uncorrected, a reactor will soon reach destructive (i.e. meltdown) levels under these conditions. In addition to damage to the reactor structure, the extreme heat can result in the production of hydrogen in the reactors. This began at Fukushima in Reactor 3 and was followed by explosions of these gases. Then, one after another, units 1, 2, and 3 over heated and started to melt down. Reactor 4 did not fully melt down but was severely damaged nonetheless. In summary, three of the reactors were left totally inoperable, one de-fueled, and the remaining two in a *cold shut-down* state. Evacuation of surrounding areas out to 20 km from the complex was ordered. Consideration was later made of a 30 km radius (Fig. 7.16), although this would have led to even more disruption of daily life and the Japanese national economy.

The Fukushima event was not nearly as destructive as the one at Chernobyl (Steinhauser et al. 2014). Though multiple reactors melted down in the Japanese accident, the discharge of radioactivity was four orders of magnitude higher at Chernobyl. Most of the radioactivity that did reach the outside from Fukushima was deposited in the Pacific Ocean. Thus, the area of contamination was a small fraction of that at Chernobyl. There were no prompt fatalities and no unambiguous fatalities from subsequent radiation exposure – although these effects may not be present for another decade or more. This is in contrast to the earthquake and tsunami themselves, which killed approximately 16,000 people.

Still, Fukushima raises the question again: *how safe is safe enough?* Is the only lesson, to locate nuclear power plants high above sea level so as not to be flooded over by a tsunami? Such a question is not unprecedented. For example, in the design of bridges and dams almost anywhere in the world, design judgments have to be made as to how high the structure should be to avoid being washed out by infrequent but high floods. Civil engineers have long dealt worldwide with such questions – they have found that reasonable compromises can be struck between safety and cost. In the case of bridges, such compromises came to be called the *100-year bridge* design, meaning that it could be *expected* that the structure would be washed out by flooding only once in a hundred years. This is, of course, a *probabilistic* statement, meaning, considering the random variations in rainfall from year to year, how high must the bridge footings be built so that rainfall in a given year in the future will not

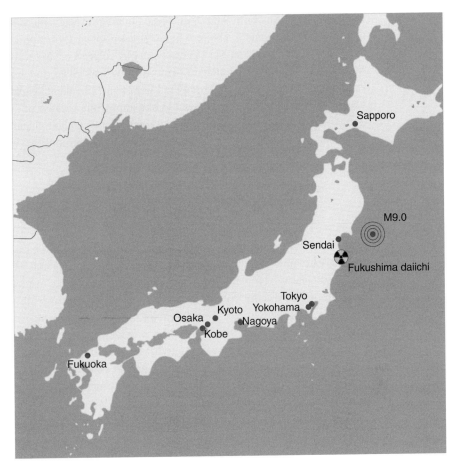

Figure 7.16   Fukushima accident overview map.

cause flood waters to overflow them. A typical length of time chosen for bridges was 100 years and thus the design was characterized as a 100-year bridge. For the case of nuclear plants located in tsunami-prone regions or near active earthquake zones, this translates as: how high above sea level should the plant be built so as not to be inundated within a specified time span (e.g. 100 years)?

Emergency Evacuation

The nuclear incidents beginning with TMI dramatized the need for emergency management of the surrounding population. At the time, some experts and the media raised questions about evacuation and about the need to have thyroid protection pills available if radiation were released. (Iodine tablets help prevent absorption of radioactive iodine into the thyroid.) With the massive radiation release at Chernobyl in 1986, of course evacuation became a clear necessity, although it was delayed by the Soviet government which sought to avoid making the disaster public (Aoki &

Rothwell 2013). Since 1979 the NRC has required emergency plans for all operating or planned nuclear plants. The formulation of such plans (not required at the time of the WASH 1400 study) resulted in heated public controversy at several nuclear plant sites. In the case of the Shoreham plant in New York State, for example, political leaders and citizens groups opposed the Long Island Lighting Company's evacuation plan on the grounds that safe evacuation was impossible because of the geography of Long Island. The outcome of these debates in the United States has been the establishment of *emergency preparedness* plans by the responsible federal agencies: Federal Emergency Management Agency (FEMA) and the US Nuclear Regulatory Commission (NRC).[8]

Criteria for Safety

At the heart of the issues of nuclear plant safety is the question posed above (and applicable to other modern technologies): *How safe is safe enough?* (Fischhoff et al. 1981). Nuclear safety is but one of several technological risks of modern society. In Appendix B we discuss criteria for safety and health standards on a more general basis. We look there at risk from the point of view of *costs and benefits*. Such an approach, though fraught with difficulties, provides the starting point for evaluating the societal benefits of improved safety/health standards or, conversely, the societal costs of lower standards. An economic question becomes evident, though, which overlays a moral dilemma. A standard of safety cannot simply be cost effective, but must have some *qualitative standard* that must be met.

That standard, however, is inevitably difficult to put into a clear, unambiguous form. For example, NRC policy (as noted in the *New York Times*, January 10, 1983) on nuclear plant accident prevention states: "individual members of the public should be provided a level of protection from the consequences of nuclear plant operation such that individuals bear no significant *additional risk* to life and health." Such a policy is grounded in the concept of *acceptable risk*, which has its origins in the earlier safety studies (for example WASH 1400). But of course, there are questions about the acceptability of the risk, its significance, and what measures are required to provide adequate protection.

After the official inquiry into the TMI accident by the Kemeny Commission (1979), the NRC and the US nuclear industry began a reexamination of nuclear safety. As part of this evaluation, there was a reconsideration of the NRC safety goals. This led to an adoption of a *Safety Culture* program,[9] which was dedicated to giving "the necessary full attention to safety matters," which became the Commission's official policy: *safety first*. A similar policy has been advocated internationally. As Pfotenhauer et al. (2012) argue:

> There is ... an urgent need for international regulatory oversight that separates nationalist goals for advanced technologies from the safe operation and maintenance of such technologies.

Such a system, perhaps desirable in the abstract, would challenge national sovereignty over very costly and sensitive technology, and so is unlikely to exist except in its voluntary formulation – as we discuss in the next chapter.

References

Aoki, M. and G. Rothwell. 2013. "A Comparative Institutional Analysis of the Fukushima Nuclear Disaster: Lessons and Policy Implications." *Energy Policy*, 53, 240–247.

Cantelon, P.L., H.G. Hewlett, and R.C. Williams (eds.). 1991. *The American Atom: A Documentary History of Nuclear Policies from the Discovery of Fission to the Present*, 2nd edition. University of Pennsylvania Press, Philadelphia, PA.

Collier, J.G. and G.F. Hewitt. 1987. *Introduction to Nuclear Power*. Hemisphere Publishing, New York.

Fischhoff, B., S. Lichtenstein, P. Slovic, S. Derby, and R. Keeney. 1981. *Acceptable Risk*. Cambridge University Press, Cambridge, UK.

Ford, D. 1982. *The Cult of the Atom: The Secret Papers of the Atomic Energy Commission*. Simon & Schuster, New York.

Grenon, M. 1981. *The Nuclear Apple and the Solar Orange: Alternatives in World Energy*. Pergamon, New York.

Kaku, M. and J. Trainer. 1982. *Nuclear Power: Both Sides – The Best Arguments For and Against the Most Controversial Technology*. Norton, New York.

Kemeny, J.G. (Chairman) 1979. *The Accident at Three Mile Island: Report of the President's Commission on TMI*. US Government Printing Office, Washington, DC.

Krohn, W. and P. Weingart. 1987. "Commentary: Nuclear Power as a Social Experiment – European Political 'Fall Out' from the Chernobyl Meltdown." *Science Technology and Human Values*, 12, 52–58.

Lamarsh, J.R. 1983. *Introduction to Nuclear Engineering*, 2nd edition. Addison-Wesley, Reading, MA.

Lifton, R.J. 1976. "Nuclear Energy and the Wisdom of the Body." *Bulletin of the Atomic Scientists*, September, 16–20.

Pfotenhauer, S.M., C.F. Jones, K. Saha, and S. Jasanoff. 2012. "Learning from Fukushima." *Issues in Science and Technology*, 28 (3). Online at http://sahalab.bme.wisc.edu/wp-content/uploads/2012/09/Learning-from-Fukushima.pdf.

Segre, E.R. 1965. *Collected Papers of Enrico Fermi, Volume 2*. University of Chicago Press, Chicago, IL.

Steinhauser, G., A. Brandl, and T.E. Johnson. 2014. "Comparison of the Chernobyl and Fukushima Nuclear Accidents: A Review of the Environmental Impacts." *Science of the Total Environment*, 470–471, 800–817.

Stobaugh, R. and D. Yergin. 1983. *Energy Future*, 3rd edition. Vintage Books, New York. (See Chapter 5, Nuclear Power: The Promise Melts Away.)

Tenner, E. 1996. *Why Things Bite Back: Technology and the Revenge of Unintended Consequences*. A.A. Knopf, New York.

Weart, S. 1988. *Nuclear Fear: A History of Images*. Harvard University Press, Cambridge, MA.

NOTES

1 Historical material is taken from Grenon (1981), Kaku & Trainer (1982), Stobaugh & Yergin (1983), and Cantelon et al. (1991).

2 The PWR and BWR technology is discussed later in this chapter.

3 Data are from the Nuclear Energy Institute at www.nei.org.

4 Deuterated water (or heavy water) is water with a larger than normal percentage of deuterium, an isotope of hydrogen with a nucleus containing both a proton and a neutron. It occurs naturally in water but when it is used as a moderator most of the hydrogen atoms of the water (H_2O) are of the heavy type.

5 This is sometimes referred to as technological bias, whereby people are expected to listen to experts – and experts are confused when in fact people do not.

6 The report was entitled, "Chernobyl's Legacy: Health, Environmental and Socio-Economic Impacts."

7 The history and impact of the three cases of nuclear power plant disasters is derived from articles and reports from the *Bulletin of the Atomic Scientists*, *Science, The New York Times*, and the *IEEE Spectrum*, as well as the report of the Kemeny Commission (1979) on Three Mile Island, "Health and Environmental Consequences of the Chernobyl Nuclear Power Plant Accident," by the Interlaboratory Task Group on Health and Environmental Aspects of the Soviet Nuclear Accident, USDOE, and "Fukushima Nuclear Accident Analysis Report," The Tokyo Electric Power Company, Inc. Other sources are cited in the text.

8 The NRC is the successor agency to the AEC.

9 "Safety Culture," September 16, 2010, Statement at the Public Meeting Between NRC and Stakeholders Regarding Safety Culture. Issued July 28, 2010.

The Nuclear Fuel Cycle, Weapons Proliferation, and Nuclear Power: The Next Generation

Introduction

When the prospect of nuclear power was first presented to the public in the late 1940s, it was portrayed as an energy panacea: so abundant it would melt the snow as it fell; so cheap it would be too cheap to meter; so powerful it would propel aircraft the size of ocean liners from continent to continent. The wonders were to be without end. But things did not turn out so well. So far from the energy cure-all, by the 1980s it seemed an overly expensive and dangerous technology. Environmentalists as well as consumerists fought against the expansion of nuclear power in the United States. While a hundred or so nuclear power plants continued to operate, construction of new facilities stopped.

But then in the 2000s came the concern about climate change, and the recognition that nuclear power did not create any greenhouse gases, caused far less damage than did coal or oil, and that with fairly abundant resources, could potentially replace coal and even natural gas in the production of electricity. Leading environmentalists switched their position. James Hansen, one of the most prominent scientists in the climate debate, and three of his climate scientist colleagues issued a letter to those who supported action on climate change but opposed nuclear power. The letter said, essentially, that nuclear power was necessary to blunt the effects of climate change. They wrote:

> With the planet warming and carbon dioxide emissions rising faster than ever, we cannot afford to turn away from any technology that has the potential to displace a large fraction of our carbon emissions. Much has changed since the 1970s. The time has come for a fresh approach to nuclear power in the 21st century.[1]

This was not received well by many environmentalists, but others agreed. It was not going to be possible to replace coal power plants solely with wind and solar for electricity generation. As we will discuss in Chapter 11, many alternatives to fossil fuels lack the "density" of those fuels and cannot replace them except at enormous cost over enormous amounts of space. Nuclear fuel, on the other hand, is a dense (see Chapter 11) fuel and therefore power plants take up a relatively small amount of space. The argument for nuclear power seems compelling unless there is a deliberate wish to de-industrialize the world (which a few – but only a few – environmentalists actually have professed[2]). Otherwise, nuclear power will have to play a major part in electricity production in the twenty-first century if the world is to maintain a high standard of living along with a low output of CO_2.

But, as the last chapter indicated, nuclear power is not without its problems. We focused on safety issues in Chapter 7. In Chapter 9, we will examine the problem of the cost of nuclear facilities. Here we look at a few other problematic aspects of nuclear power and at some advanced technologies that hold the potential of solving these as well as the safety issues of the previous chapter.

The Nuclear Fuel Cycle and Weapons Proliferation

Nuclear fuel undergoes a cycle of *extraction, preparation, use,* and *disposal* to supply the energy to run nuclear reactors. Throughout the course of that cycle there are hazards that threaten health and property, but fuel-cycle problems are due mainly to two factors: First, nuclear fuel fabrication and use are not only potentially hazardous in the present, they may also continue to be hazardous for generations to come. Second, the nuclear fuel cycle generates elements that can be used in the making of nuclear weapons.

A diagram of the steps in the nuclear fission fuel cycle for commercial power plants is shown in Figure 8.1. In the cycle of uranium use, we can distinguish the following steps:

1. Mining of uranium ore in the form of uranium oxide (U_3O_8).
2. Conversion to gaseous form (uranium hexafluoride or UF_6).
3. Enrichment of the ^{235}U concentration.
4. Fabrication of the processed uranium into reactor fuel rods.
5. Reactor use for power-producing fission.
6. Storage (and possibly reprocessing) of the spent reactor fuel, including uranium, plutonium, and other radioactive waste products.
7. Disposal of radioactive and fissile products of spent fuel.

Some of these steps were discussed in Chapter 7. The environmental impact of the nuclear plant, and the *front-end* steps (1–5) of the cycle were the prime concern in that chapter. Here we will examine the *back end* of the cycle, the question of what happens to spent fuel that comes out of the reactor. We will consider, too, the dangers the waste poses, and how the problems might be solved.

Management and Disposal of Nuclear Wastes

Nuclear reactor (LWR) fuel is considered *spent* (or "used up") when about 75 percent of the ^{235}U has been fissioned. During its cycle of use as a fuel, however, uranium (both ^{235}U and ^{238}U) produces many other elements through fission and other transmutation processes. One of these elements, plutonium, can itself fission and release usable energy, but the remainder of the reactor products are *nonfissionable* and contribute nothing to power generation, at least in current designs.

Yet many of these products are extremely toxic and their handling, storage, and ultimate disposal are therefore difficult. Not only are they highly radioactive, but they

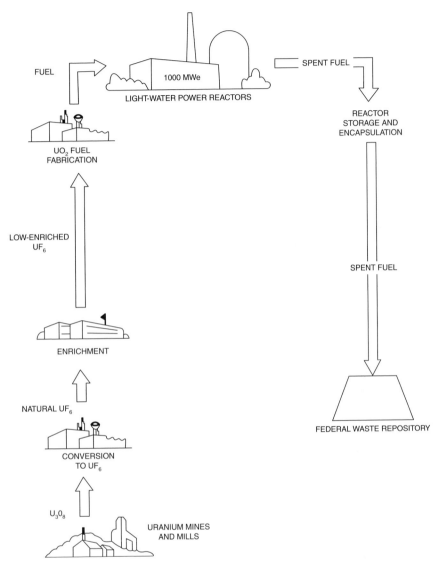

Figure 8.1 The nuclear fuel cycle.
Source: US Nuclear Regulatory Commission, 1976, NUREG–0016, October, Washington, DC.

continue to generate heat. Both the *reactivity* and the *heat* decline through the processes of radioactive decay. However, the time required for decay varies greatly among the various elements found in spent fuel. Tables 8.1 and 8.2 show the various fission products and actinides[3] that would be typical for a ton of spent LWR fuel. (The actinides follow *actinium* in the periodic table and include plutonium isotopes.) The radioactive half-lives (the decay times to one-half the radioactive emission levels) of the elements shown range from 285 days to 17 million years, with the important fissionable plutonium isotope ^{239}P reaching its half-life only after 24,400 years, a time greater than all of recorded human history. Table 8.1 includes the amount of radioactivity measured in curies (Ci) for each element over time intervals

Table 8.1 Typical yields of actinides[a]

Element	Half-life (yr)	0.3 yr (g)[b]	0.3 yr (Ci)[b]	10 yr (g)	10 yr (Ci)	500 yr (g)	500 yr (Ci)	10,000 yr (g)	10,000 yr (Ci)	100,000 yr (g)	10,000 yr (Ci)
^{237}Np	2.1×10^6	760	0.59	760	0.59	786	0.61	810	0.63	790	0.61
^{238}Pu	86	5.8	105	5.5	100	0.1	1.8	—	—	—	—
^{239}Pu	24,400	27.5	1.7	27.5	1.7	32	2.0	59	3.5	4.5	0.3
^{240}Pu	6,580	8.5	2.0	19.2	4.5	38.4	8.8	13.9	3.2	—	—
^{241}Pu	13.2	4	464	2.4	273	—	—	—	—	—	—
^{242}Pu	379,000	2	0.009	2	0.009	2	0.009	2	0.009	1.7	0.007
^{241}Am	462	54	189	55.6	198	29.5	103	—	—	—	—
^{243}Am	7,370	82	17.0	82	17.0	77	16.0	31	6.5	—	0.001
^{244}Cm	17.6	30	2,570	19.7	1,700	—	—	—	—	—	—
Total		974		974		965		916		796	50
Actinides, including daughter products (approx.)			200,000		10,000		900				

[a] Includes intermediate and long half-lives in the waste stream from processing 1 metric ton of light water reactor fuel irradiated to 33×10^9 watt days (thermal) per metric ton of fuel.
[b] g: grams per metric ton of fuel. Ci: curies.
Source: Reproduced with permission from *The Nuclear Fuel Cycle*, The Union of Concerned Scientists, The MIT Press, Cambridge, MA, 1975.

Table 8.2 Typical yields of fission products[a]

Element	Half-life	Curies remaining			
		10 years	100 years	500 years	1000 years
^{144}Ce/^{144}Pr	285 days	300	—	—	—
^{106}Ru/^{106}Rh	367 days	1100	—	—	—
^{155}Eu	1.8 years	160	—	—	—
^{134}Cs	2.1 years	8300	—	—	—
^{125}Sb/^{125}Te	2.7 years	980	—	—	—
90Sr/^{90}Y	28.1 years	1.2×10^5	1.32×10^4	0.6	—
^{137}Cs/^{137}Ba	30 years	1.6×10^5	2.1×10^4	2	—
^{151}Sm	90 years	1100	520	30	0.4
^{99}Tc	2.1×10^5 years	15	15	15	15
^{93}Zr	9×10^5 years	3.7	3.7	3.7	3.7
^{135}Cs	2×10^6 years	1.7	1.7	1.7	1.7
^{107}Pd	7×10^6 years	0.013	0.013	0.013	0.013
^{129}I	17×10^6 years	0.025	0.025	0.025	0.025
Total curies (approx.)		300,000	35,000	53	22

[a] Includes intermediate and long half-lives in the waste stream from processing 1 metric ton of light water reactor fuel irradiated to 33×10^9 watt days (thermal) per metric ton of fuel.
Source: Reproduced with permission from *The Nuclear Fuel Cycle*, The Union of Concerned Scientists, The MIT Press, Cambridge, MA, 1975.

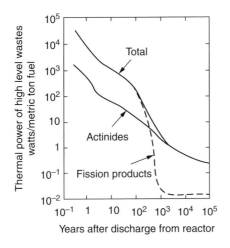 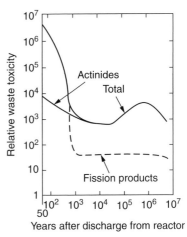

Figure 8.2 Radioactive decay of high-level nuclear waste from light water reactors. (a) Thermal power per ton of reactor wastes versus years after discharge. (b) Relative toxicity (compared to uranium ore) versus years after discharge.
Source: Adapted from Ford (1975).

as long as 100,000 years. (Thermal power decay over time per ton of spent fuel is shown graphically in Fig. 8.2a.)

If, during this tremendous time span, there is an uncontrolled release into the biosphere of these waste materials, the public will be endangered. The hazard may be measured by the relative toxicity of the wastes compared to the toxicity of uranium ore, which is radioactive in its natural state. We see in Figure 8.2b that it takes about 1000 years for the reactor waste to decay in toxicity from over a million times to a few thousand times that of natural uranium ore. After 1000 years, the decay never drops much below a relative factor of 1000. Indeed, it actually rises again after 1,000,000 years or so because of the spontaneous creation of *daughter elements* from the radioactive decay process.

At the present time, in the United States nuclear plants temporarily store, rather than permanently dispose of, nuclear wastes, and as of 2014 there were about 74,000 metric tons of nuclear waste stored in this fashion – an amount increasing in the United States alone by about 2000 metric tons per year. This situation exists because there are substantial doubts about the feasibility of the schemes proposed to date for permanent, or long-term, disposal. In the United States, for example, all spent fuel is stored at reactor sites in ponds that resemble swimming pools (the water absorbs the radiation and dissipates the decay heat). There could be alternative interim methods of dealing with waste; for example, it could be stored in central depots or spent fuel could be reprocessed to take out fissionable elements for use again in fuel rods. Several European countries, notably France, as well as Russia, India, and Japan, reprocess spent fuel rods, extracting the plutonium and using it in newly fabricated fuel rods. The United States began developing reprocessing facilities, but President Jimmy Carter opposed it because reprocessing would have meant large amounts of fissile plutonium traveling around the country. Consequently, in the United States (and in other countries without repro-cessing) the wastes only accumulate, raising the question of long-term disposal.

Long-term disposal must meet acceptable criteria of safety, given the dangers of accidental radioactive releases into the environment (Kasperson 1983). Due to the extraordinarily long lifetimes of fission products, the disposal of these wastes presents unprecedented technological and ethical problems (Spash 1993, Murdock et al. 1983). Technologically, the method of disposal must ensure a high degree of isolation for many *thousands* of years, thus requiring containment materials and disposal facilities that are known to be stable for periods of this magnitude. Furthermore, unless the disposal is terminal (i.e. requires no further human intervention), society may be faced with the problem of guarding or monitoring the wastes into an indeterminate future, a time frame beyond that of any civilization in history.

United States government policymakers – Congress and the Department of Energy – have turned to the concept of *geological isolation* as the sole objective for radioactive waste disposal. They have adopted a policy to achieve reliable, long-term safety against accidental radioactive release of the most dangerous nuclear wastes called *high-level wastes* (HLW).[4] If they can do this, then their obligation to public safety will be met. By geological isolation, they mean the placement of wastes in rock or sedimentary formations in the Earth's crust that have remained intact – that is, have not been subject to seismic or tectonic movements – for millions of years.

Such formations exist both on land and beneath the oceans, although how easy it is to identify and use them has remained unclear. In the United States, for example, critics have argued that sites under consideration as disposal repositories may not be as stable in the past as the government has claimed. Nevertheless, the plan called for the placement of wastes deep within formations through superdeep drilled holes or in mined vaults, or by rock melting. This last technique would use the decay heat of the wastes themselves to melt the rock until the wastes were deep within the Earth.

Attempts to develop an acceptable disposal technique have proven difficult.[5] Yucca Mountain in Nevada was designated as the repository for nuclear waste from civilian power plants by Congress in 2002, but although money was appropriated the repository was not built. It was contested by many groups and opposed by most of Nevada's politicians; indeed, when the project was abandoned in 2011, federal government officials said the rationale was political not safety related.

If our society decides to go to great lengths to protect even the generations of the not-so-distant future, we will have to pay more now. The cost will be reflected in the overall generating cost of nuclear power, because waste disposal will become part of the cost of the power. This is not always so explicit in the debate over waste, which tends to focus on safety. But safety has a cost that in this case will affect the cost of the energy we use.

Nuclear Power and Nuclear Weapons Proliferation

The problems of long-term disposal are greatly magnified when we consider not just the *accidental* release, but also the *intentional* release of radioactive material. By intentional release we mean the retrieval of wastes for the *fabrication of weapons*. It might seem unlikely that wastes placed deep in the ground could be easily recovered, but there is no reason to believe that, over the thousands of years involved, retrieval could not be simplified. Wastes could be used to create explosive nuclear devices, or they could be fabricated more easily into *radiological weapons*, where the release and diffusion of toxic particles – such as in an aerosol sprayer – could lead to widespread death and social upheaval.

Perhaps the most dangerous substance in high-level waste (HLW) is plutonium (^{239}Pu). Unfortunately, separation of plutonium from spent fuel rods is a relatively straightforward chemical procedure, and plutonium is produced constantly in nuclear reactors. About 8 kilograms of ^{239}Pu is bred in a typical LWR during every month of operation. A few micrograms of plutonium ingested or inhaled by a human are fatal; 15 kilograms or less may be sufficient to fabricate a crude fission bomb. Because plutonium is fissionable, some have suggested reprocessing spent fuel rods to extract the plutonium for reuse in reactors. However, reprocessing would also present an additional short-term danger: large quantities of separated plutonium that could be readily assembled into a weapon.

As we stated at the outset, although intended for peaceful purposes, nuclear power technologies are tied inextricably by physics with nuclear weapons technologies. As a result, the world is confronted with the threat of nuclear weapons proliferation.

The chain reaction described in Chapter 7 is the basic process for both *nuclear power and nuclear weapons.* This holds true at two levels:

1. The chain reaction is the means by which energy is released for both peaceful and violent purposes.
2. The chain reaction is a prime means of breeding (see Chapter 7) plutonium for *either* weapons or nuclear fuel.

Historically, of course, nuclear fission was first applied in the development of the atomic bomb during World War II. But the first controlled chain reaction preceded the bomb. In 1941, physicist Enrico Fermi demonstrated a carbon-moderated *pile*, a primitive version of a reactor. After the first experimental proof of the chain reaction, US physicists worked on developing the means of suddenly assembling within a bomb a *supercritical mass* to create a rapidly growing chain reaction and a violent release of energy. In constructing atomic bombs, scientists used two different elements – *enriched uranium* (^{235}U) or *plutonium* (^{239}Pu). Both worked, and one of each type was dropped on Japan to end World War II.

Scientists later demonstrated that the same chain reaction in its controlled form could produce nuclear power. In other words, current processes of nuclear power and its fuel cycle are *physically indistinguishable* from those of nuclear fission weapons (i.e. the *chain reaction*). Inevitably, then, in the process of spreading nuclear power around the world, we increase the prospect for nuclear weapons proliferation as well.

Both weapons and power depend on having fissionable isotopes (either ^{235}U or ^{239}Pu) in sufficient concentration to sustain a chain reaction. The achievement of separating the fissionable isotope in sufficient concentration was the *technological hurdle* to be overcome during the creation of the atomic bomb.

In Chapter 7, we discussed briefly the enrichment of natural uranium ore. Natural concentrations of (fissionable) ^{235}U in uranium are too low to sustain a chain reaction in an LWR and the largest constituent (^{238}U) is nonfissionable. The ^{235}U concentration can be raised by enrichment through a process called *diffusion* or through other high-technology means. The diffusion process, which was first developed on a mass scale at Oak Ridge, Tennessee, during World War II, represented a major industrial effort of great sophistication that few other countries had duplicated at that time or since. The enrichment process remains a technological hurdle to any country that is trying to achieve a chain reaction with uranium.

Actually, there is an alternative to enrichment. The CANDU or heavy water reactor permits a chain reaction in natural uranium ores. The deuterated water achieves its moderation of neutron energies more efficiently by capturing fast neutrons, thereby leaving slower ones to participate in the fission chain reaction.[6] In World War II, Germany attempted to build a fission weapon by producing heavy water to breed plutonium. Had the Axis powers overcome this hurdle first, they would have had the means to produce sufficient plutonium for weapons and, conceivably, could have won the war.

But the elimination of the enrichment process does not solve the problem of weapons proliferation, because heavy water reactors as well as LWRs breed plutonium. Since the separation of plutonium from spent reactor fuel is a relatively simple process, a country would have no further technological hurdles to obtain *weapons-grade* materials (that is, materials with high concentrations of fissionable nuclei).

Today, either of these two technologies, enrichment or heavy water, is a key to nuclear energy for either peaceful or nonpeaceful purposes. Several developed nations – France, the United Kingdom, the United States, China, and Russia – have long had the capability to manufacture both nuclear power reactors and nuclear weapons. Other countries, namely Pakistan, India, North Korea, and Israel, have either shown weapons capability or are believed to possess it. Iran is believed to be developing weapons capability. Finally, several developed countries – Canada, Germany, Italy, Switzerland, the Netherlands, and Sweden – have highly developed nuclear power capabilities, some even for export, but have so far chosen not to assemble nuclear weapons.

In 1968, following several years of negotiations and debate, 89 countries signed the Treaty on the Non-Proliferation of Nuclear Weapons (or Non-Proliferation Treaty, NPT).) In addition, there is a nuclear nonproliferation agreement, the Treaty of Tlatelolco, signed by 24 Central and South American nations, in 1972. This treaty declares the southern portion of the Western Hemisphere a "nuclear-free" zone. The NPT includes the major nuclear powers and now has 190 signatories. Each nuclear weapons nation that signs the treaty agrees not to supply weapons or the means to fabricate them to non-nuclear nations and all nonweapons states agree not to receive or acquire such devices or technology. In order to monitor compliance with the treaty, all signatory states have agreed to a system of international inspection of nuclear facilities.[7]

This inspection system was set up by the International Atomic Energy Agency (IAEA) with the following objectives: "Timely detection of diversion of significant quantities of nuclear materials from peaceful nuclear activities to the manufacture of nuclear weapons and deterrence of such diversion by early detection."

The key phrase in this statement is *timely detection.* Timely detection refers to a country's imminent ability to assemble a critical mass of fissionable material. The imminent capability of a country or clandestine organization to do so depends on whether it has overcome the major *technological hurdles* to creating a weapon. If, for example, it has uranium in enriched concentrations or reprocessed plutonium, then the only remaining question is whether it has sufficient weapons grade material to fabricate one or more weapons. How much are the "significant quantities" that are required?

1. For plutonium (^{239}Pu):
   a. Ideally, it would take only 4 kilograms of *pure* ^{239}Pu inside a neutron-deflecting shell of beryllium to make a critical mass. The diameter would be the size of a baseball.
   b. More practically, the presence of fission poisons (^{240}Pu) requires a larger mass, but the presence of oxygen creates pressures that somewhat reduce the required mass. Thus, 15.7 kilograms of plutonium dioxide (PuO_2) is a critical mass.
   c. The IAEA considers 8 kilograms of ^{239}Pu a "significant quantity."
2. For highly enriched uranium (^{235}U):
   a. An ideal, 100 percent enriched, uranium would require 15 kilograms for a critical mass and would be the size of a softball.
   b. At 20 percent enrichment of ^{235}U, the critical mass of the uranium mixture is 250 kilograms.
   c. The IAEA considers 25 kilograms of 25 percent ^{235}U uranium a "significant quantity."
3. For uranium ^{233}U, the fissionable isotope of thorium, the IAEA considers 8 kilograms a "significant quantity."

The scope of the IAEA's safeguarding responsibilities is great though their actual authority is not. But the IAEA is responsible for over 1250 nuclear installations, including enrichment facilities and reprocessing operations. The Agency, based in Vienna, employs more than 850 staff in the Department of Safeguards alone. There are 182 countries with "Safeguard" agreements, and 193,500 "significant quantities" of fissile materials to monitor. In all there are an estimated 1440 tons of enriched uranium and 500 tons of plutonium, enough materials to make an estimated 60,000 nuclear weapons.[8] Of course, much of the material is in the United States and Russia; the two countries, combined, produced an estimated 125,000 nuclear weapons during the Cold War. A major concern exists with "bulk-handling facilities," mainly fuel fabrication, reprocessing, and enrichment facilities, where most of the material under IAEA supervision resides.

The basic approach of the IAEA is to verify quantities of nuclear materials at each facility it monitors. A system of material accounting is operated by each participating country. The IAEA maintains a running evaluation of discrepancies in these accountings, called *material unaccounted for* (MUF), which would indicate a possible diversion for weapons purposes. Ideally, the system works to produce timely warnings that a significant amount of potential weapons material is unaccounted for. (A *timely warning* to the international community assumes that it will take a minimum of 7 to 10 days to fabricate the materials into an explosive device.)

Critics have pointed out institutional and political problems with the NPT. These relate to the overriding issue of *national sovereignty*. In fact there are four nations –

India, Pakistan, Israel, and South Sudan – that have refused to sign it. The first two have tested nuclear weapons; the third is believed to have the capability to produce them.

More problematic is the fact that signatories can evade detection. For example, a host country can delay on-site inspections to defeat timely warning. More importantly, countries determine which facilities the IAEA is permitted to inspect, thereby allowing the possibility of separate clandestine weapons materials projects (research reactors producing plutonium weapons using noninventoried uranium ore). This was the charge made against Iran, a country believed to be developing nuclear weapons capabilities.[9] The IAEA is also limited in its ability to enforce compliance with the treaty, to prevent a signatory nation from denouncing the treaty, or to compel any nation to sign. Some have felt in recent years that the NPT is something of a dead letter, especially after the failure in 2015 to ratify the 5-year review report (Dhanapala & Duarte 2015). Some remain optimistic about the NPT (for example, see Ruzicka & Wheeler 2010). But it seems that the treaty and IAEA are still the main restraint on weapons proliferation – a risk that remains, especially so long as LWRs are producing large quantities of plutonium.[10]

Next Generations of Nuclear Power Technology[11]

The various dangers of nuclear power – in use, in the creation of lethal wastes, in the proliferation of nuclear weapons – must be balanced against the perceived need for power generation that does not produce greenhouse gases. Nuclear power does this – and does it without requiring a major transition of electric power systems (as would be required by a system using intermittent sources of electricity, wind and solar, see Chapter 11). Nuclear power plants today provide continuous power and many have been running without major problems for several decades, and for the most part have created no widespread health problems. Indeed, for baseload power, nuclear has generally performed well.

Nuclear power also is the CO_2-free means of producing electricity with what is termed a small "footprint." As Gwyneth Cravens (2007) notes, "A nuclear plant producing 1000 megawatts [a standard size of a baseload power plant] takes up a third of a square mile. A wind farm would have to cover over 200 square miles to obtain the same result, and a solar array over 50 square miles."

These characteristics have led scientists, governments, and private entrepreneurs to look closely at alternative nuclear power designs that in the next 10 years might create a new generation of nuclear power plants. Ideally these systems would (1) be inherently safe (or at least safer than any existing plants) due to safety features that will prevent meltdowns (or explosions as at Chernobyl) without the need for human interventions or electric power; (2) produce little or no waste that can be exploited as a weapon; in fact, some designs may consume the tons of existing waste as fuel; and (3) be cost effective to build and operate, unlike so many LWRs of the past half century that experienced long delays and massive cost overruns. One reason for cost overruns in the United States was that all LWRs were custom built and so gains from scale economics in production of components were not obtained. Furthermore, unique design meant that learning with respect to construction issues at a given plant was not necessarily transferred.

In some areas, progress has already been made. So-called "third generation" nuclear plants are similar to those depicted in Chapter 7, but there are some important differences. We noted one such design, the Westinghouse AP1000, which is an example of generation III nuclear technology. These designs incorporate what is termed passive safety features. That is, in the event of a LOCA, changes in pressures or temperatures within the reactor and/or the containment structure cause emergency coolant to fill the chamber without human intervention. Nor do they require emergency backup electricity as the safety systems at Fukushima did, which led to the meltdown of the reactors. Instead, emergency action is based on some combination of gravity, natural convection, and/or response to high temperatures. These forces initiate the processes that will shut down the reactor and reduce temperatures and pressures.

ESBWR

Another example of a generation III nuclear power design is the Economic Simplified Boiling Water Reactor (ESBWR). The ESBWR received Nuclear Regulatory Certification in 2015 and may soon be installed at the Detroit Edison Fermi power plant. The ESBWR has passive safety features, using gravity for the reactor core and a separate passive system to cool the containment structure. The manufacturer (GE Hitachi Nuclear Energy) claims that if there is a major problem in the reactor, it can sit for a full week or more without human intervention or electricity.

It is expected to be "economic" for two reasons: first, the design has been simplified, eliminating (according to the company that makes it) the need for 25 percent of the pumps, valves and so on that were part of previous BWR designs. Second, the reactor is modular so that parts are constructed off site according to a standard design. That presumably will mean a more rapid construction process and costs that should fall over time because standardization should lead to gains in production efficiency. The reactor also permits longer stretches between refueling and is expected to produce electric power for 60 years.

It is, however, conventional in many respects. It still uses water as its main working fluid and coolant; its basic operating system is the same as BWRs for the past 60 years; waste will still be of the same basic variety as most other nuclear power plants; and exactly how much it will actually cost to build and operate remains to be seen. As for safety, only significant operating experience can assure that it is truly "safe," although its systems would seem to be much safer than earlier nuclear power plants.

There are other generation III designs for pressurized water reactors (e.g. the AP 1000) and even heavy-water reactors that employ some of the features, especially passive safety and modular construction, of the ESBWR. A few of these are under construction and their performance in terms of cost and safety will be of considerable interest to those who hope to see an expansion of nuclear power.

Small Modular Reactors

Modular construction, as noted with respect to the ESBWR, potentially lowers costs. Moreover, if a system is small enough – much smaller than any version of the ESBWR – it should be easily transported fully assembled. Instead of lengthy, costly on-site construction, a small reactor could be put on a truck or train, driven to the site, put in place and, not all that long after, started. Small modular reactors (SMRs) are rated at only 50–300 MW, but can be clustered to provide the same level of power as the large LWRs.

SMRs, it should be noted, come in various configurations, some using the basic principles and design of LWRs, while others are gas cooled or cooled with more exotic means, for example liquids of chloride, fluoride, or sodium. According to the IAEA, as of 2015 there were "more than 45 SMR designs under development" worldwide and four under construction. One US design, for example, would offer a prefabricated reactor with a containment system that could be transported by rail, truck, or barge to the plant site, with construction completed in 36 months – one-third to one-half the time for conventional LWRs.

A particular design element is that the reactor vessel is to be placed within a containment vessel, which is filled with water. The entire unit is then positioned below ground. The reactor functions as a normal PWR with water as coolant and moderator in the reactor vessel. But in the event of a problem with the reactor, the water in the outer vessel passively removes the heat, according to designers, indefinitely.

Some SMRs are close to commercial application and they could become widely adopted if they in fact prove both safe and economical. There remain, however, questions related to waste disposal especially – questions that may continue to hold back commercialization of these (and many other) types of reactors.

There is, it should be noted, a more advanced, generation IV, set of nuclear designs – reactor concepts that are still in the development stage – which hold out the greatest hope for the future of nuclear power. Below are just a few of the hoped for generation IV reactor designs. Designs, like generation III models, all include passive safety features, but these reactors can potentially avoid problems of weapons proliferation and long-term waste disposal.

Molten Salt Reactors

Water is the most extensively used moderator/coolant in most currently operating or licensed nuclear plants. But an alternative seems to hold out promise. Molten salt reactors (MSRs) are nuclear reactors in which the nuclear fuel is in a *liquid* form and is combined in a liquid medium of chloride or fluoride salt instead of solid fuel rods. Since the fuel/salt is liquid, it is not only the fuel but also the moderator and the coolant. In other words, the liquid produces the heat and can also transport it away from the reactor to the power plant. The salts remain in a liquid state at higher temperatures and lower pressures than those found in conventional PWRs. Depending on the composition of the mixture, it can remain in a liquid state at up to 1400 °C.

The MSR systems with circulating fuel salt have much lower inventories of fissile plutonium, as well as no requirement to fabricate or handle solid fuel, which lowers costs. Refueling also will reduce net costs because it can take place even as the reactor is continuing to operate – that is, while the reactor is still providing electricity to the grid. Actinides are less readily formed when a fissile form of thorium (^{233}U) is used instead of ^{235}U, and many designs are based on thorium and the thorium cycle rather than the standard nuclear fuel cycle from ^{238}U. The thorium cycle would breed fissionable products in the reactor fuel. These can be removed from the mixture, but in some concepts fissile and nonfissile, heat-producing radioactive materials stay in the fuel mixture and continue to "burn," contributing to the heat production. Other designs will use high-level waste (HLW) itself for fuel. Essentially, once the fuel is burned, there would be little HLW remaining that would require monitoring for generations – again a cost saving. What waste products emerge will have a much shorter lifetime than the residues from LWRs, so that geological time frames for disposal will no longer be needed.

These characteristics make MSRs a promising technology. But some questions remain. Most importantly, molten salts are highly corrosive and so how will materials hold up under long-term use? Clearly, MSRs are still some way from commercialization – and like other new energy technology concepts may never get there. But there are many people who see MSRs as the way to provide safe and greenhouse-gas-free power in the decades ahead.

High-Temperature Gas-Cooled Reactors

High-temperature gas-cooled reactors (HTGR) are intended, like SMRs, to be modular and buried underground. The innovation here is that the reactor is cooled by helium. It is designed especially for safety. For example, if temperatures rise above those for normal operating conditions the reactor automatically shuts down as control rods fall into the reactor core entirely by gravity, and shut-down occurs without the need for human intervention or for the injection of water or other fluid coolant. Because helium, the coolant gas, is inert there is no possibility of fire or explosion, yet there would be no danger even if normal air or water entered the containment structure since shut-down and passive insertion of the control rods would immediately occur. It is expected that an HTGR would not cause disastrous conditions even in the event of natural events such as earthquakes or flooding – a concern especially after Fukushima.

Fast Neutron Reactors

Many of the generation IV nuclear reactor designs are what are termed "fast neutron reactors" (FNR). In a standard LWR (or any variants that use similar technologies) water acts as a moderator to slow neutrons that are emitted during fission. But in an FNR, neutrons move unmoderated at a speed three orders of magnitude faster than those in an LWR. Instead of water, most FNRs use liquid metal coolants, often liquid lead or a liquid lead–bismuth combination. The basic concept is that fast neutrons can utilize the most common isotope of uranium (nonfissile) ^{238}U and so potentially provide fuel for several centuries. Indeed, in its "breeder" configuration (FBR), it will potentially create more fissile fuel than it uses.

The typical LWR that we saw in Chapter 7 (and even most of the newer designs described above) is what is termed a *burner* reactor. In a burner reactor the fissionable nuclei are split, liberating energy for a power plant, and the nuclei are thereby *spent*, i.e. used up in much the same way as a combustible fuel is burnt up. The breeder, on the other hand, creates *new* fissionable nuclei, as a by-product in the process of fission energy production, and does so faster than these nuclei are used up to generate power, thus creating more nuclear fuel than it consumes. Breeding occurs through the bombardment of ^{238}U nuclei by fission-liberated neutrons. The fissioning of either ^{235}U or ^{239}Pu in any nuclear reactor creates high-energy neutrons, some of which will collide with ^{238}U nuclei and be absorbed. The ^{238}U nuclei will then be transmuted into ^{239}Pu (after 2.3 days) by the process of radioactive decay.

The FBR could increase the amount of usable uranium fuel, breeding plutonium that could be separated from the fuel rods and used as fuel in standard "burner" reactors. Or, to put it more concretely, if there are 200 years or so of burner uranium fuel (that is, given that uranium contains a small, natural amount of ^{235}U that is enriched to provide the heat to turn the electric generator), a breeder reactor would extend that 100-fold or fuel for 20,000 years. An FBR reactor under development in India would use thorium, which would extend for many centuries India's large thorium resources.

The fast breeder reactor (FBR) is actually far from a new idea, and was touted in the 1969s and 1970s by the head of the Atomic Energy Commission, Glenn Seaborg, who believed it was the answer to America's apparently dwindling supply of oil and natural gas. But in early designs it was recognized that fast breeder cores held far greater potential for destructive accidents than moderated reactors. An early US experimental breeder reactor (EBR-I) suffered a partial meltdown of its core in 1955 when it was inadvertently allowed to slip into a condition whereby a rise in core temperature caused an increase in the rate of reaction (i.e. an increase in neutron flux). This effect, called a *positive temperature coefficient*, occurred when thermal expansion pushed the fuel rods closer together. Another experimental FBR, the Enrico Fermi I, also suffered a partial meltdown in 1966 when the flow of sodium coolant was accidentally blocked by a loose metal plate. A radiation emergency was declared at the plant near Detroit, but almost all of the radioactive contamination was kept within the containment shell (Collier & Hewlett, 1987).

During this period of FBR development, it was soon recognized that breeder reactors of sizes larger than the EBR-I or the Enrico Fermi I could conceivably suffer an even more violent criticality called a *core collapse* or a *core disruption accident* (CDA) (Collier & Hewlett, 1987). A CDA can occur, it is theorized, when a blockage of coolant flow causes melting of fuel rods in one portion of the core, thus allowing a collapse of the fissionable mass into a lower portion of the core.

One of the major concerns relating to breeder reactors is that they would involve large quantities of fissile materials, especially plutonium, which would expand the international nuclear weapons proliferation problem.[12] But some FNRs are not built to provide excess plutonium but rather are designed to convert natural uranium to the fissile form through neutron bombardment which then would burn in the same core. One such idea is a gas-cooled, small modular FNR that would have a "convert-and-burn" design. This concept would extend the nuclear fuel cycle so that the reactor would only need to be fueled once and would then provide heat and electric power for 30 years.

All of the current FNR/FBR concepts are stressing the need for safety and there appears to be a great deal of development work to be done. FBRs are, for the most part, not yet at the demonstration stage and so if they are to provide electric power it may well be a decade or more in the future before they are ready for commercial application. Then again, if evidence of catastrophic climate change grows, the timetable to commercialization of these new designs might well be accelerated.

References

Brinton, S. 2015. "The Advanced Nuclear Industry," Report. At www.thirdway.org/report/the-advanced-nuclear-industry.

Cravens, G. 2007. *Power to Save the World: The Truth About Nuclear Energy*, Knopf, New York.

Collier, J.G. and G.F. Hewitt. 1987. *Introduction to Nuclear Power*. Hemisphere Publishing, New York.

Dhanapala, J. and S. Duarte. 2015. "Is There a Future for the NPT?" Arms Control Association, at www.armscontrol.org.

Greenwood, T., H.A. Feiveson, and T.B. Taylor. 1977. *Nuclear Proliferation: Motivations, Capabilities and Strategies for Control*. McGraw-Hill, New York.

Häfele, H. 1981. *Energy in a Finite World: Paths to a Sustainable Future*. Ballinger, Cambridge, MA.

Kasperson, K.E. (ed.). 1983. *Equity Issues in Radioactive Waste Management*. Oelgeschloger, Gunn & Hain, Cambridge, MA.

Lipschutz, R.D. 1980. *Radioactive Waste: Politics, Technology, and Risk*. Ballinger, Cambridge, MA.

Murdock, S.H., F.L. Leistritz, and R.R. Hamin (eds.). 1983. *Nuclear Waste: Socio-economic Dimensions of Long-Term Storage*. Westview Press, Boulder, CO.

Ruzicka, J. and N.J. Wheeler 2010. "The Puzzle of Trusting Relationships in the Nuclear Non-Proliferation Treaty." *International Affairs*, 86 (1), 69–85.

Shrader-Frechette, K.S. 1993. *Burying Uncertainty: Risk and the Case Against Geological Disposal of Nuclear Waste*. University of California Press. Berkeley, CA.

Spash, C.L. 1993. "Economics, Ethics, and Long-Term Environmental Damages." *Environmental Ethics*, 15 (2), 117–132.

Yager, J.A. 1974. *Energy and U.S. Foreign Policy*. Ballinger, Cambridge, MA.

Resources for the Current Debate on Nuclear Power

Debate: Does the World Need Nuclear Energy? Mark Z. Jacobson vs. Stewart Brand At www.ted.com/talks/debate_does_the_world_need_nuclear_energy?language=en

Articles (pro): "Environmental Heresies," *Technology Review*, May 2005.

"Why we still need nuclear power," *Foreign Affairs*, Nov/Dec 2011.

"Why Scientists are Calling for Nuclear Power to Save Biodiversity," Climate Change National Forum, August 2015.

Articles (con): "Reject Nuclear Power – Here's Why," Global Research, February 2013.

"A Plan to Power 100 Percent of the Planet with Renewables," *Scientific American*, November 2009.

"Do we really need nuclear power?" *New Republic*, March 2011.

NOTES

1 Open letter, dated November 3, 2013, signed by Dr. Ken Caldeira, Senior Scientist, Department of Global Ecology, Carnegie Institution; Dr. Kerry Emanuel, Atmospheric Scientist, Massachusetts Institute of Technology; Dr. James Hansen, Climate Scientist, Columbia University Earth Institute; Dr. Tom Wigley, Climate Scientist, University of Adelaide and the National Center for Atmospheric Research.

2 There are many papers in the scholarly literature calling for "sustainable de-growth." Economic growth, it is argued, is unsustainable and so world industrial economies can either crash or figure out a way to live more simply and in harmony with nature.

3 The actinides belong to a group of isotopes following actinium (number 89) in the periodic table and include the transuranium elements (which themselves include the various plutonium isotopes).

4 High-level waste (HLW) is officially defined as "the waste streams that result from the reprocessing of spent reactor fuel" (Lipschutz 1980). However, in practice, HLW primarily consists of spent fuel rods themselves. In any case, it is distinct from materials that have been contaminated but were not initially radioactive – for example, reactor pipes.

5 See Shrader-Frechette (1993) for "the case against geological disposal of nuclear waste."

6 The production of heavy water was itself a difficult technological hurdle.

7 See *Arms Control and Disarmament Agreements, Texts and Histories of Negotiations*. US Arms Control and Disarmament Agency, Washington, DC, 1980 edition.

8 These data are from 2014 and can be found on the IAEA's website, at www.iaea.org/safeguards/basics-of-iaea-safeguards/safeguards-facts-and-figures.

9 And the fear is that should Iran develop nuclear weapons, Saudi Arabia and other Middle Eastern nations might feel compelled to do likewise.

10 It should be noted that one major incentive for non-nuclear nations to agree to the NPT was the assurance of the transfer of nuclear *power* technology to the signatories. These nations did not want to remain nuclear have-nots as a result of not signing the treaty. This telling fact illustrates well the political nature of the issue (see Yager 1974 and Greenwood et al. 1977).

11 The information in this section is taken from Brinton (2015); the IAEA, "Small and Medium Sized Reactors (SMRs) Development, Assessment and Deployment," at www.iaea.org; and the World Nuclear Association, "Advanced Power Reactors," at www.world-nuclear.org/information-library/nuclear-fuel-cycle/nuclear-power-reactors/advanced-nuclear-power-reactors.aspx.

12 The prospect of massive amounts of plutonium did not trouble everyone. Some in fact thought that energy issues would be solved by developing a "plutonium economy." This idea has mainly vanished from energy policy discussions – even among those who advocate expanded nuclear power development (see for example, Häfele 1981).

9   The Economics of Electric Power

Introduction

Since the 1970s, public discussion of electric power generation has dealt most often with the issues of safety and environmental impacts. But electric power generation always has had significant economic importance as well. Consider that in 2014, in the United States alone, over 4 trillion kilowatt hours (kWh) of electricity were generated and sold to millions of residential, commercial, and industrial customers – all of whom have come to depend on reliable *low-cost* electric power. These customers paid a total of nearly $400 billion for that electricity during that period.

Moreover, the cost of equipment to generate electric power is great and has continued to rise over the years. By 2012, the installed electric generating capacity in the United States was over 1100 million kilowatts, representing a truly enormous investment.[1] As of 2012, according to the Energy Information Administration, a large coal-fired plant (in the range of 1,000,000 kW = 1000 MW = 1 GW) would cost about $3000/kW to build. Nuclear plants cost almost double that amount. Even relatively low-cost gas-fired baseload plants (that is, ones to compete with coal and nuclear) cost around $1000/kW. Put in terms of total outlay, a 1000 MW coal plant would cost around $3 billion; a nuclear plant approximately $5.5 billion, and a combined-cycle gas plant about $1 billion. Furthermore, any fossil- or nuclear-powered generating plant requires millions more per year for fuel and maintenance.

In spite of the huge costs involved, until recently the economics of electric power were based on some very straightforward premises: First, the industry was considered a *natural monopoly*. A natural monopoly is an industry where there are pervasive economies of scale, which means that average or unit costs of production (kWhs for electricity generation) continually fall the more a firm produces (e.g. Fig. 9.1). Since lower average costs allow a firm to charge lower prices, one big firm can always underprice any combination of smaller firms. Because the industry had been assumed to be a natural monopoly, typically only one electric power producer has been allowed to exist in any given city or region of the United States. This company is typically either a government-owned entity or, more commonly, an investor-owned utility (IOU). IOUs are regulated by various governmental authorities, but most directly through state *public utility commissions* or PUCs. The task of PUCs had been to allow an electric company to make a "fair" return, while protecting consumers from the monopoly power the companies typically have been given.

Because of the presumed economies of scale of electric power companies, a second premise has been that large monopoly firms should build large central (*baseload*) power plants, and complement them with plants that can come in quickly

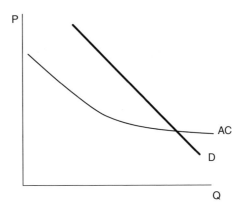

Price versus quantity (P vs. Q) for a natural monopoly.
Note: Average costs (AC) fall the greater the quantity (Q) produced. In this instance, one firm should provide all of the market demand (Q). The price (P) that a single firm can profitably charge will be lower than that which any combination of smaller firms can charge.

when the load is especially heavy – for example on a hot day in the summer. Indeed, if the economies of scale argument were always true for the production of electricity, then the *bigger-the-better* reasoning would prevail. With large plants and single companies, it should be expected that more power can be provided at the lowest possible cost. That would mean that consumers could spend more on material goods and less on electricity. And businesses and industries dependent on electric power would be able to produce more at lower cost, thereby benefiting the economy as a whole, or so the reasoning goes.

By the 1970s, the issue of what kind and size of power plant to build had become more complex. Although economies of scale were still thought to prevail, it appeared that some types of large power plants were not as cost effective as others. With the sudden rise in oil and natural gas prices (and then significant fluctuations), the imposition of various environmental regulations, and cost overruns in nuclear power construction, many types of power plant began to display very troubling cost characteristics. If a utility chose one technology, say coal-fired steam generation, the cost of production could be much higher (or lower) than for some alternative, such as natural gas, depending on such factors as construction costs, environmental regulations, and the price of fuel.

United States utilities had not worried about these possibilities before the 1970s. First, most energy resources had been cheap and abundant. Also, the cost of power seemed to fall because of size alone, and overall costs per unit of electricity were falling. Second, PUCs routinely let electric companies charge a price for their electricity that would allow them to recoup all of their investment costs within reasonable times. If one plant cost a little more to build, prices would be raised incrementally in order to limit the impact on customers, and to provide continuing profits for the electric company.

But unexpected escalating fuel and construction costs changed both the economic and political climate. By the 1980s, PUCs did not grant rate requests as easily as before, especially when it was deemed that a utility had constructed a facility unwisely. But even when electric utilities could charge more to cover higher costs, because the price of electricity sometimes rose considerably, the quantity demanded fell, and economic development regionally and nationally diminished.

Moreover, in the 2000s, as concerns about climate change grew, 29 states passed Renewable Portfolio Standards (RPS) that required a certain percentage of new electric capacity be free of carbon dioxide emissions through the use of renewable fuels. This generally meant wind-generated or solar-generated electric power or power generated from the burning of biomass. These technologies clearly cost more than the old ones but they were often subsidized by various government programs, and PUCs have allowed cost recovery for renewable energy – although sometimes at the expense of consumers, particularly low-income ones. But renewables present their own set of economic and (related) technological issues that are still being confronted in 2017 as this edition of *Introduction to Energy* is being written.

In general, electric utilities have sought facilities that are either less costly to build and operate or that meet some subsidized, mandated goal (or both). But what kind of facility would this be? The answer to this question was not clear cut even before RPS mandates. Take lowest cost: To decide which kind of power production facility is lowest cost, power providers have to consider costs, not just the construction costs but also costs that extend over the entire lifetime of the plant – 20 to 40 years into the future. Such forecasting to date has provided, at best, uneven results, and every conceivable cost factor, from the future cost of fuel to the cost of environmental regulation, has been the subject of dispute. Some costs, such as the disposition of spent nuclear fuel, have become almost entirely conjectural. Thus, the choice of which type of power plant to build has become a difficult one, and there has been a great need for companies to simply get the *best cost* estimates possible for all types of power plant. Below, we will consider how power companies calculate the "*life-cycle*" costs of a power plant.

In the 1970s and 80s, the debate about cost centered on the relative cost of large baseload plants of at least a several hundred megawatts capacity, and specifically it centered on the comparison of the costs of nuclear versus coal-fired generation. While the issue was still in some doubt into the early 1980s, by the middle of the decade, it was widely acknowledged that coal had a definite cost advantage even with the added burden of pollution abatement equipment.

But electric power cost estimation has become far more complicated in recent years. While coal is generally still regarded as superior (from a cost standpoint) to older types of nuclear generation, combined-cycle natural gas-fired technology (see Chapter 3) has lately been considered as superior from a cost standpoint to both coal and nuclear plants. Moreover, there are now other alternatives – even to large central power stations themselves – not only from renewables but also from improved energy efficiency by end users. *Not* building a new electric power plant may have cost advantages over any new central power generating stations; that is, it may be cheapest of all to provide incentives to users to consume less.

Indeed, so much had changed by the mid-1990s that few new baseload stations of any kind were being built in the United States. To meet demand, in some cases, older plants were simply being refurbished and upgraded – termed *repowered* by the electric industry (Moore 1995). But also new capacity was being provided by companies that were not traditional electric utilities. This group of "*nonutility generators*" (or *NUGs*) was taking advantage of new laws that require utilities to accept new operators onto the distribution grid, and pay fairly for power generated by *anyone*, not just designated public utility companies.

The growth of NUGs points up how fundamentally the economics of electric power has changed over recent decades. It is no longer accepted that bigger is necessarily better. Few believe anymore that electric production should be considered a natural monopoly (Grossman & Cole 2003).[2] The small independent providers using relatively small electric generators – often driven by gas turbines – are able to produce electric power profitably and may well be able to improve on the costs of large companies and gigantic central stations.

Nonutility generators have introduced a form of competition into an industry long thought to be by definition noncompetitive. But this competition was only the beginning. In some American states, notably Texas, there is competition between the utilities themselves in which they can make pitches to consumers for additional services or power generated from various sources (for example, "green" power) as well as competition on price. Because of competition as well as new regulations, especially those concerning CO_2 emissions, the electric power industry has been undergoing some stresses. We will consider later in the chapter the future of electric power economics.

Calculating the Cost of Electric Power

In the 1970s and 80s, huge costs for nuclear power plant construction garnered headlines in the press. But plant construction is only one component of the cost of electric power. Indeed, in the 1960s, although nuclear power plants were more costly to build than coal plants, they appeared then to have a cost advantage over coal. The point was not what type of plant is costliest to build but rather what will be more economical over the *lifetime* course of operation of a plant. Although construction costs must be factored into the determination, the overall problem for power planners is to estimate how much it will cost to produce a unit (kWh) of electric power and how much electricity any facility will produce over the course of a year.

More specifically, the *life-cycle cost* is calculated by combining the initial cost of construction – the capital cost – with ongoing and future charges: operations and maintenance, fuel, and decommissioning. There is another factor to consider as well, one that is technical but, as we will see, has important *economic* consequences: the proficiency with which the plant operates. This is measured typically by the plant's capacity or availability to generate its designated power output as continuously as possible and is quantified as a *capacity factor* or an *availability factor*. All of these life-cycle components must be considered in determining the cost of generating electric power, as expressed in terms of *cents per kilowatt hour* (¢/kWh) of electricity delivered. But, it should be recognized, each of these factors may be subject to great uncertainties for forecasting, even only a few years into the future.

The capital costs – construction and equipment – would seem on the surface somewhat less problematic to predict because the time horizon for construction expenses is shorter than that for the fuel costs over the plant's lifetime. For a small gas turbine facility, designed to provide *peak* or *intermediate* load, the time horizon for design and installation of equipment can be relatively short; the equipment is modular and can be ordered and shipped in a matter of weeks. But baseload plants typically take at least several years to go from the design to the operational stage. Nuclear plants, in particular, have become notorious for extremely long construction phases; some have taken more than a decade to complete. But even a new baseload coal or a large combined-cycle gas facility would still need as much as 5 years to go from the drawing board to operation.

Within a 5-year span on any given project, uncertainties abound and factors can change so much that capital cost estimates can prove wildly inaccurate only a few months after they are made. New environmental and safety regulations, for example, can necessitate design changes and add millions of dollars to the capital cost and years to construction time. Labor and material costs can also change suddenly. And above all, interest rates – the cost of borrowing money – can change suddenly and drastically.

Interest costs are important to this discussion because capital costs are customarily not one-time outlays of huge amounts of cash. Rather, money is borrowed and paid back over time. In the case of a new baseload plant, money will likely be paid back over decades. For a typical plant, in fact, amortization of the debt becomes an ongoing expense, expressible on a ¢/kWh basis as a *capital fixed charge*. Of course, that fixed charge is affected by the cost of borrowing money.[3]

To raise the necessary financial resources, a utility will get loans from banks and/ or sell debt to investors. Debt instruments include bonds sold in the bond markets. (They may also sell new issues of stock – shares sold on the stock exchanges to raise money.) Bonds and bank loans must be paid back with interest by the utility over varying terms from a few months to 30 or 40 years. So a utility that builds a coal power plant costing $1 billion or more early in this century, will have to pay interest charges well into mid-century as well as paying off the debt principal at some specified point in time. Of course, if interest rates are around 15 percent, the cost of a plant will be much greater than if rates are around 8 percent. In fact, a 15 percent rate would mean a *71* percent increase in total costs of construction – interest and principal over 30 years – compared with an 8 percent rate.

An electric power producer can know the rate it must pay for debt *only* at the moment it borrows. (Most utility debt has a fixed rate of interest, so that the company can know the cost for the entire term of that loan.) If the company could borrow the full cost of construction and equipment at one time, the company would have a better defined idea of the capital fixed costs over the life of the plant. But borrowing all the money at once is neither feasible nor sound. For one thing, the company would have to know exactly the final cost of construction, which generally is not possible. Also, it would be pointless for a company today to pay interest on $1 billion when it only needs, say, $100 million for this year's construction work, and then the remainder incrementally over the remaining years until completion. That means that interest costs become increasingly subject to forecast and uncertainty.

Future demand for electric power also must be estimated. Whatever demand growth is expected firms will need to pay for whatever new capacity they believe they will need. In the 1970s, incorrect forecasts about electricity demand led several utilities to overbuild; some ended with as much as 40 percent overcapacity. Low demand estimates, on the other hand, can result in insufficient capacity to meet demand for electricity. Demand and interest rates are also tied together. Incorrect estimates of interest rates can, and do, force prices higher (or allow them to drift lower), which in turn throw off the estimates of the quantity of energy that consumers will demand.

There are many operational charges and ongoing costs incurred by any power facility, but they can generally be grouped into two categories: *fuel* and *operations and maintenance* (O&M). Fuel costs are perhaps the least predictable variable of all. Who would have guessed in the 1960s that the price of oil would increase by 1000 percent in the 1970s? Or that it would then decline by more than half during the 1980s? Or that in 2008, when oil was $145/bbl, that 8 years later it would be about 75 percent lower? Or that natural gas prices would fall by even more in that same time period.

Costs of O&M, on the other hand, are perhaps the least contentious cost variable, although even here there may be some disputes. For example, the future costs of environmental regulation for coal plants or the maintenance costs of aging nuclear plants certainly are subject to uncertainty and debate, often in the political arena.

Finally, we must consider the operating proficiency of the power plant, as measured by the *capacity factor* (CF). By CF, we mean the quantity of electricity *actually* produced annually, compared to the amount that would have been produced if generators were run at full power continuously. Ideally, a baseload plant would operate continuously to supply that irreducible low level of electric demand that never ceases (see Chapter 3). However, all operating machinery must be shut down periodically for maintenance. Also, unscheduled interruptions will occur randomly due to mechanical failure and accidents. The better the design and quality of operation, the higher the fraction of hours in the year the plant will operate, but of course it can *never* achieve 100 percent. Obviously, a generator that runs at a high CF will produce more revenue than a generator with a low CF.

We calculate the CF of a plant by dividing the number of *actual* kilowatt hours produced in a year by the number of kilowatts the plant is *rated* to put out (in other words its full capacity) times the number of hours in a year:

$$CF = \frac{\text{actual kWh} / \text{yr}}{\text{rated kW} \times 8760 \text{ hr} / \text{yr}}$$

Using the CF, we can calculate the electric rate (¢/kWh) necessary to pay for the actual energy (kWh) generated, in order to pay off the capital fixed charges. The lower the CF, the higher the electric rate required. Or, to put it another way, if a plant operates at low capacity, it will not produce enough revenue to pay for itself unless the rate is increased. As we will see, this is an important issue for intermittent sources of electric generation such as solar and wind.

Actually, with respect to baseload plants, it may be better to use a slightly different measure than the CF in determining the effective operation of a plant. Although utilities report CF as a measure of actual use, such reports may be misleading. For example, if a power provider has overestimated demand and therefore must leave some of its generating capacity idle, the capacity factor for the idle plant (that is, actual use) will be too low. But in such a case, the idle facility was working well and could have been otherwise utilized. An alternative measure, the *availability factor* (AF), gives the hours that an electric generating facility is ready for service if needed. The AF is calculated simply by dividing the hours available by the number of hours in a year:

$$AF = \frac{\text{actual hours available} \, / \, \text{yr}}{8760 \, \text{hr} \, / \, \text{yr}}$$

The AF is an important indicator of a given type of plant's *dependability* and should be a factor in any discussion of the economics of electric power. At the same time, it should be noted that the *capacity factor* is the measure typically used in calculation of power plant costs.

Comparing Power Plant Costs

Table 9.1 gives life-cycle costs for three power plants if newly constructed, all baseload – coal-fired, combined-cycle gas-fired (CCG), and advanced (generation III) nuclear. All use heated water to produce steam that will turn a generator, although the CCG system includes a gas turbine, which improves its overall efficiency. Shown here are representative comparisons for plants if built in the mid-2010s. The *capital costs* for the coal plant are much higher than the CCG plant because the former requires extensive pollution abatement equipment (Chapter 6). Gas-fired plants, furthermore, have substantial advantages with respect to pollution. As noted in Chapter 6, they produce no sulfur dioxide and very limited amounts of nitrogen oxides. Moreover, per kilowatt hour, gas-fired facilities create less than half as much carbon dioxide (the principal greenhouse gas) than a typical coal burning plant. The capital cost for the combined cycle plant is around 30 percent that of a baseload coal plant, although it is generally assumed that fuel costs will be higher (and more uncertain) for the natural gas plant. Nuclear is the most costly, although with a design such as the ESBWR (Chapter 8) there is the expectation that costs may fall through scale economies, learning-by-doing, and process efficiencies as production of the modular components evolves.

All three technologies have proven to be highly reliable, and the projected capacity factors of the technologies (by the EIA) are 80 percent or better for all three. That said, it should be noted that in practice these CF averages are not always attained. Indeed, although coal and CCG potentially could achieve the rated CF, they are sometimes deliberately idled, since RPS mandates require that renewable power take precedence. If, on the whole, that means too much electricity is entering the grid, coal or gas plants are the ones taken offline.[4] Coal generally has an *availability factor* of

Table 9.1 Life-cycle cost of new baseload power (cost figures are in 2013 dollars; projected to 2018)			
	Coal	CCG	Nuclear[b]
Capital cost[a]	$2900/kW	$900/kW	$5000/kW
Capacity factor	85%	87%	90%
Capital fixed charge	6¢/kWh	1.8¢/kWh	8.3¢/kWh
O&M[c]	3.4¢/kWh	5.9¢/kWh	2.4¢/kWh
Total	9.4¢/kWh	7.7¢/kWh	10.7¢/kWh

[a] Capital costs are approximate based on various sources and reflect likely designs given; a common interest rate is used although the rate might be higher for a new nuclear plant, given regulatory uncertainty.
[b] Generation III nuclear design.
[c] Including both fixed and variable operations and maintenance costs; it also includes estimates of fuel costs and the costs of transmission and connection to the grid.
Source: Adapted from USEIA, 2013 for "new central station generating technologies" due to come online 2016–2022 depending on the technology.

over 80 percent, and gas turbines combined with a basically conventional steam system have had similarly excellent reliability. Nuclear power plants are seldom idled, and after years of relatively poor performance and low CF, nuclear plants in the United States typically are at a 90 percent CF, near the limits of their availability.

The capital fixed charges reflect the differences in construction costs of the different types of plant. Note that, while the capital cost of a CCG plant is about 30 percent of the coal plant, the higher CF means that the unit will provide proportionately more revenue per year. These are figures, of course, for new plants. According to a study in 2015 by the Institute for Energy Research, existing facilities were operating at considerably lower cost per kWh. Coal facilities on average produced power at a total cost of less than 4¢/kWh, while existing CCG and nuclear plants produced power at slightly less than 5¢/kWh and 3¢/kWh respectively. Of course, many of these were built without some of the safety and pollution control equipment required today. As for the numbers in the table, it is important to keep in mind that these are projections. New environmental regulations, for example, might force higher or lower charges, thereby altering the ratio of costs. Also, interest rates might change, changing the costs for both. But for a power company planning to build a baseload plant, officials *must* take into account the *expected* capital costs over the life of the plant.

With concerns about climate change and with necessity of fulfilling RPS mandates, it is important not only to analyze costs of longstanding traditional technologies but also to examine the cost characteristics of renewables, which some argue can replace fossil-fired and nuclear power even for baseload (for example, Jacobson & Delucchi 2010). Table 9.2 shows the same kind of analysis of life-cycle cost for onshore wind, solar photovoltaic, and geothermal electric generation.

It would seem from looking strictly at the numbers, that wind especially would be an economically viable alternative to coal and nuclear power. (Geothermal, too, but it is very limited to areas where the Earth's heat is close to the surface, as in volcanic zones.) But this conclusion does not take into account some important elements, which if added to cost pushes wind above both coal and new nuclear power. It is

Table 9.2 Life-cycle cost of new utility-scale renewable power (cost figures are in 2013 dollars, projected for 2018)

	Wind	Solar PV	Geothermal
Capital cost	$2000/kW	$3200/kW	$2400/kW
Capacity factor	34%	25%	92%
Capital fixed charge	7¢/kWh	13¢/kWh	7.6¢/kWh
O&M[a]	1.6¢/kWh	1.3¢/kWh	1.3¢/kWh
Total	8.6¢/kWh	14.4¢/kWh	8.9¢/kWh

Assumptions here are generally the same as in Table 9.1.

[a] There are of course no fuel costs, but values do include the costs of transmitting and connecting to the grid, which are likely to be higher with renewables because their characteristics restrict their geographic placement.

Source: Energy Information Agency

important to keep in mind that wind is an intermittent source of energy and therefore there must be backup, which means additional costs (estimated at around 0.2¢/kWh). There is also a subsidy of about 2¢/kWh, raising the cost of wind to over 10¢/kWh, and a tax credit simply for employing wind that lowers the capital fixed charge, although both the subsidy and the reduction in taxes means that costs are essentially shifted from electric rate payers to taxpayers. Connection and transmission are also more costly. In general, there are estimates of the life-cycle cost of wind power of 10.9¢/kWh or more (Simmons et al. 2015).[5] There are those who dispute these figures, and alternatively it can be argued that the benefit from a non-CO_2 source of energy is worth taxpayer dollars (though sometimes the same people argue it is not worth taxpayer or ratepayer dollars to advance non-CO_2 nuclear power). We will return to this discussion in Chapter 11.

It should be noted that the Energy Information Administration has given a rough projected range of costs (for 2018) over the menu of electric generation possibilities, which we have put into Table 9.3. Ranges are probably more useful than averages or illustrative examples (as we use in the previous tables). The point is that costs vary even for conventional technologies. For example, a coal-fired power plant established at the site of a surface coal mine will have lower costs than a similar coal plant that is a hundred miles away and requires a continuous stream of freight trains in order to provide the resources for generation.

Demand-Side Management and the Smart Grid

In the past, the question of which power plant an electric company was to build was of great importance because utilities continually needed to expand capacity to meet growing demand. The growing economy needed more or bigger power plants. Or so the thinking went until the 1980s. As we saw in Chapter 4, in the 1980s, in the US energy demand increased much more slowly than the increase in the economy as a whole.

Table 9.3 Cost* of electric power, 2018 (projected)	
Type	Cost/kWh (cents)
Gas	7.6–9.3
Coal	8.3–11.4
Hydro[a]	7.8–10.8
Geothermal	8.1–10
Wind (on- and offshore)[b]	7.3–29
Biomass	9.8–13
Solar thermal	19–41
Nuclear	10–11.5
Photovoltaic	11.2–22

* Constant 2011 dollars.
[a] Hydro is assumed to have seasonal storage capability.
[b] How often the low end of the estimates are achieved is not clear.
Source: Adapted from data in USEIA 2013.

Of course, as we discussed in Chapter 4, conservation potential remains even as the economy expands. We can use less while getting more from the energy we use, and in the process cut down on the need to build new power plants.

But, of course, conservation has a cost, too. To use less energy but get the same or greater benefits, consumers would have to buy more efficient appliances; industrial users would have to invest in more efficient machines; and commercial users would also need new equipment.

But how does the cost of that improvement in technology for users compare with the cost of generating more electricity? Is the added cost of a device that requires one fewer kilowatt hours generated less than or greater than the cost of producing one more kilowatt hour? The electricity saved has been termed *negawatts* (Lovins 1989), and the comparison is between the cost of negawatt hours and kilowatt hours.[6] Overall, the conservation approach is called *demand-side management* or DSM (World Bank 2005, Strbac 2008).

This effort began in the 1980s when a few state utility commissions, to promote conservation generally, started to penalize utilities that did not have a program to encourage efficiency. But these efforts put utilities in the awkward position of being forced to discourage customers from using their product.

A boost to DSM came in 1989 when the National Association of Regulatory Utility Commissioners recommended that regulators allow utilities that adopt active DSM programs to be compensated for them. In other words, if a utility started and managed a program to discourage people from using more electricity, they could build the costs of the program into their rate base. They could get paid for negawatts as well as (generated) kilowatts.

DSM programs are intended to encourage consumers to adopt energy-efficient technologies. This is typically accomplished by offering rebates to consumers who purchase efficient equipment. However, in some cases utilities have subsidized appliance dealers to lower the cost of efficient equipment, have provided financing to consumers, have leased equipment, and have even given electricity-saving devices away. Along with the new technologies, DSM programs emphasize marketing and education so that people know their options and the benefits of saving electricity.

Consider how this *should* work – ideally it would provide benefits to all. The Niagara Mohawk Power Corporation in upstate New York (now a part of the Niagara Hudson Power Corporation) embarked on a program in the 1980s to provide consumers variously with (i) a low-flow shower head, (ii) a compact fluorescent bulb, and/or (iii) insulation for water heaters and pipes. According to projections, the equipment should have saved on average 960 kWh per customer per year. At a residential rate of 7.5¢/kWh, the utility loses $72 per customer per year, but saves money by needing less generating equipment online, estimated at approximately $40 per customer per year. Customers, however, pay the utility the difference ($32) plus $6 per year (for 8 years) for the equipment. In theory, the utility makes a profit and the customer saves $34 per year (see Fickett, Gellings & Lovins 1990).

But, as Palensky and Dietrich (2011) observe, DSM should involve more than simply changing inefficient appliances for more efficient ones. It also should include load shifting, that is, changes in "time of use" (TOU). An electric system is taxed more heavily at certain times of the day or night depending on location, climate, and so on. But utilities provide power 24 hours a day, 7 days a week, 365 days a year. During much of that time the system has relatively low usage. While hot evenings in July may require more than baseload power (it will often include so-called "peaking" generators, usually gas turbines that can start quickly and be shut down equally fast), it would be beneficial if on those evenings customers limited usage to essentials, such as air conditioning and lighting. Clothes or dishes, on the other hand, could be washed and dried at any time – say, after 10 PM. In fact, the absolute peak in annual demand may be less than 50 hours total. If usage (load) could be shifted from those peak times to off-peak hours, less of that peaking reserve would be necessary and energy and money could be saved. It makes sense in that context to signal to consumers when it is wise to use which sort of appliance – signaling, as economics would argue, most effectively by raising the price of power (per kWh) in the evening (referred to as time-of-use tariffs) and lowering the price later.

But for much of the past 35 years there has been a basic problem with DSM: In order for such a program as the one described above to work best, consumers need information about price changes and it needs to be communicated in real time. It is far more likely that people will change their TOU when they see price changes happening in front of them instead of just seeing numbers on a monthly bill.

As of 2012, 40 US states had active DSM programs (and there were similar programs in Europe), but they were largely based on a model whereby utilities come up with ways for consumers to save energy, such as the Niagara Mohawk program. As one expert noted, there has been "a relatively slow uptake of DSM" (Strbac 2008). But TOU requires a different sort of system. In the new model, consumers make decisions based solely on a real-time appreciation of price, which would be based on TOU. With the development of smart grid technology (discussed in Chapter 4), it is possible for real-time, two-way communication between consumer and provider. Strbac (2008) envisions an "electronic energy market, supported by the internet." According to a 2010 study by the consulting firm McKinsey, a system like this based on smart grid technology could reduce US peak demand by 20 percent and cut overall energy consumption by more than 9 quads annually. As noted in Chapter 4, there are some pilot programs in smart grid technology but it has moved slowly. For DSM to be effective, it seems such technological change will be required.

The New Electric Power Market

In the past few decades, the market for electric power has been undergoing significant change. In general, production and distribution of electric power has become more competitive and deregulated (or at least re-regulated), and in fact in several states there is retail competition. A customer can switch from one utility to another if he or she is dissatisfied or learns that there are attributes (such as green power) available from one power provider but not another.

What was this new market? And why did it develop? The changes in the electric power market began in the 1970s. At that time, the US government sought to encourage greater energy efficiency, the use of renewable energy resources, and the expansion of power production. One result of this effort was the Public Utilities Regulatory Policy Act (PURPA) of 1978. PURPA required utilities to buy power from independent providers. At the time, the most likely sources were seen to be "co-generators," generally industrial plants that burned energy for heat – heat that after its primary utilization could be captured to make steam and run a generator (see the *technical efficiency* section in Appendix C). In fact, a number of major cogeneration projects were developed as a result. Other early independent providers included small hydro facilities reconstructed along old mill runs and dams.

By 1992, independent power producers (IPPs or nonutility generators, NUGs) had over 55,000 megawatts in generating capacity nationwide – or about 7 percent of total US generating capacity – in operation. This power was sold on a wholesale basis to the utilities, not to individual customers. The 1992 Energy Policy Act encouraged further independent production, including the creation of independent generating subsidiaries of existing utilities. Indeed, by that time, the amount of new installed capacity that was brought online by independent producers was equal to that of the new capacity installed by the electric utilities themselves.

Most of the IPPs have used smaller generating units, on average no more than 25 megawatts. While IPPs initially focused on hydro and existing heat sources, they began in the 1990s to build facilities for low-cost gas combustion turbines. Modular in design and relatively inexpensive to build and maintain (requiring two-thirds fewer employees to operate than a steam system), the price of these turbines became increasingly competitive, prepackaged for as little as $500/kW (Bayless 1994). These units, operated primarily for peak and intermediate loads, have proven to be profitable, and as a result more and more of them are being installed.

The success of small producers has only reinforced what was becoming clear in the industry: bigger was not necessarily better. It was not necessarily more profitable. The monopoly character of the industry was certainly open to question – as some economists had argued for years (Moorhouse 1995, Grossman & Cole 2003).

What had really changed? To some extent the change emerged from the technology of electric power generation itself. Although capital cost and fuel cost tended to fall with bigger units, it was noted that reliability of large steam power plants suffered. In the early 1970s, as one book noted, forced outages of steam generators over 600 MW were double those of generators under 600 MW (see Berlin, Cicchetti & Gillen 1974). As the Electric Power Research Institute (EPRI) noted, the movement from large plants to small gas turbines could be explained by the fact that centralized economies of scale were overtaken by "distributed economies of precision."

But the market also had changed because of advances in other kinds of technology. Through computerization and telecommunications, it had become much easier to monitor and report transactions between suppliers and customers. As a result, a power provider could keep track of customers spread through a wide geographic area, even when they are interspersed with customers of other providers.

While the development of independent power providers was increasing anyway, in April 1996 a Federal Energy Regulatory Commission (FERC) ruling opened the market to even greater competition. This required that the *entire* electric power transmission system be made available to *any* producer, enabling a utility in one part of the country to buy power from any provider – whether a utility or NUG. Although the ruling applied only to wholesale sellers and buyers of power, it was expected to force electric prices lower as utilities sought low-cost power from regions of the country that can produce power cheaply. Regional differences were striking. For example, at the time the ruling was announced, utilities in the northeastern states charged a retail price for electricity that was more than double the price charged by southern utilities. Such disparities reflected different costs, often due to the ease of access to resources or local environmental constraints.

But the FERC ruling was only one more step in what many believed would be a complete opening of the electricity market, thereby ending monopoly control of utility systems. Indeed, at almost the same time as the FERC ruling, several states, most notably Texas, took the final step to make electric power competitive: they allowed retail customers – including average consumers – to buy directly from whichever power provider they chose. The process came to be called *retail wheeling*. Wheeling refers to the use by any producer of a local utility's transmission and distribution system to deliver the power to their customers. In other words, a consumer in Houston is able to buy power from a long list of providers, some headquartered in Houston but others in Dallas, Austin, and even outside the state. One company boasts that it "is one of the largest providers of energy and energy-related services in North America, with customers in all 50 states, 10 Canadian provinces, and Washington, DC." Whichever choice a Houstonian makes, transmission access is assured. Modern technology permits firms anywhere to interact quickly with any of their customers.

Much of the competition takes place on price; low-cost producers are able to undersell the high-cost generators. But there are other dimensions on which competition may take place. For example, some providers might offer more in customer services such as maintenance, repair, warranties, and planning. DSM programs with smart grid technology could provide a competitive advantage for some, but customers would pay directly for cost-reducing programs that would fit their individual needs. The character of the market, if it became competitive, would require more in the way of advertising, marketing, technical support, and so on. But if it was developed effectively, it could keep prices low while providing better and more varied services (Douglas 1994).

Despite the apparent benefits of a competitive retail market, some experts were cautious about retail wheeling. One problem was that high-cost producers, such as the utilities that had built large nuclear plants, would lose the most customers – leaving them with plants that were unpaid for and so costly still as to threaten the companies' solvency. These, so-called *"stranded investments,"* could not only hurt utilities, but could leave some consumers with very high retail prices as low-cost suppliers reached capacity.

And in at least one case, California, an attempt at deregulation was so badly constructed that 5 years after a bill that allowed for wholesale competition passed the state legislature, the state faced a power crisis, with blackouts of cities and bankruptcy of one utility (Grossman 2003). The response was to return to the old system, but the real lesson was that deregulation had to be undertaken carefully. The California experience slowed the deregulation movement, and as of this writing fully competitive retail electric power markets exist in only 17 states and the District of Columbia.

Still, competition in the electric power market is, it seems on balance, beneficial to consumers as it has been in what was once another paradigm of natural monopoly, the US telephone system. Telephone deregulation in the 1980s did lead to lower prices, more choice, and technical advancement. There were some problems over the years with the deregulated phone system as there likely will be if there is complete deregulation of the market for electric power. Overall, deregulation will not be the panacea some of its supporters had foreseen, but neither will it cause the chaos that the detractors feared. In any event, an increasingly competitive market for electric power is likely to be a fact of life in the twenty-first century.

References

Bayless, C.E. 1994. "Less is More: Why Gas Turbines Will Transform Electric Utilities." *Public Utilities Fortnightly*, December 1, 21–25.

Berlin, E., C.J. Cicchetti, and W.J. Gillen. 1974. *Perspective on Power*. Ballinger Press, Cambridge, MA.

Douglas, J. 1994. "Buying and Selling Power in the Age of Competition." *EPRI Journal*, June, 6–13.

Fickett, A.P., C.W. Gellings, and A.B. Lovins. 1990. "Efficient Use of Electricity." In *Energy for Planet Earth*, W. H. Freeman and Company, New York.

Grossman, P.Z. 2003. "Does the End of a Natural Monopoly Mean Deregulation?" Chapter 10 in P.Z. Grossman and D.H. Cole (eds.), *The End of a Natural Monopoly: Deregulation and Competition in the Electric Power Industry*. JAI Press, Amsterdam.

Grossman, P.Z. and D.H. Cole (eds.). 2003. *The End of a Natural Monopoly: Deregulation and Competition in the Electric Power Industry.* JAI Press, Amsterdam.

Jacobson, M.Z. and M.A. Delucchi. 2010. "Providing all Global Energy with Wind, Water, and Solar Power." *Energy Policy*, Part I, 39 (3), 1154–1169; Part II, 39 (3), 1170–1190.

Joskow, P.L. and D.B. Marron. 1991. "What Does a Negawatt Really Cost? Evidence from Utility Conservation Programs." *The Energy Journal*, 13 (4), 41–74.

Lovins, A. 1989. The Negawatt Revolution – Solving the CO_2 Problem. Retrieved from www.ccnr.org/amory.html.

Moore, T. 1995. "Repowering as a Competitive Strategy." *EPRI Journal*, September-October, 6–13.

Moorhouse, J.C. 1995. "Competitive Markets for Electricity Generation." *The Cato Journal*, 14 (3), 421–442.

Palensky, P. and D. Dietrich. 2011. "Demand Side Management: Demand Response, Intelligent Energy Systems, and Smart Loads." *IEEE Transactions on Industrial Informatics*, 7 (3), 381–388.

Simmons, R.T., R.M. Yonk, and M.E. Hansen. 2015. *The True Cost of Energy: Wind Power.* Report, Strata Institute of Political Economy, Utah State University, Logan, UT.

Strbac, G. 2008. "Demand Side Management: Benefits and Challenges." *Energy Policy*, 36, 4419–4426.

World Bank, 2005. *Primer on Demand-Side Management.* CRA No. D06090, Prepared for the World Bank by Charles River Associates, Oakland, CA.

NOTES

1 "Updated Capital Cost Estimated for Utility Scale Electricity Generating Plants," USEIA 2013.

2 An argument can be made that the transmission system and national grid are natural monopolies still.

3 One might think self-financing – that is, paying from cash reserves – would be a big improvement over borrowing. But one must factor in the opportunity cost of that cash – that is, what else could be done with the money that might provide a better return?

4 They may continue to burn fossil fuels because intermittent sources such as solar and wind may diminish unexpectedly so that power from fossil-fired plants must be added quickly. For that reason even if 10 percent of all electricity is renewable it might not mean that CO_2 emissions have been reduced by that percentage.

5 This is for wind turbines erected on land; those erected offshore are far more expensive as Table 9.3 shows.

6 Also, see the discussion in Joskow and Marron (1991).

PART III

ENERGY TECHNOLOGY AND THE FUTURE

10   Alternative Technologies

Introduction

> In the future, we will need alternative energy technologies to replace our present conventional sources of energy.

Few would disagree with this statement. Indeed, many people would say that alternative technologies are needed right now. Yet even though people generally agree on the usefulness and desirability of alternatives, they may well have different opinions about the extent to which (finite) societal resources should be expended now in the development of such technologies. Promoters of alternative technologies express few doubts of the need for new means of providing the engines driving the economies of the twenty-first century. Typically, they represent alternatives as technically feasible, economically viable, and environmentally safe – if not immediately then in the near future. Such promotion, often by the technical innovators themselves, is not necessarily intentional misrepresentation, but more likely reflects the innovators' hopes and aspirations for their projects.

However, as we will see in this chapter, the development of new energy technologies, or any new technology, is an uncertain process. From the conception of the project to its final adoption by society, the uncertainties may encompass not just *whether* the concept will work but, just as importantly, *when* and *if* widespread adoption can be feasible. We will see, in fact, that the time scale (or term) of expected adoption of an evolving technology is an important consideration in judging its chances of success; the longer the term for the prospects to be realized, the greater the uncertainty of its ultimate success.

Finally, we should note that alternative technologies frequently promise great advances: freedom for the future from dependence on unreliable sources of energy, the reduction in the harmful impacts of current technologies, and the elimination of the risk of accidents. Such claims should be regarded with some skepticism. Not only must a technology function, but we must ask, at what cost – both financial and environmental? In some cases, the cost of development and adoption of an alternative technology could be so great that public welfare would be harmed in the process. Synthetic fuels fit into this category, according to critics. Solar and wind technologies seem to offer better prospects, but at what cost to develop them?

The Evolution of New Technologies

Stages

The development of a new technology, from its initial conception to its widespread adoption, is both a technological and a social process. There are many historical examples of the evolution of technologies.[1] Before the twentieth century, development evolved through stages – from *invention* to *adoption* – without necessarily the benefit of developmental subsidies from government or big business. Development essentially was determined by the market. So, for example, starting in the nineteenth century, steel manufacturers gradually discarded the Bessemer process in favor of the open-hearth process of steel production because it was in their economic interest to do so. The new process worked better and gave its users and producers advantages over those who still relied on the old Bessemer process, although at least initially those firms using the older process profitably were not persuaded to make a major change. In fact, many years passed before the new technology was overwhelmingly predominant. The same kind of market-driven evolution – sometimes called the *natural* pattern of industrial evolution – has given us many other now familiar technologies, including many energy technologies. Fossil-fired electric generation and the refining and processing of crude oil into gasoline are two examples of energy technologies that have followed the natural pattern.

The natural pattern generally has proceeded as follows:

1. *New knowledge* – the establishment within society of basic scientific principles.
2. *Development of technical capability* – proof of the principle for technological application.
3. *Prototype construction* – the first working model designed for practical operation.
4. *Commercial introduction* – low-cost design and manufacture for mass markets.
5. *Widespread adoption* – promotion and wide utilization by society.

Although many present-day energy technologies have followed this natural pattern, others have not. Rather they have developed with the help of programmatic government sponsorship, especially during and after World War II. The most notable example is nuclear power. It has had government assistance all the way, even into the process of commercialization, and probably would not have been developed at all were it not for government support (Camilleri 1984).[2] Alternative energy technologies for the future will likely also require significant government sponsorship to advance in the marketplace. Indeed, few have been able to sell in markets without subsidies or government mandates also continuing. Nevertheless, in the twenty-first century, many individual inventors and entrepreneurs (e.g. Microsoft founder Bill Gates) have invested in various new energy technologies.[3]

While government-sponsored technology development may not be market driven, it nonetheless is intended to parallel a natural innovation process. The stages of government-sponsored programs have come to be:

1. *Research* – proof of the scientific principle for technological potential.
2. *Development* – engineering design for technical feasibility.
3. *Demonstration* – the proof of operational feasibility and economic competitiveness for the technology.
4. *Commercialization* – the process of dissemination of information, building of experience, and overcoming various barriers to widespread adoption of technology; the stage ideally leads to the privatization of the production process.

These stages of programmatic development have been called *RD&D*, standing for *research, development, and demonstration*. In reality, with respect to energy technologies there has not been a true commercialization stage, which is a relatively recent goal, that has proceeded by market forces alone. For this reason, this last stage is not included in the jargon of government-sponsored development.

Time Scales

There are currently many alternative energy technologies under development – some at every stage of the natural and programmatic stages of evolution. Accordingly, each technology has a different time scale of expected adoption. These times scales are commonly grouped into the following categories:

Near term – at the *commercialization* stage; the new technology has been judged feasible but is as yet underutilized (impact expected within 5–10 years); often economic viability is either limited to niche markets or is not yet fully realized.

Medium term – at the *development* (engineering) stage; proof of technical feasibility and prototype *demonstration* have been achieved; proof of *economic viability* is uncertain but it is hoped that with continued operational experience (10–15 years before expected impact) viability will be achieved.

Long term – in the *research* stage; scientific/technical feasibility has yet to be proven; development of technical workability and economic viability remains unclear (15–25 years before expected impact).

Examples of each category are given in the following lists. Each time-scale category carries a different implication of uncertainty. Technologies with near-term prospects, for example, are the most certain to proceed to wide-scale use, whereas the outcome for those in the medium term is significantly less certain.

Near Term
1. *Industrial conservation* and *technical efficiency* – continuing efforts of technical demonstration of products that repay buyers with lower energy costs. Examples are reduction of heat losses, materials recycling, combustion efficiency, heat recuperation and cogeneration.
2. *Solar thermal* – domestic hot water (and also swimming pool heaters); US market penetration is relatively small (300,000 units or 0.5 percent of all hot-water systems) and most exclusively in the south and southwest. Much greater degree of penetration in Mediterranean countries such as Israel and Cyprus.

3. *Solar-direct conversion* (*small-scale*) – rooftop solar photovoltaic (PV) systems; also remote photovoltaic conversion, for rural and wilderness sites located outside the electric grid. About 800,000 US homes have rooftop solar (qualifying for incentives from the US federal and often state governments).

4. *Solar-direct conversion* (*medium-scale*) – domestic and commercial uses of photovoltaic conversion providing intermittent moderate electric loads, mostly in Europe, though increasingly also in the United States. Economic battery storage needed for further market penetration.

5. *Solar-thermal power* (*medium-scale plants*) – sizes range up to 100 MW, in Europe with price support. Economic heat storage needed. A few recent projects in the United States are in the range of 250–400 MW. All are heavily dependent on subsidies.

6. *Wind farms* – Many are in operation, particularly in superior wind regions; in some places they are feeding significant generation to electric utility grids. Zoning and environmental site disputes need improved means of resolution in some regions for further market penetration, and large-scale storage is needed to account for intermittency and to fully realize potential.

7. *Geothermal power generation* – currently available only in particular sites; the result is that only 0.3% of the US national electric capacity is geothermal, but more significant national capacities are possible elsewhere in the world.

8. *Small hydroelectric generation* – accessible resource currently only at limited available sites.

9. *Enhanced oil recovery and hydraulic fracturing* – extending and adding to recoverable resources; active and commercially viable with high oil prices.

10. *Super- (and ultra-) critical coal combustion* – well into the commercial development phase (see Chapter 3), although coal development in the United States has been slowed over concerns about climate change, and US market penetration is low.

11. *Resource recovery* – heat recovery from municipal solid waste; currently being pursued in localities nationwide in the United States and abroad, with government assistance.

12. *Electric automobiles* – marketed with large government support; greater penetration will require large investment in infrastructure, especially charging stations. Low gasoline prices as well as expensive battery packs are also likely to slow commercial development.

13. *Biofuels* – ethanol and biodiesel are mandated as additives into US gasoline, and some engines run on mixtures of up to 85 percent ethanol (E-85). Most is derived from corn or soy and all depend on government mandates and/or subsidies.

Medium Term

1. *Synfuels* – coal-derived and biomass-derived, especially ethanol from cellulosic feedstocks.

2. *Alternative fossil fuels* – oil shale. As noted in Chapter 5, some shale formations contain hydrocarbons in the form of kerogen, locked in shale rock; as much as 2 trillion barrels of oil shale are estimated in one location alone, the Green River Formation, in the Western United States.

3. *Solar industrial process heat* – processing industries, low to medium temperatures.
4. *Utility-scale renewable generation* – larger-scale solar or wind electric generation feeding utility grids (dependent on very large-scale storage, on the order of gigawatt-weeks).
5. *Resource recovery energy* – reduction of solid waste to synfuels.
6. *Advanced storage* – advanced batteries, hydrogen storage, thermal storage.
7. *Alternative conversion* – fuel cells for transportation, electric backup, electric power generation.
8. *Breeder reactors* – uranium-cycle or thorium-cycle fission; both types are in pilot stages but neither are at present under development in the United States. (See Chapter 8.)
9. *Superconducting transmission lines* – more efficient electrical transmission.

Long Term
1. *Nuclear fusion* – power reactors.
2. *Hydrogen economy* – large-scale transmission and distribution of energy using the medium of hydrogen, for all purposes including transportation.
3. *Universal, dispersed-site generation* – widespread use of small-scale and/or renewable source generation in autonomous or electric grid usage.

Any technology that reaches the commercialization phase will have passed from stage to stage, progressing down from long- to medium- to near-term prospects as the result of scientific or engineering breakthroughs or changes in market conditions. From the mid-1980s through the late 1990s, we saw examples of breakthroughs leading to a change in the time scale of development of a technology. In the case of superconductors, materials that offer no electrical resistance (to direct currents), the scientific principles had been known for decades. Yet despite the fact that superconductivity could mean, among other advances, low-loss electrical transmission, it had virtually no technological potential, because the previously known superconductive compounds had to be operated at such low temperatures that the capital and operating costs were prohibitive for commercial use. By the end of the twentieth century, however, researchers in the United States and Switzerland found compounds that proved to be superconducting at temperatures dramatically higher than ever before. Suddenly many practical uses, such as in magnetic resonance imaging machines (MRI), could then be considered and engineering development for such uses started. As a result, superconducting devices have moved into the development stage. Superconducting transmission lines are not as far advanced as, for example, superconducting components of medical equipment. However, superconductors can be considered as having moved from the *long-term* to the *medium-term* time scales for expected adoption to energy-related applications.

The case of superconductors also demonstrates the way sponsorship of development changes. When it was strictly a long-term prospect, *superconductivity* research was relegated to the laboratory usually under the sponsorship or direct control of the government. Much of the energy-related research for the government has been carried out at a group of national laboratories (under the auspices of the US Department of Energy) located in Argonne, Illinois; Brookhaven, New York; Los Alamos, New Mexico; Oak Ridge, Tennessee; and Sandia, New Mexico.

Once breakthroughs have been achieved at the national labs, one might expect venture capital would be invested by entrepreneurs into product-development companies. In reality, the major work typically has remained in the laboratory. However, with improved prospects of some new energy technologies, it is hoped both government and business will step up funding of research leading to the development phase. In theory, once the development phase has been achieved, work (along with risk) would shift entirely to the private sector, and there are examples of this in fields outside of energy. But this has never been entirely true of what are the most prominent alternative energy technologies: solar, wind, nuclear, and biofuels. None has been cut loose entirely from subsidies or mandates that have been established at the federal and/or state levels.

Near-Term Technologies

The near-term technologies are those that have reached the stage of *commercialization*. Technical feasibility has been proven, demonstrations (or pilot units) have operated successfully, and costs are deemed close to competitive with those of conventional technologies. Still these technologies have not penetrated the relevant market to their full potential. In many cases, penetration is nowhere near full potential. Here we explore some of the reasons for this lack of penetration.

The Cost Barrier to Adoption

The commercial viability of a new energy technology depends primarily on its cost effectiveness. Since it is to substitute for an existing technology, it must be comparable or superior in terms of costs before it can have wide success in the market. Even though some near-term alternatives may approach cost competitiveness, they may still fall short. In other cases, there may be some change in the energy market or a small cost-reducing advance in the technology itself to make the alternative cost competitive. So, for example, an increase in the price of oil has made *enhanced oil recovery* technology economically viable, though it may not remain so in times of low oil prices. In some cases, the government might step in and offer incentives that will at once make an alternative technology cost effective. For example, rooftop solar PV systems have become more common because various subsidies and other incentives have substantially lowered the price homeowners must actually pay and raised the return on investment they receive. The danger of course is that, in these cases, government incentives do not actually make the alternative cost competitive, and if the incentives are withdrawn the marketability of the alternative disappears. This was the case with solar thermal water heating in the 1980s.

The general economic situation can also be a deterrent to investment in alternatives. During the early 1980s, for example, there were high interest rates and a recession. Both discouraged any major capital investments by private businesses. And so while heat recuperators, fuel-efficient boilers, and cogeneration units were available, and could at least in some cases have been cost effective, adoption proceeded at a slower rate than it might have in a period of economic growth and lower interest rates. A deeper recession occurred in the first decade of the twenty-first century, prompting the US Department of Energy to initiate the *Energy Efficiency & Renewable Energy* (EERE) program to foster solar energy, wind power, hydrogen fuel cells, biomass energy, and geothermal energy technologies, as well as investments in efficiency. It should be noted that just a year or so before this program was launched oil hit an all-time high price of $147/bbl and natural gas was also at a record level; fears of peak oil and gas were ubiquitous.

Finally, there may be problems for consumers attempting to assess the relative costs of technologies. The cost of an alternative (or of conservation) may be an added *capital* cost, whereas the savings may be in terms of *operating* costs, particularly fuel costs. It may be difficult, however, to persuade an individual to spend a large amount now to save small increments over time, especially when the savings are not entirely apparent because the future prices of factors are unclear. For example, we may project that the cost of natural gas will average $10 per 1000 cubic feet or more over the next decade, and therefore calculate how much gas a solar hot-water system will save. But the savings will depend on gas prices, and those prices will be increasingly uncertain the further into the future the analysis is taken. In some cases, it may take many years for an investment in an alternative to pay off – that is, for the buyer to make up the extra capital cost in saved fuel costs. The payback time also depends on where one was living, the subsidies that were available, and whether one bought a straight solar thermal system or a hybrid solar thermal/PV system. As of 2016, it was estimated that the payback period in sunny, arid Arizona would be about 7 years.[4] Elsewhere the average payback time could be quite a bit longer – long enough to discourage many from buying.

The Social Process of Adoption[5]

Cost effectiveness is not the only reason some near-term alternatives have failed to gain widespread adoption. Even with financial incentives, adoption can be slow. In 1978, the US government offered tax incentives to make solar hot-water heating a viable alternative and the technology made some market headway before the support was withdrawn. But even before the incentives were removed, solar hot-water heating had not achieved the hoped for market penetration, even in the Sunbelt states where it seemed to have financial advantages. The problem lay mainly in the basic *social* process by which a new technology is adopted.

The behavioral patterns of people in adopting a new technology[6] have been studied by sociologists (see, among many, Rogers & Shoemaker 1971, Rogers 1995, Lynn et al. 1996, Hall & Khan 2003). The process leading to widespread adoption is called the *innovation process*. When tracked in time, it can be shown typically to follow the S-shaped pattern in Figure 2.2a. The S-shaped curve is familiar, of course, from our discussion of natural resource extraction and turns out to be applicable to human technological behavior in general. An S-curve for the innovation process would show the cumulative number of adoptions of a new technology versus time. In the beginning phase, the increase in adoptions would be exponential, where the rate of rise at a given point in time would be proportional to the number of adoptions at that point. The last phase would represent market saturation, indicating the limits of adoption, as has occurred in the previous century with many household electrical appliances, and common electronics.

In each phase of this time pattern of innovation, individual adopters have distinct characteristics. In the very earliest phase, for example, the adopters tend to be risk takers who are eager to be seen as trend setters. The *early adopters* of all-electric automobiles have been mostly wealthy and well-educated people who could risk a financial failure of investment in order to be the first with a new technology. The decision to be an early adopter is not, however, strictly a matter of economic class, because many people in the same socioeconomic group are not so adventuresome. Early adopters are then both of a particular social class and of a particular personality type.

After the early adopters, we next see the *early followers*. These people are innovative, but not quite as willing to take the initial risk on an unknown technology. With rooftop solar panels (PV) for example, early followers may be influenced merely by the fact that their more innovative neighbors have already purchased solar systems, but they would not necessarily wait to see what that neighbor's experience would be. They would, however, tend to be influenced more by their neighbors than by advertising.

This means of building confidence in a new technology is an example of what sociologists call *networking*. The term networking here refers to interpersonal communications and influence, where the participants are most probably unaware of their contributions to the process of innovation and the diffusion of a new technology into society.

As the process continues, we encounter the *late followers*. Here, we are definitely out of the exponentially rising part of the S-curve of adoption. If we were to move into the future to even later adoption, we would begin to reach the saturation range of the curve. Rooftop PV has been installed in more than 700,000 homes and businesses as of 2015, but clearly the technology has not yet approached the late-follower stage, much less the saturation point. There are over 100 million homes in the United States, so the percentage with rooftop solar is a tiny fraction of the possible number of private homes. Add businesses to the possible buyers, and one can readily see rooftop solar is a long way from reaching saturation.

But other technologies have in recent years more quickly scaled the upper portions of the S-curve. The cellular telephone is a notable example (Agar 2013). It was invented in 1973 but the exponential growth phase was not reached until the 1990s. By the 2000s (or about 30 years after its invention) it reached a saturation level, with over 5.5 billion cellphones in use in the world as of 2011. Of course, the cellphone is relatively cheap (now) and represents an advance over landlines in terms of convenience. But in the 1980s, it was then still in the early adopters phase.

Cost and convenience were probably the most important characteristics leading to the success of the cellphone, and it also depended on the expansion of applications. But success depended, too, on other characteristics that overcame resistance to mass adoption. First, social communications mattered; initially interpersonal networks and later the general media helped overcome psychological barriers, particularly fears about the new technology's complexities. The cellphone market also began to exhibit what social scientists call *trialability*, which may be required for the late-follower stage. Trialability is, in essence, the estimation of whether the technology is going to be workable and convenient for any given individual, especially a late follower. A late follower will want to know: will there be organizations and facilities to repair the equipment and will repair service be readily available? Will repair costs be reasonable? Is the product guaranteed against malfunction or theft? And is there a dependable performance rating of normal operation?

Repairs, guarantees, and ratings are *institutional* barriers to adoption. In other words, because no organizations preexist in society to facilitate the use of the technology, they have to be established before widespread adoption is likely to take place. Not only must repair businesses and a credible system of performance testing and ratings be set up (by government or industry), there must also be some institutional system (public or private) to provide consumer protection to identify and end dishonest business practices. Solar hot-water systems needed the equivalent to the nameplate ratings and warranties of a conventional hot-water boiler. The failure of the manufacturers of domestic solar equipment to provide such institutional guarantees, in the opinion of observers, contributed to the low penetration of the potential market in the late 1970s and early 1980s (see issues of *Solar Engineering & Contracting* magazine from that period). By the turn of the twenty-first century, consumer's guides and warranties had become available although solar heating systems remain in the early stage of adoption.

Legal Barriers

Legal questions may also produce barriers to adoption of new technologies. In the 1970s, entrepreneurs who saw the potential for dispersed-site generation of electricity found legal barriers to the marketing of their energy. They proposed to use a number of different alternative technologies, including cogeneration, small hydroelectric, small-scale solar and wind generation. But regardless of the kind of technology proposed, local electric utilities refused to purchase the electricity. And because the utilities controlled the power distribution grid, and because the dispersed-site innovations had to tie into that grid, the utilities effectively blocked this form of alternative energy development. They were supported by public utility laws (mostly at the state level), which gave them monopoly control. In other words, the problem for the dispersed-site innovators was not technical, financial, or sociological; it was strictly *legal*. The utilities had the right by law to refuse to open the market.

The utilities were acting primarily in the interests of their stockholders; however, legislators, who were expected to consider the perspective of the public interest overall, came gradually to a different view. At both the state and federal levels, lawmakers recognized that public utility statutes, drawn up in an earlier era to enhance the expansion of the utility grids, were now being used to prevent adoption of alternative energy sources. The result was legislation in 1978 at the federal level, the Public Utilities Regulatory Policy Act (PURPA), which required the utilities to purchase electricity at full *avoided costs*. Some states, as well, required the utilities to make such purchases at what would be considered a *fair price*. This all led to moves, at both the state and federal levels, to begin restructuring the electric industry, thus permitting (nonutility) independent power producers to compete in open markets for the sale of electric power. With the passage of the Energy Policy Act of 1992 and its provision for competitive power markets, the opportunities for nonutility generators grew.

Old laws have also affected the adoption of modern solar energy systems. The issue revolved around the rights of an investor in a solar energy system to have an unobstructed flow of sunlight. This question of solar access deals with disputes that can arise between owners of adjacent properties, when buildings or vegetation on one property cause a shading of the solar collector of the other. There have been many court cases in which the question revolves around one individual's wish for unobstructed solar access sometimes demanding that another individual trim his or her trees or cease a building project that would block the first party's sunlight. There has been no uniform judicial outcome. In the 1982 case *Prah v. Maretti*, for example, Maretti was enjoined from a building project blocking Prah's sunlight, but in *Zipperer v. County of Santa Clara* (2005) the court ruled the plaintiffs could not demand that the county cut back trees that were blocking solar access.

It should be noted that there is an old doctrine of *ancient lights* in British common law, the tradition from which US law developed. In the old common law, the householder had essentially a legal, property right to sunlight, a property right that extended to the heavens. This law is untenable in an age of airplanes (otherwise a homeowner could sue for trespass every time a plane passed overhead), and the old doctrine was simply concerned with the property owner's level of illumination rather than collection of solar energy. Ancient lights is an absolute doctrine, however, and also conflicts with other constitutional rights because it would limit the right of the owners of other properties to develop them (e.g. grow trees).

Workable legal doctrines on solar access have been explored to enhance the adoption of this technology, while at the same time preserving constitutional rights (Rule 2010, Susman & Lund 2011). Toward these goals, perhaps the most workable doctrine is called a *solar easement*. A solar easement is analogous to a land easement in which a strip of land owned by one party may be used by another (usually to gain access to another piece of land) under an agreement or contract. The agreement usually involves an exchange of money in return for use of the easement. In this case, the solar collector owner would negotiate an easement of sunlight whereby his neighbor would forbear obstruction of his or her access to it. States, notably Massachusetts and New Mexico, have statutes that include easement rules to foster solar development. But the legal issues have not been entirely resolved and it may well take experience with various statutes and many court precedents before there is general agreement on what legal formula works best. In the meantime, the legal uncertainty may pose an institutional barrier for solar development.

There can also be jurisdictional questions that slow the diffusion of innovations. For example, "smart" grid technology (see Chapter 4) may face the problem of satisfying federal, state, and even local statutes before it can be widely utilized (Tabors et al. 2010). In the meantime, smart grid technology has been slow to diffuse with only a few pilot projects in the United States as of this writing.

Medium-Term Technologies

Medium-term technologies are those that are technically feasible but still some time away from widespread adoption. None have been developed to the point where widespread commercialization is possible. During this period, working models can be used for limited applications but, for the most part, are simply prototypes. The most important distinction, however, is that production or usable applications of these working models is not economically viable.

Some medium-term technologies have the technical potential to substitute for the conventional fuels used by society – making them potentially of major importance. For example, some synfuels and alternative fossil fuels could provide *direct substitutes* for petroleum and natural gas, but have not been utilized because they are not price competitive. That is, the physical and chemical properties of the substitutes would be indistinguishable from the conventional fuels and they could be adopted without technical changes if the price were right. That is, there would be no need to design new capital equipment in order to distribute and use these fuels, and they could be sold in the same markets as the fuels (e.g. gasoline) they replaced. Solar technologies, on the other hand, offer what might be termed an *indirect* substitute. They would generate heat or electric power through the conversion of sunlight and, to a greater or lesser extent, supplant the markets for conventional fuels; in this case, however, new capital equipment would be required.

The available energy that some medium-term technologies could supply to the economy would range from the *huge* to the *inexhaustible*. For example, synfuels can be derived either from coal or from biomass: the former, as we saw in Chapters 2 and 5, is an immense resource, while the latter, if properly managed, would be renewable and so made effectively inexhaustible. Others are comparably large and not fully exploited – as noted oil shale deposits in the Western United States contain more than 300 years of domestic supply at the current rates of use.

The solar heat technologies have the potential to provide vast amounts of industrial heat. Indeed, since they use sunlight, they offer the possibility of tapping the superabundant and inexhaustible solar-energy source, the Sun. But in the medium term, solar industrial heat technologies, unlike near-term examples such as rooftop PV, are not near large-scale commercialization. As with many medium-term technologies, there is a need to overcome several barriers to adoption, such as high investment costs and analyses of long-term operational characteristics, before economic and/or technical viability can be achieved and more widespread adoption can take place.

Feasibility of Mass Production

Medium-term technologies are those that have been proven technically feasible at least at the laboratory level and appear likely to become cost effective in the not too distant future. To become widely adopted, any technology must be capable of cost-effective *mass production*. Cellulosic ethanol plants, for example, will have to be able to process thousands of tons of feedstocks per day cost effectively to become a significant substitute for gasoline. In 2007 (as noted in several earlier chapters), Congress launched the United States on a major biofuel development program. The goal of that program was to produce 36 billion gallons of ethanol per year by 2022, which would replace 20–25 percent (approximately) of US (oil-based) transportation fuel with domestically produced alcohol-based fuel. Of the 36 billion gallons, 15 billion were to come from ethanol made from corn – requiring a traditional distillation process. But another 16 billion gallons were supposed to be made from cellulosic materials, plants such as switchgrass or waste products such as wood chips, requiring more complex processes (discussed in Appendix C) that have not yet been proven capable of mass production. President George W. Bush had intimated in a State of the Union address that this technology would likely be commercially viable by 2012.[7] Only it was not. Not in 2012 and not in 2015 when according to the statute 3 billion gallons of cellulosic ethanol was supposed to be available. In reality, the statute called for production that was more than three orders of magnitude greater than actual production. There was no significant cellulosic ethanol production to speak of.

Solar industrial process heat presents a different kind of obstacle to mass production. In order for the technology to be cost competitive, the solar equipment *itself* must be inexpensively manufactured using mass-production techniques. Until such large-scale production techniques are developed, solar industrial process heat will not be able to move from the medium term to the commercialization phase. Still, the technical developments required for such mass production do not seem great. Major breakthroughs do not appear necessary. Rather, a series of practical engineering innovations and design improvements are needed.

To say that no major breakthroughs are required does not, however, guarantee early adoption or even a smooth path toward eventual market competition. Ordinary engineering development requires effort and time to progress and be analyzed. One synthetic fuel demonstration project, for instance, took over 5 years from the start of design to complete construction of the prototype plant. Once built, prototypes need years of operation and testing to demonstrate no major flaws in the basic design. Since flaws are not uncommon in prototype development, it is typically several years and can be much longer before a technology in the prototype stage moves to commercialization.

Technical Breakthroughs

There is uncertainty even in the seemingly most straightforward technology development programs – if not for the ultimate success of the final designs, at least for the amount of time it will take for the program to succeed. When some sort of breakthrough is required, the uncertainties are magnified and the time frame made increasingly uncertain.

Photovoltaics – the direct conversion of sunlight into electricity – represent a technology where breakthroughs have been needed, and throughout the 1970s and into the 1990s they seemed near but never really occurred. Even more gains have been made over the past decade in semiconductors for photovoltaic conversion, although further gains are still needed at the time of writing (see Chapter 11).

It is noteworthy that photovoltaic (PV) conversion (of sunlight to electricity) *is* a working technology but has had limited applications due to its initial cost (including grid connectivity and backup) and its low conversion efficiency. Production of thin-film PV cells is worldwide, with China the leader in quantity of production. However, at present only a small fraction of *gross electric power* generation worldwide is derived from PV arrays, since they have not yet been deployed for major production. But the potential of PV has been evident, and some have envisioned large arrays of solar conversion cells powering industry and providing major additions to electricity supply, especially in sunbelt areas. This has not happened by the middle of the second decade of the twenty-first century. Engineering efforts have focused both on improving efficiency and reducing the cost of production; but most importantly, intense research efforts continue with the goal of low-cost electrical storage to reduce the effects of the sunlight's intermittency on the grid. Improvements have lowered the cost per watt of peak power output, but further improvements are needed.

Photovoltaic cell manufacturers have aimed for advances in basic cell materials (semiconductors) and in the techniques for handling them. Since only a very thin layer near the surface of a semiconductor is effective in photovoltaic conversion, semiconducting thin films have been deposited on a host crystal to form the operational PV cell. This also serves to lower the cost of the cell, since a single-crystal material, as grown, would be very expensive. These cells have mostly used various forms of crystalline silicon (Si). However, this has still resulted in relatively high fabrication costs, considering the large surface areas of solar-cell arrays required for significant electric output. Amorphous silicon (a-Si) or compounds such as cadmium telluride (CdTe) have been substituted as the semiconducting layers. These can be deposited in thin films on flat sheets of less expensive materials. Further R&D work needs to be done to create new cells with efficiencies of over 30 percent, close to the theoretical limit of conversion. This has been achieved with very costly specialized materials and equipment, and 20 percent has been achieved with more basic materials. But, as of 2015, commercial solar cells produced efficiencies in the range of 11–15 percent.

Of course, greater power output per cell decreases the need to reduce manufacturing costs and the new cells may indeed bring PV technology closer to large-scale application. But such a new development would only be a step in that direction. Although higher efficiency cells promise an increase in revenue from electricity production, they have not been fully realized in commercial production. Such cells, though efficient, will require intricate production engineering, and even then, projections are that without further refinements the cells will not be cost competitive with conventional sources for utility-scale deployment. Development of manufacturing techniques and site testing will then be needed before they can be widely utilized at utility scale.

Alternative Technologies and Energy Markets

To say that photovoltaic or any other new technology is economically viable is to say that it can supply its product at a price that is competitive in the market for that product (e.g. electricity). If the product is an *exact substitute*, such as synthetic fuel oil derived from coal as noted above (see also Appendix C) is for petroleum, then that price must be equal to or less than that of the product it is to displace from the market. If some properties of the new product differ somewhat from the existing product (e.g. ethanol, which has less energy per gallon and has properties that can corrode engine parts) then the price must be such that the consumer would still want to purchase it instead of the existing product.

The market price of any product must, of course, be greater than the cost of production in order that the producer will not be selling at a loss. The production of any product – and hence its market price – must include costs for investment, operation, and the input factors of production (see Chapter 4). In the case of cellulosic ethanol, the cost of the input feedstocks must be low, and then the remaining problem is that the initial capital costs and the cost of processing must be low enough that the output can be priced comparably to the gasoline it is meant to replace. For the user, there is no other measure than price, since it is presumed that the product will directly substitute to operate existing capital equipment – as some engines are capable already of utilizing high percentages of ethanol fuel. However, experience to date shows that either the price of conventional oil will have to rise greatly or the technology of synfuel production will have to improve before these substitutions can approach cost competitiveness.

With solar technology, on the other hand, economic viability is determined when the market price of new equipment (including the interest cost to borrow the money) results in net savings over conventional energy (electricity) sources for the user. A solar user has limited operating costs – i.e. solar has no fuel costs and low operating expenses. But this user has to be convinced that the initial capital cost represents a saving over conventional sources, where the ongoing expense of fuel and maintenance may be greater. Solar technologies, and any indirect substitutes, have a greater hurdle to overcome because users must make a large outlay of money in the present for an estimated, but ultimately uncertain, amount of future savings.

Market uncertainty, of course, also will affect the willingness of investors or entrepreneurs to back development ventures that are often needed to make a technology viable. Even a huge rise in oil prices might not encourage investment in new technologies for several years, and will then only if it becomes clear that the rise was permanent, not just a temporary spike. Such investment behavior has been especially true in the early years of the twenty-first century, when oil and natural gas prices soared, then dropped, and then rose part of the way, fluctuated, plateaued, and then dropped again. Investors, who had invested quickly in cellulosic ethanol plants and solar equipment manufacturers, lost money – although in all of these cases taxpayers also lost since most of these new ventures were heavily subsidized.[8]

Two alternatives, oil shale and liquid coal-derived synfuels, have been undercut several times by market conditions in the past. Oil shale research, conducted by the US Bureau of Mines, was initiated in 1916. World War II accelerated this research and by 1944 some small pilot projects had been completed and the construction of larger prototypes was being considered. But this was a period of increasing availability of large amounts of low-priced oil principally from the Middle East. The government had passed the Synthetic Fuels Act (SFA) in 1944 to develop oil shale along with coal-derived and other alternative fuels during World War II, but the SFA projects were discontinued by the early 1950s (Grossman 2013).

This start–stop process was repeated in the 1970s and 1980s. First, concerns about energy supplies led to government-sponsored development, as well as investment by private industry, in alternative fuel pilot projects. In 1980, Congress passed a major coal to oil (or gas) synfuels bill that was projected to cost $88 billion.[9] But oil prices, expected to more than double by 1990, fell instead, and as a result the US government dropped its support and the program essentially ended (Grossman 2013). South Africa continued to develop synfuels in the decades following World War II, so that by 1983 a prototype synfuel oil plant was in operation producing 58,000 bbl/day (see Appendix C). However, the South African government has had to provide ongoing subsidies because the output has not been cost competitive under current (2016) conditions in the oil market.

Environmental Unknowns

The uncertainties of eventual mass use of some *medium-term technologies* for alternative fuels have included concern over unforeseen risks to the environment. Such concerns have tended to be ignored by proponents in their enthusiasm over the potential benefits of these new technologies. But environmental questions must be addressed before societal resources – perhaps in the billions of dollars – are expended in new development. Depending on the technology in question, the environmental impacts may include air pollution, waste disposal, occupational hazard, and water pollution. Environmental issues are especially important with respect to coal-based synfuels, biofuels, and alternative fossil fuels.

A good example in history can be found in oil-shale processing. Oil shale, as noted earlier in this chapter (and in earlier chapters of this book), is a solid hydrocarbon, kerogen, locked up in shale rock deposits. To get the oil out of the rock, the shale must be mined, crushed, and heated, called the *retorting process*, and then the oil must be refined (discussed in more technical detail in Appendix C). There may be many environmental risks throughout the process. Mining may pose occupational hazards similar to coal mining, including pneumoconiosis and possible gas explosions. The sheer volume of material to process might also prove environmentally detrimental. In order to produce 25 gallons of oil – less than one barrel – a ton of shale must be processed; 1 million bbl would leave almost 2 million tons of waste rock to be disposed of. The processing operation would also require large amounts of water, but most of the shale is found in relatively arid country. Not only does that leave a problem of water supply, it also would probably induce a conflict over water resource rights. In addition, the shale is extremely alkaline and water runoff could irreparably damage nearby rivers and aquifers. Clearly, such impacts would have to be minimized before oil shale development could proceed. The potential environmental hazard not only adds safety questions, it adds economic ones as well. The environmental cost, when added to the cost of production, could keep oil shale from economic viability even in the event of a dramatic increase in the price of oil. As of the mid-2010s, the world was facing an oil glut and low prices, and fracking for shale oil (not to be confused with oil shale) boosted US reserves, so that there is little impetus for exploiting the vast reserves of oil shale. Should supplies tighten this could change, but would likely require that producers provide solutions to the difficult environmental issues oil shale raises.

Public Conflicts

Protest against a massive oil shale operation is virtually certain. Oil shale projects of the late 1970s provided examples of the kind of opposition such projects probably will face in the future. Critics leveled two charges, in particular, against the shale operations. They claimed that shale disposal was leaching harmful chemicals into drinking-water supplies, and they raised concerns over the safety of shale workers. The disputes pitted the oil companies, such as Union Oil and Occidental Petroleum, that operated the projects, against environmental groups, public interest groups, labor unions, and federal regulatory agencies, including the National Institute for Occupational Safety and Health (NIOSH) and the Mine Safety and Health Administration (MSHA).

The period leading to adoption of massive production of *any* alternative technology, when it comes, is likely to be stressful. Present-day protests against fracking, for example, have led to outright bans or moratoria against it in a few states, most notably New York, and there are protest movements against it elsewhere even where it is permitted.

There are also active protest movements against wind energy projects, especially offshore projects, which have at times been scaled back or canceled outright. In Vermont, a bill introduced in the state legislature in early 2016 would ban any large-scale wind farm. The objections to wind focus on noise pollution, bird mortality,[10] and aesthetics, all of these important in a state such as Vermont, which is an important bucolic tourist destination. Thus, a vast expansion of wind power is likely to lead to increasing opposition.

No doubt a great expansion of nuclear power plants, notwithstanding a claim that they are inherently safe as well as CO_2-emissions free, would face intense opposition due to fears of accidents and of nuclear material proliferation (Price 1990). Of course, some environmentalists have become strong supporters of nuclear power (Chapter 8), but others remain vehemently opposed and maintain that the electric system can become completely renewable (and non-nuclear) by 2050. An attempt to do this, however, would require massive expenditures, the huge expansion of wind facilities, and would also surely engender wide opposition.

Long-Term Technologies

Proof of Principle

Long-term technologies have the greatest uncertainty of all. Not only do we have the uncertainties of commercial potential and widespread adoption, as we do in the medium-term and near-term time scales, but also we have to answer a more basic question: Will the proposed technology (or technological system) work?

The proof required goes beyond *technical feasibility*, which should be understood in the sense of engineering and design. Coal-derived synfuels, for example, have been produced by the processes described in Appendix C and so we can say the technology is "*feasible*." Some prospective synfuels, however, may not turn out to be efficiently scalable. Nor may they be demonstrated to create fuels at competitive costs. Prototype plants alone do not fully determine feasibility since the problems of large-scale operations may not have appeared in demonstration projects. With such prospects, they will have cleared the initial research stage of technological evolution, but are unable to move beyond that point. But in this section, we are dealing with technologies that have not even passed the research phase, but their promise is so large that research continues in the hope that prototype engineering analysis can begin.

Nuclear fusion research provides the clearest illustration of the distinction between the proof of principle and technical feasibility. In the fusion program, the specific proof of principle has been that a *net yield of energy* can be achieved from controlled nuclear fusion reactions. This net yield, furthermore, must be in a form that would produce significant amounts of useful energy for power production. But proof of principle has yet to be achieved, as we will discuss in more detail in Appendix C; if it can be achieved, fusion will solve energy dilemmas presumably forever.

The first major objective in the fusion research program overall has always been to conduct a laboratory experiment that shows a net *breakeven* of energy; that is, the energy output from the fusion reactions should at least equal the energy input required to ignite the reaction. In 2014, an experiment, using inertial fusion technology (Appendix C), briefly achieved not just breakeven but also apparently a net energy yield. But the energy gain lasted only an infinitesimal 150 picoseconds (or 150×10^{-12} s). Until a net gain can be sustained (called *ignition*), it will not be proven that fusion can be a useful source of energy. Moreover, most of the fusion research effort has focused on a different (and to many scientists, ultimately more promising) means of achieving a fusion reactor capable of generating electric power. Magnetic fusion technology (Appendix C), despite many attempts going back to the 1950s, has yet to achieve breakeven, much less a net gain of energy or ignition.

The lack of success should not be attributed to the lack of funding of fusion research programs. With the possible exception of nuclear fission, fusion research funding has had more continuity in support, since its initiation, than any other energy R&D program in the United States. Also, the scientific talent working in national laboratories, corporate research laboratories, and universities has been of top quality – both in the United States and worldwide.

The underlying reason for the limited success of fusion experiments has appeared to lie in basic issues of plasma physics (discussed in Appendix C). Such an assessment was not only accepted within the fusion research community, but was also the focal point of critics of the program's direction. They contended that the program should have concentrated from the beginning more on basic physics before attempting to construct breakeven experiments and steps beyond (Lidsky 1983). Similar debates had taken place at the Joint European Torus (JET), an internationally sponsored fusion research project, over where the proper emphasis should have been.

Support of Research

Technologies in the research phase do in fact need financial support, even though they have highly uncertain prospects of success. Actually, it is precisely because of their uncertain prospects that at least some of the funding is likely to be public. Typically, few private investors are willing to put their money into ventures that seem to have little hope of a short- (or even medium-) term payback. Consequently, large research projects, such as fusion, have been funded primarily by governments; but governments, too, will often decline to expend their resources on the basis of mere promises of success from researchers.

The problem with government research funding, when results are slow in coming, lies in assuring its continuity on a long-term basis, to allow the possibility of ultimate success. The ongoing operation of research projects requires career commitments from the scientists in them, starting with years of graduate training in universities and continuing into professional specialization within the projects themselves. Large experimental projects, such as those related to fusion, also require cadres of highly trained technicians and design engineers. The construction of experimental devices demands contracting with specialized suppliers for custom-built equipment that often requires years to complete. Disrupted or unpredictable funding in this complex management of large research efforts has impacts far exceeding those of simpler organizations. Thus, we could say that research projects have virtually no chances for success without long-term continuity of support.

The long-term continuity of government funding is a political issue in representative democracies, where government expenditures are determined by legislation. Long-term commitments of funding are very difficult to legislate, because legislators are usually elected on a short-term basis and their perspectives are thus limited. Also, there are problems of accountability on the part of large projects that are given long-term commitments of funds for purposes that cannot always be clearly understood by the legislators.

These difficulties were evident in the attempt by the US Congress to set up a long-term fusion research, development, and demonstration program. The Magnetic Fusion Energy Engineering Act of 1980 (MFEE) laid out a fast-paced $20 billion program, but one geared to tangible results rather than scientific inquiry. It aimed first for breakeven experiments and was to end with a demonstration prototype of a fusion-power reactor, all within 20 years. Although the legislation had apparently achieved a step toward assured long-term funding, it did not start at what the critics felt was the necessary beginning – the basic science questions.

The legislative success was, in any case, short lived. Following soon after its passage in 1980, the need to "do something" on energy dissipated with falling oil prices and a large surplus of supply. In the meantime, it was soon clear that the MFEE would not achieve its programmatic timetable. Fusion funding was cut, although it did continue at a level to sustain some research (Grossman 2013).

Basic research has long been recognized as a public good produced by the scientific community and so worth government funding (Arrow 1962). In the past, the scientific community has had the freedom to set out research objectives, allocate resources and select those who will carry out the research projects. It is generally acknowledged that the scientists are seeking the fundamental principles and true essence of how things work in nature. Society has understood, implicitly, that in the long run the benefits from this unfettered activity will be more than its costs. While this is an idealization, it nonetheless has been an underlying understanding in Western societies since the earliest days of European science.

In recent years, this understanding seems to have been unravelling. Legislative bodies, especially the US Congress, have been more reluctant to allocate long-term support for research. Basic research has increasingly come under attack as too expensive and unnecessary and has repeatedly been cut (though not entirely cut out) from federal budgets. The cuts in nuclear fusion and high-energy physics research are prime examples. Some of these policy shifts have resulted from the short time perspectives of legislators, who are constantly under the pressures of the next election. But budget cuts have also emerged as a consequence of ideological battles over budget balancing and government priorities. And out of this turmoil, society seems to be losing its vision of the future where "science [is] the endless frontier" (Bush 1945). Still, since the mid-2000s the United States has actually increased fusion funding, providing research grants for domestic labs as well as for the multinational, multibillion dollar International Thermonuclear Experimental Reactor (ITER) to which the United States contributed about $200 million in 2014.

Long-term choices, like fusion, could fulfill society's almost unimaginable hope. If they work, they promise lasting solutions to our energy dilemmas, perhaps for the rest of human history, transforming society in a fundamental way. But we must recognize that it is uncertain that these technologies will ever fulfill our visions, even if billions of dollars are expended in their development. Indeed, the uncertainties are so encompassing that even if the technology is proven feasible, we cannot be sure now that it will not beget a new set of unforeseeable problems; in other words, we may find only after decades of effort that it is not *the* solution after all.

Such promise and uncertainty leaves a society such as ours with large questions: should we invest massively in these hopes or in more probable, limited realities? Should we make a great effort to see if we can obtain lasting solutions to all energy problems, or should we focus our efforts on medium-term and short-term technologies, even though they may only help us forestall the need for long-term solutions?

Of course, there is no need to make *either/or* decisions. Society may, and probably will, continue to seek ultimate answers even as it pursues stopgap measures. The mix of choices and the emphasis in energy policy may finally depend less on grand questions and more on immediate concerns. For example:

- How can we best manage fossil-fuel resources?
- Which new technologies will assist in creating a more equitable world distribution of energy resources?
- Are there attainable alternative technologies that will reduce the spillover effects of energy conversion, such as pollution and potential hazards of radioactive waste?
- Which technologies will best help eliminate the possibilities of major ecological disasters, such as climate change, deforestation, or nuclear proliferation?
- Which energy technologies can help the environment and also sustain worldwide economic growth?

New energy-conversion technologies offer different answers to these questions. If we consider the resource question, for instance, we can readily see that if a policy focused on development of synfuels and alternative fossil fuels, it would only extend the life on fossil resources. A more direct solution would be the adoption of *renewable resource* technologies such as biomass fuels (see Appendix C). Biomass fuels may be truly *renewable* – resources that replenish themselves. Over time, we might see systems of sustainable harvesting of energy crops. Once *biomass management* is accomplished, fossil fuels could be displaced for use in transportation, and one energy dilemma, the dilemma of fossil fuel spillovers, eliminated.

Any renewable or superabundant source could also solve the present problems of world resource distribution (see Chapter 5). Thus, if fusion could be successfully developed and made available, it would eliminate the problem of energy deficiencies in resource-poor regions of the world. The ending of energy resource deficiencies would be a major step toward economic development and the reduction of poverty. Fusion (as well as solar) offers the potential of powering development in a sustainable way. Present trends are producing deforestation, erosion, and pollution in many areas of the developing world. In the industrialized world, solar, fusion, and, to some extent, biomass energy technologies could reduce spillovers. Solar energy would appear to be the most environmentally safe. There are few spillovers apparent in solar production and operation. On the other hand, many industrial countries such as the United Kingdom and Canada are geographically and climatically poor candidates for widespread use of solar energy.

Eliminating the possibility of major ecological devastation may require the adoption not just of renewables or superabundant sources but only those alternatives that do not involve combustion. Biomass, for example, though renewable, poses a threat in two ways: (a) deforestation (which is already a concern in developing industrial countries such as Brazil) and (b) continuing emissions of carbon dioxide.

Our discussion comes down to this in the end: there is tremendous promise in developing long-term energy technologies but there probably is no panacea that solves every conceivable dilemma related to the production and consumption of energy. We can only try to develop what we can, in the hope that we will provide adequate answers to the questions posed above.

"Energy Justice": Present and Future

Underlying much of the discussion on alternative technologies and the direction of energy policy within and between nations are moral questions. Of course, policymakers must weigh the relative benefits from funding one program over another or one kind of technology over another or, even, technological development itself versus alternative uses of scarce financial resources. The choices must necessarily take into account ethics as well as technology.

But it is not only policymakers who need to take the ethical dimension of energy use into account. As we are all consumers of energy, our own behavior will have an ethical dimension. Part of the solution to the world's dilemmas connected to energy, it is argued, will involve our recognition of our own complicity in the environmental and social problems energy entails.

What exactly would this mean? For some observers, the moral dimension of energy implies maximal precaution and forbearance – a return to a more simple lifestyle, with low-impact technologies. Some environmentalists and scholars advocate the concept of "de-growth," which would require the developed world "to depart from the promethean economic growth paradigm and to embrace a vision of sustainable de-growth, understood as an equitable and democratic transition to a smaller economy with less production and consumption" (Martínez-Alier et al. 2010). Production would be small scale, environmentally benign, low energy, and materials intensive (Lovins 1977). For this to happen, however, there must be "systemic political, institutional and cultural change" (Kallis 2011). De-growth takes one outside a capitalist model and its advocates seem to believe there are ways to achieve such a radical reordering without massive disruption to existing social structures and economic relations.

Other experts advocate government policies that are more in keeping with the mixed (that is, market economies with strong government-provided social welfare components) capitalist traditions to protect the environment and more equitably distribute energy resources. Here, the emphasis is usually on government efforts to curb excessive energy production and consumption. Among the policy measures might be higher prices (i.e. taxes on energy consumption) and redistributive benefits (i.e. subsidies for conservation). As the Worldwatch Institute has argued (2013), governments have a major "role in curbing consumption excess, primarily by removing incentives to consume – from subsidized energy to promotion of low-density development."[11]

Ultimately, the goal of policy, from this viewpoint, would be justice – justice in the sense described by John Rawls (1971), whose work we first noted in Chapter 6. The basic principles of justice are, first, the equal right of all in a society to "the most extensive basic liberty," and second, equal fairness in treatment of all. If there are inequalities, then those with the least should have maximal opportunity to improve their situation, the "maximin" principle.

Operationalizing energy justice is especially complex given its global dimension. That is, consumption of energy in the United States has consequences felt through both space and time. With our (large) demand for fossil fuels, the price of energy resources rises; if the fuels are fossil or nuclear they are finite so less is available for others by definition; and finally, consuming fossil resources means that pollution increases (even if we are greatly curbing it). Depending on the nature of that pollution, our actions are potentially impacting our own people, people in other nations, people in the future, and the ecosystem of the globe now and tomorrow.

Benjamin K. Sovacool (2013) has outlined what he has called the eight principles of "energy justice."[12] These principles reflect ethical imperatives related to energy in the present along with the obligation of the present to future generations. As Sovacool notes, a choice to do nothing, or more to the point observe business as usual, is itself a choice that will impact the world – its people, climate, and environment. These principles provide at least a starting place for discussion of the ethical dimensions of energy.

The first principle is "*Availability.*" All people in the present should have access to "sufficient," what Sovacool calls, "high quality" energy resources and technologies. This is quite close to the views of many other scholars. Macer (2013), for example, argues "Access to energy is important for a reasonable quality of life. Many poor people [as we noted in Chapter 5] are dependent on traditional biomass fuels (wood, dung and so on) for their heating and cooking needs." A solution would be for them to have reliable continuous access to electricity. But the "solution" as Sovacool sees it is a combination of local "distributed generation (including renewables), energy efficiency and carbon taxes," which may seem a good few steps removed from the needs of the world's poorest.

One concern behind the concept of availability is the problem of market disruption, such as the oil embargo of 1973–4 or more recently in the threat by Russia to withhold supplies of natural gas from European customers.[13] In both instances, the real lesson is never to rely on a single supplier (or a few suppliers) for a key energy product. The Arab oil embargo had an impact in large part because the United States and other industrial nations received a disproportionate share of its oil from the Arab states of the Middle East. As the United States diversified its suppliers (it was importing supply from more than 60 different nations by the early 2000s) and received a large percentage of its oil from neighboring Canada and Mexico, the potential impact of an embargo has very much diminished.

"*Affordability*" is Sovacool's next "principle." He argues, "[A]ll people including the poor, should pay no more than ten percent of their income for energy services." This point is important given the fact that energy poverty is itself a serious problem and is usually coincident with general impoverishment. But the point extends beyond the developing world. The relatively poor in rich countries (in the United States for example an individual with income $11,670 or less is considered poor) are often hard-pressed to pay the power bill or fill up the gas tank. Sovacool argues that greater energy efficiency, presumably devices or structural improvements that are subsidized, will mean less of an impact on family budgets. Of course, in the developing world, even a few cents per day would be a burden to many and so electricity will have to be heavily subsidized, if not offered free to the poor, in order for all to benefit from its services.

Sovacool's third principle, "*Due Process,*" is one that seems in keeping with democratic traditions, but at the same time could lead to conflicts among these principles. By due process, Sovacool means that people should be involved in making energy policy decisions that affect their communities. People must be able to give "fair informed consent." But what if a community rejects programs that are suggested to solve availability? Most notably, what if communities reject renewables? There are many wind projects in the United States and other industrial nations that communities are seeking to block. Other efforts, carbon taxes come to mind, never have had much support in the United States, thus putting due process in opposition to availability.

Of course, one might argue that the key phrase above is "informed consent," that if people are aware of the facts as they really are they will endorse renewable energy and carbon taxes; but that leads either to an argument for paternalism over democracy, or for the rule of knowledgeable scientists and technocrats over that of citizens. Indeed principle four is "*Information*," but this is framed more in the context of transparency with respect to policy and curbing corrupt behavior of decision makers.

"*Prudence*" also potentially clashes with other principles. Prudence in this model means sustainability of natural resources; or that "countries have sovereign rights over their natural resources, that they have a duty not to deplete them too rapidly, and that they do not cause undue damage to their environment or that of other states beyond their jurisdiction." While this view is widely held, it is difficult to say at what point depletion becomes "too rapid." This was certainly an issue – a belief – that the United States faced in the 1970s, but the policy analyses were basically incorrect, or at least conflated ultimate recoverable resources with falling reserve levels.

Equity to current generations and to future ones is embodied in the next two principles: "*Intragenerational Equity*" and "*Intergenerational Equity*." "Intragenerational Equity" means that all people today are "entitled to a certain set of minimal energy services, which enable them to enjoy a basic minimum of wellbeing" (Sovacool and Dworkin 2015, p. 440).[14] It is a matter of distributive justice that in the modern world people should have access by right to food, shelter, unpolluted air, water, and, as others note, energy.

Our choices today on energy will of course have implications far into the future, affecting the welfare of generations to come, regardless of what we do. If we invest in one alternative over another, that choice will be a given for future generations; they must either accept it or undo it, where the cost of doing either may not now be known. If we choose no investment in alternatives at all, that too becomes a reality for the future. No present commitment on energy technology exists in isolation. We may leave the future with secure resources or not, cheap resources or expensive, growing spillover problems or diminishing ones, a permanent solution for our energy needs or only a greater need to find one.

But of course, this is a central conundrum posed elsewhere in this book. There are always *trade-offs*. If we ask people to sacrifice for the good of the future we are reducing their well-being today. If we adopt the view that we *will* have sufficient resources to carry us and our descendants far into the future (at least until something inexhaustible such as fusion is harnessed), then we are meeting intragenerational needs possibly at the expense of the intergenerational ones. That is, an intragenerational policy should be oriented to exploit resources for the present good. It is true that outcome of investment in alternatives is uncertain, but then so is the decision *not* to invest. Investment may fail and there may be hardship, but the supposition that no investment is necessary may be wrong. Then generations of the future will pay the price. Because they have had no voice in the decision today, the decision may be both incorrect and unjust.

The basic principle of intergenerational equity has long been recognized. The EPA, for example, has stated that future generations should bear no greater risks than those acceptable to the current generation. But how are we to judge? To Sovacool, the answer lies in an acceptance of the principle (his last) of "*Responsibility*." In his view this is broadly encompassing "responsibility of governments to minimize environmental degradation, a responsibility of industrialized countries responsible for climate change to pay to fix the problem (the so-called "polluter pays principle"), a responsibility of current generations to protect future ones, and a responsibility of humans to recognize the intrinsic value of non-human species, adhering to a sort of 'environmental ethic.'"

Sovacool's principles are intended to provide an outline for energy justice, but in the end whether such principles are embraced comes down to the will of millions of individuals – individuals who do not necessarily have similar views of the world and its prospects. But people might agree that we need to have a moral framework with respect to energy production and consumption, and that the absence of some framework does seem to speak of injustice.

References

Agar, J. 2013. *Constant Touch: A Global History of the Mobile Phone*. Icon Books, London.

Arrow, K.J. 1962. "Economic Welfare and the Allocation of Resources for Invention." In R.R. Nelson (ed.), *The Rate and Direction of Inventive Activity: Economic and Social Factors; a Conference of the Universities-National Bureau Committee for Economic Research and the Committee on Economic Growth of the Social Science Research Council*, Princeton University Press, Princeton, NJ.

Bhatia, R. and A. Pereira (eds.). 1988. *Socioeconomic Aspects of Renewable Energy Technologies*. Praeger, New York.

Bush, V. 1945. "As We May Think." *The Atlantic*. At www.theatlantic.com/maga zine/archive/1945/07/as-we-may-think/303881/.

Camilleri, J.A. 1984. *The State and Nuclear Power*. University of Washington Press, Seattle.

Girifalco, L.A. 1991. *Dynamics of Technological Change*. Van Nostrand Reinhold, New York.

Grossman, P.Z. 1992. "Alternative Energy Technologies: Market Solutions." In *Working Papers*, 23. Center for the Study of American Business: Washington University (St. Louis).

Grossman, P.Z. 2013. *U.S. Energy Policy and the Pursuit of Failure*. Cambridge University Press, Cambridge.

Hall, B. and B. Khan. 2003. "Adoption of New Technology," UC Berkeley Working Paper No. E03-330.

Kallis, G. 2011. "In Defence of Degrowth." *Ecological Economics*, 70, 873–880.

Kranzberg, M. and C.W. Pursell (eds.). 1967. *Technology in Western Civilization: Volume 2, Technology in the Twentieth Century*. Oxford University Press, New York.

Lidsky, L.M. 1983. "The Trouble with Fusion." *MIT Technology Review*, October. At http://fusion4freedom.us/science/The-Trouble-With-Fusion.pdf.

Lovins, A.B. 1977. *Soft Energy Paths: Toward a Durable Peace*. Ballinger, Cambridge, MA.

Lynn, L.H., N.M. Reddy, and J.D. Aram. 1996. "Linking Technology and Institutions: The Innovation Community Framework." *Research Policy*, 25, 91–106.

Macer, D. 2013. "Ethics and Energy Consumption," *World Social Science Report 2013*, International Social Science Council, UNESCO and the OECD co-publishers.

Martínez-Alier, J., U. Pascual, F.-D. Vivien, and E. Zaccai. 2010. "Sustainable De-growth: Mapping the Context, Criticisms and Future Prospects of an Emergent Paradigm." *Ecological Economics*, 69, 1741–1747.

Price, J. 1990. *The Antinuclear Movement*, Revised edition. Twayne Publishers, Boston, MA.

Rawls, J. 1971. *A Theory of Justice*. Harvard University Press, Cambridge, MA.

Rogers, E.M. and F.F. Shoemaker. 1971. *Communication of Innovation: A Cross-Cultural Approach*, 2nd edition. The Free Press, New York.

Rogers, E.M. 1995. *Diffusion of Innovations*, 4th edition. Free Press, New York.

Rule, T.A. 2010. "Shadows on the Cathedral: Solar Access Laws in a Different Light." *University of Illinois Law Review*, 2010, 851–896.

Smallwood, K.S. 2013. "Comparing Bird and Bat Fatality-Rate Estimates among North American Wind-Energy Projects." *Wildlife Society Bulletin*, 37 (1), 19–33.

Sovacool, B.K. 2013. *Energy & Ethics: Justice and the Global Energy Challenge*. Palgrave, New York.

Sovacool, B.K. and M.H. Dworkin. 2015. "Energy Justice: Conceptual Insights and Practical Applications." *Applied Energy*, 142, 435–444.

Susman, H.E. and M. Lund. 2011. *Lex Hellius – The Law of Solar Energy: A Guide to Business and Legal Issues*, 3rd edition. Stoel Rives Initiative, LLP, San Diego, CA.

Tabors, R.D., G. Parker, and M.C. Caramanis. 2010. "Development of the Smart Grid: Missing Elements in the Policy Process." In Proceedings of the 43rd Hawaii International Conference on System Sciences, Honolulu, HI.

NOTES

1 Examples of analyses of technology development and the social process may be found in Kranzberg and Pursell (1967), Girifalco (1991), and Agar (2013).

2 Some would argue that even now nuclear power is not commercially viable. That is, it cannot win in the marketplace without ongoing government support (Grossman 1992).

3 See also Bhatia and Pereira (1988).

4 Seven-year estimate is from Southface Energy Institute (2016), at www.southface.org/learning-center/library/solar-resources/solar-thermal-costs-paybacks-and-maintenance.

5 Much of this discussion is taken from Rogers (1995).

6 A *new technology* includes *any* innovation in the methods of accomplishing practical ends, whether it involves a working mechanism or not.

7 George W. Bush, State of the Union address, January 31, 2006.

8 Two cases in particular stood out. The first, Solyndra, made a novel kind of solar cell but it could not be produced at a commercially viable cost. The US government – and thus taxpayers – had guaranteed $535 million in loans, which became a taxpayer liability. The second, Range Fuels, was created to produce cellulosic ethanol, which it never could do in quantity. It went bankrupt in 2011 costing taxpayers over $150 million.

9 Adjusted for inflation this would have been about $250 billion in 2015 dollars.

10 Though the exact numbers are not known, it is estimated that wind turbines kill hundreds of thousands of birds and bats in the United States every year. This includes protected species such as eagles. Smallwood's (2013) estimate is that overall 888,000 bats and 573,000 birds die annually from wind turbines – numbers that are likely to grow as more wind turbines are constructed. This issue is also addressed in Chapter 11.

11 "The State of Consumption Today," At www.worldwatch.org/node/810.

12 This section draws heavily on the book by Sovacool, *Energy & Ethics* (2013). See the references.

13 Russia threatened this in response to protests throughout Europe to Russia's actions in the Ukraine.

14 Most of this section is taken from Sovacool (2013), but the book makes a peculiar switch of intra- and intergenerational equity, describing the former as equity across time, which is more typically described as intergenerational. The paper published with Dworkin (2015) uses the more typical forms.

Energy Transition

"A non-fossil future is highly desirable and eventually inevitable, but a civilization built on fossil fuels cannot make that transition easily or speedily." Vaclav Smil, author of more than 30 books on energy and the environment.

"[We took on the challenge] to determine how 100 percent of the world's energy for all purposes could be supplied by wind, water and solar resources, by as early as 2030. Our plan is presented here." Mark Z. Jacobson and Mark A. Delluchi, writing in *Scientific American*, November 2009.

Introduction

Beginning in the early 1990s, Germany decided as a matter of public policy to transition away from fossil fuels and nuclear power and embrace renewables, mainly solar photovoltaics (PV) and wind power, for the generation of electricity. This policy has come to be termed the *energiewende*, meaning energy transition.

The German turn to renewables was strengthened in 2000 by the Energy Sources Act, which set as a goal that 80 percent of all electricity would be generated from renewables by the year 2050. There were interim goals established later: 38 percent in 2020; at least 50 percent by 2030. The goal for the overall renewables component in the total consumption of energy (including transportation) was set at "above 60 percent in 2050" (Hirschhausen 2014). In other words, successive German national administrations have committed the country to a major shift in its energy technology; put a little differently, the German government intended to make non-fossil, non-nuclear electricity generating sources "commercial," simply by making them mandatory. This agenda is being pursued even though it was clear that from a market standpoint these technologies, while improving, were still not commercially viable without government support.

But could a complete transition to renewables even in the electricity sector actually be achieved by policy? And if so, at what cost? As the quotes above indicate, many experts believe a transition is possible and even "inevitable." But they are divided on whether it can happen quickly, that is, hastened by major government-driven initiatives, or if it must be allowed to happen gradually primarily through market forces. Experts are even divided as to whether the German model is a success or not. It is argued by some that "[by chasing] the goal of generating 80% of its energy from renewable sources by 2050, Germany has aggressively pursued a Green dream with unsustainable subsidies that have produced an unstable system described by [the *Financial Times*]... as 'a lesson in doing too much too quickly on energy policy.'"

Others, such as Jacobson and Delluchi quoted above, believe that Germany's *energiewende* is not ambitious enough, that Germany (and the United States too) could achieve 100 percent of their energy by 2050 from renewables: wind, water, and solar (WWS) alone.

Of course, part of the rationale for an all-renewable power system is to reduce CO_2 emissions and, in doing so, protect humans from the effects of climate change. Germany projects that the *energiewende* will reduce its emissions of carbon dioxide by 85–95 percent over the levels of 1990. In this regard Germany is not off to a good start. Because of the simultaneous shut-down of several of its nuclear power facilities, Germany has resorted to domestic coal for much of its electricity generation. Consequently, CO_2 emissions, which had declined from 1990 through 2009, rose in 2010 and stayed at a higher level even as more renewables were added to the grid. In Chapter 10, we discussed the general obstacles to the adoption of new technologies. Here we will discuss the specific issues connected to the two major alternative technologies of the *energiewende* – technologies most renewable advocates endorse – solar photovoltaics (PV) and wind. We will also consider the question of how such an energy transition might be implemented.

Intermittency and the Problem of Storage

Solar PV and wind have one crucial problem: intermittency. The Sun does not always shine even in Arizona or Saudi Arabia – not every hour of the day of course, but also not every day. The wind does not blow strongly or consistently 24/7/365, not even in the windiest locales on Earth. Cloudy days reduce or eliminate solar electric generation. Calm days leave wind turbines unable to generate a single watt. Moreover, wind can blow strongly one minute and die down a minute later. It can also blow *too* strongly, damaging wind turbines.

The central problem is that electricity needs to be available every minute of every day so that supply always equals demand. That means that suppliers have to plan to have enough power to meet the expected demand (or *load*), and they also must have electrical capacity in reserve in case of unexpected increases in demand. With conventional generation, there can be unexpected outages so that more power will be needed quickly, but more often the issue is that demand will spike, for example, on an unusually hot day when thousands of air conditioners are running full blast. Thus, there must be reserves to meet those contingencies as well. Often the reserves will be in the form of gas turbines because they can start up and begin supplying power quickly. A coal plant on the other hand, cannot make a cold start. The working fluid must be heated to make steam to run the generator. This process takes time; hours in fact. Still, sometimes a coal-burning plant will be kept offline but in readiness (burning fuel and reaching a high temperature, and incidentally producing CO_2 emissions) just in case more power is needed.

However, the situation changes with wind and solar. Of course, sometimes the wind will blow for hours and the Sun will shine from morning until night producing more electricity than is actually needed, which often means a coal-burning or nuclear power plant is taken offline.[1] A system of solar and wind can go from peak power to nothing and back up again. Load balancing becomes a delicate problem. As one system operator noted, supply can be erratic, with "40, 50, 60 percent change in the output in a very short amount of time."[2] If load is not balanced there may be spot outages or even a full-scale system-wide power failure.[3]

At the same time, if the excess electricity that is at times produced could be stored at low cost, the usefulness of solar and wind generation would be greatly enhanced. Electricity generated on a windy night or a weekend day (when demand tends to fall) of bright sunshine could be put away and restored on a calm or cloudy day.

There are of course ways to store electricity but they are either not effective or too expensive or (most often) both. Moreover, the scale has to be much greater than is typically available. It is true that battery arrays can store up to several megawatt hours, though at high cost. But to be truly effective a battery or other storage option would have to accumulate gigawatt days of electric power, not just megawatt hours, and be able to do so at relatively low cost. Advanced batteries can cost anywhere from about $150 per kilowatt hour[4] to more than $1000, but even at the low end this is far too high to provide backup for renewables. (We discuss some of the technical issues of advanced batteries in Appendix C.) At present the highest volume storage is in the form of *pumped hydro*. With this type of storage, excess electricity is used to pump a body of water from one place to a reservoir at higher altitude. When electricity is needed, the water is let out, using gravity to turn a generator, with the water collected in the original reservoir. Pumped storage facilities can provide up to a gigawatt or more and can provide power for a day or sometimes longer. But pumped storage is expensive (costing over $5000/kW but providing power only intermittently) and it cannot work without a significant amount of water. That is a critical problem because many of the best locales for wind and solar are in arid regions. Consequently, much of the backup for solar and wind is fossil-fuel fired or, when available (as it is sometimes in Germany), hydro.[5] Unless it is hydro, backup even defeats the goal of reducing emissions substantially. In general, until some form of battery or other advanced storage system becomes cost effective and of utility scale, it will be difficult for wind and solar to overcome the limitations of intermittency.

Grid Parity

The economic goal of solar and wind electricity is what is called "grid parity," meaning that the cost of each solar/wind kilowatt hour is approximately equal to the cost of gas, nuclear, or other conventional power source. It should mean, though it is not always framed this way, that parity is reached when different technologies have comparable cost characteristics without subsidies. This becomes a tricky exercise since some calculations include a hypothetical cost of climate change from carbon-based fuels, raising the cost not only of coal but also natural gas. It is also difficult because the calculations rarely are of directly comparable technologies; that is, it is misleading to compare "dispatchable" forms, such as gas-fired electricity, with unsteady renewables, which are not dispatchable. In any case, there are many articles

Table 11.1 Estimated levelized cost of electricity (LCOE) for new generation resources, 2020 ($2013)

Plant type	Capacity factor (%)	Levelized capital cost	Fixed O&M	Variable O&M (including fuel)	Transmission investment	Total system LCOE	Total LCOE including subsidy
Dispatchable technologies							
Conventional coal	85	60.4	4.2	29.4	1.2	95.1	
Advanced coal	85	76.9	6.9	30.7	1.2	115.7	
Advanced coal with CCS	85	97.3	9.8	36.1	1.2	144.4	
Natural gas-fired							
Conventional combined cycle	87	14.4	1.7	57.8	1.2	75.2	
Advanced combined cycle	87	15.9	2.0	53.6	1.2	72.6	
Advanced CC with CCS	87	30.1	4.2	64.7	1.2	100.2	
Conventional combustion turbine	30	40.7	2.8	94.6	3.5	141.5	
Advanced combustion turbine	30	27.8	2.7	79.6	3.5	113.5	
Advanced nuclear	90	70.1	11.8	12.2	1.1	95.2	
Geothermal	92	34.1	12.3	0.0	1.4	47.8	44.4
Biomass	83	47.1	14.5	37.6	1.2	100.5	
Nondispatchable technologies							
Wind	36	57.7	12.8	0.0	3.1	73.6	
Wind – offshore	38	168.6	22.5	0.0	5.8	196.9	
Solar PV[3]	25	109.8	11.4	0.0	4.1	125.3	114.3
Solar thermal	20	191.6	42.1	0.0	6.0	239.7	220.6
Hydroelectric	54	70.7	3.9	7.0	2.0	83.5	

Source: Energy Information Administration, 2015

on the web that claim parity has been achieved for one or the other (solar or wind) at least in some locations where the Sun shines regularly or the wind blows steadily. So, for example, Deutsche Bank of Germany claims that solar has achieved grid parity with any alternative in northern Chile, especially with respect to natural gas, which is imported in costly liquefied form. But even here one must remain somewhat skeptical since transmission costs, which can be quite high, are not included. In the United States, some have argued that grid parity has been reached with regard to some wind locations, and it is maintained by the EIA that solar will be at, or even past, parity by 2020 in places in the US Southwest. But, as Table 11.1 shows, solar overall (at \$125/MWh or 12.5¢/kWh) will not be close to parity, especially as against combined-cycle gas. Wind, according to this table, would however seem to be at parity or better especially against coal or new nuclear power.

Note that the technologies are divided between those that are *dispatchable* and those that are not. Dispatchability means that a generator can be turned on and off on command and power added or subtracted as needed to balance load. Fossil-fuel and nuclear power are dispatchable; solar and wind are not. This question impacts the assessment of grid parity for wind that Table 11.1 seems to imply.

According to some energy economists (e.g. Giberson 2013, Simmons 2015), the low cost of wind in Table 11.1 does not include the price of intermittency. As we noted in Chapter 9, the need for backup adds costs to wind and solar as does the need for new, often long, transmission lines. The conclusion was that instead of 7.3¢/kWh, the real cost will be just about double and therefore also double that of gas-fired combined-cycle power. The revised figure for wind means that it is also more costly than nuclear, conventional coal, and hydroelectric power. Put another way, grid parity will be spotty at best. For the most part it will still cost quite a bit more to have a system that relies heavily on solar and/or wind.[6]

Energy Density and Power Density[7]

The concepts of energy and power *density* are significant when considering any program for the development of solar and wind electric generation. Though both the sunlight and the wind are enormous energy sources as measured over the whole Earth, they are relatively light and diffuse in any one place. They are not *dense* – like fossil fuels and nuclear fuel – and that is an important fact.

The two measures of density can be applied to the same resources but it is mainly the latter, power density, that will concern us in this section. *Energy density* is the quantity of available energy in a mass of a given size – a battery, a gallon can of gasoline, a nuclear fuel core, or even a living organism. It is measured in energy units (typically joules (J) or watts (W), or megajoules (MJ) or kilowatts (kW) – see Appendix A) per unit of volume or weight – referred to as *volumetric* or *gravimetric* energy density. For most energy resources, one can use either measure or both. That is, gasoline's gravimetric density is about 44 MJ/kilogram, and volumetric density of 35 MJ/liter. In considering electric power resource densities, often watt hours or kilowatt hours are used instead of joules, where 1 kWh equals 3.6 MJ. Thus, the gravimetric density of gasoline can be written as 12.2 kWh/kg. It should be noted that the energy densities can vary even within certain resource groups. The different varieties of coal have different calorific values. Of course there can be vast differences between resources. For example, while the volumetric energy density of gasoline is almost double that of coal, it is over 200 times that of a standard lead acid battery.

Power density – the time rate of energy transfer per unit of area, volume, or mass – is a vital metric when considering electric power sources. Power density of electric plants (including renewables) is often measured by the horizontal area required for the entire installation, that is, the power available per square meter (m^2)/kilometer (km)/mile.[8] This is often referred to as the *footprint* of an electric power facility. Nuclear reactors, for example, though typically sited on fairly large tracts of land (one to two square miles for safety reasons), nonetheless, have generation capacities of 1000 MW or more and power densities on the order of 250–900 W/m^2.

Solar and wind are *diffuse* energy sources. In other words they lack high power densities. Consider a single wind turbine: Let us say it is rated at peak power of 3 MW. It sits on a massive tower (125–150 meters high or around 400–500 feet) and has propeller-like blades turned by the wind. The turbine blades sweep an area of around 2.5 acres (1 hectare). Because wind turbines will cause air turbulence in the immediate vicinity the towers must be placed well apart. The average area of a wind farm is, according to the DOE, 85 acres per MW of rated capacity.[9] Thus, the total horizontal area of every turbine must be calculated to determine the power density of a wind farm. In general, depending on the location, but using the DOE's acreage figure, wind turbines have at full capacity a flow of just under 3 watts per square meter (3 W/m^2, sometimes referred to as the "capacity density") of area. But to calculate the *power* density, one must adjust the figure by the capacity factor, which is about 32 percent on average for wind, meaning that in this case, the power density would be about 0.9 W/m^2, about a 250th or less of a nuclear power plant.[10] Solar is somewhat better. Rooftop PV panels, for example, have a power density of 12–17 W/m^2, still well below that of conventional electric generation.

Now consider what this means in terms of a transition to an electric grid dependent largely on solar and wind. How large a wind facility would one need to generate the same amount of energy in a year as a single nuclear plant? Take an actual nuclear plant such as one in Michigan. It is sited on 2 square miles of land and produces 1150 MW, at about a 90 percent capacity factor. Its output is about 9 million MWh per year. To get the same output from a wind farm of 3 MW wind turbines might seem to be a simple problem of arithmetic: 1150 MW times 85 acres means 152 square miles of horizontal area on which one would put up 1150 divided by 3 or about 380 wind turbines.

That is very large tract of land; but wait. The capacity factor of wind-powered electric generators is only about one-third that of the nuclear power plant, so to get 9 million megawatt hours of electricity one would need about 1140 turbines and a wind farm of about 450 square miles, an area roughly the size of the cities of New York and Denver combined. Replacing three large nuclear plants with wind farms would mean using an area greater than the state of Rhode Island. To replace all existing coal-fired electric generating plants in the United States with wind farms would require, according to one estimate, an area the size of Italy (that is, about the size of the state of Arizona; Bryce 2015).

It should be noted that wind installations leave a lot of open ground for crops or cattle but they cannot be sited in populated areas, where there are natural or manmade impediments to the wind and, of course, where the winds are inconsistent. Furthermore, there has been something of a backlash against wind-energy installations in recent years. For instance, in Germany, the state of Bavaria has passed a law that effectively bans many wind projects by mandating large setbacks from residential areas.

Solar has a greater power density than wind but, unlike wind, utility-scale solar arrays do not lend themselves to other uses. In fact, utility-scale solar PV would also involve very large spaces. Below we note an array that provides 17.5 MW of peak power covering an area of 76 acres, or a bit over 4.3 acres per MW. Again, to scale up to 1150 MW would mean almost 5000 acres (or 7.8 square miles), except that capacity factors for solar are lower than for most wind facilities, so one would need about five times the area to generate the same amount of electric power per year, that is almost 40 square miles of solar panels – double the area of the borough of Manhattan in New York City.

Does that disqualify it from consideration? Not necessarily. As noted, wind farms often allow for other land use beneath the high towers and solar is distributed on existing rooftops. But low density is itself another barrier for the widespread use of solar and wind.

Below we consider the technical and economic issues of solar and wind.

Direct Photovoltaic Solar Conversion

Solar Technology

Direct conversion of sunlight to electricity is a very attractive technological prospect. It is pollution free and uses the renewable energy of the Sun. It would also eliminate

Figure 11.1   A solar photovoltaic installation.
Source: Getty Images/Westend61.

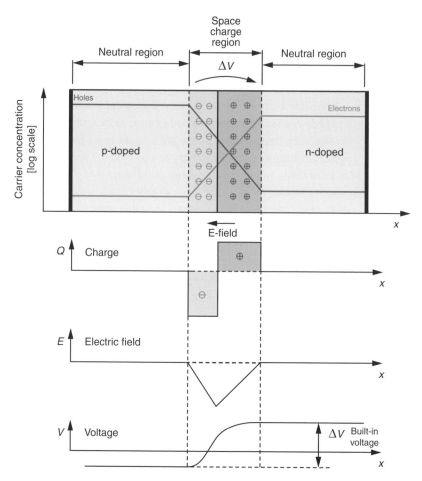

Figure 11.2 A p-n junction, in thermal equilibrium, with zero bias voltage applied.
Source: Wikipedia: http://en.wikipedia.org/wiki/Pn_junction.

all of the intermediate processes of thermal plant generation, which are necessary for all technologies collected in the form of heat.

Direct conversion can occur through several different physical mechanisms, including photochemical processes, the thermoelectric effect, and the photovoltaic effect. Of these alternatives, *photovoltaics* (PV) have received the most attention and development. The photovoltaic effect was first observed over 150 years ago by A.C. Becquerel, but was not fabricated into a useful device until the 1950s, when it was first used in the space program. Today, photovoltaic conversion is a working technology, but its major barrier to widespread adoption is its cost relative to performance.

Photovoltaic conversion cells are semiconductor, junction-type devices. *Junction* refers to the region within the material between volumes that have *doping* of oppositely charged (p- and n-type) carrier atoms, which produce an electric-potential barrier within the crystalline material (see Fig. 11.2). They are able to supply electric currents when illuminated by sunlight, converting the incident solar energy into electricity directly with

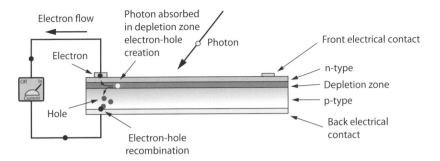

Figure 11.3 Photovoltaic cell, with electric bias applied.
Source: Images SI Inc.

efficiencies ranging up to a theoretical limit of about 30 percent (which limit corresponds to the ideal Carnot limiting efficiency for thermal conversion - see Appendix A).

Photovoltaic electricity is created in the semiconductor material when a photon of sunlight is absorbed into the semiconductor and energizes a charge carrier (an *electron* or a *hole*). This energy enables the carrier charge to overcome the electric potential barrier at the p-n *junction* within the cell. Junctions are a feature of many semiconductor devices, including transistors (Wenham et al. 2007). They are created by *doping* two adjacent volumes of the semiconductor with two different impurities (Fig. 11.3). On one side, diffusion creates an excess of negative carriers (mobile electrons), while the other side has an excess of positively charged mobile carriers (so-called *holes*). The transition region between the two sides is the *junction region* and it contains an *electric potential barrier* caused by the fixed ions of opposite-charged polarities in the two adjacent volumes.

If a mobile charge carrier (electron or hole) gains sufficient energy from an incident solar photon to overcome the potential barrier at the junction, then the carrier is in a position to do useful *electrical work*. This follows from the physical principles outlined in Appendix A and has its mechanical analog in hydropower, where a mass of water is recognized to have gravitational potential energy if it is raised to a higher height.

The efficiency of PV conversion depends on two major factors: optical losses and the *height* (electric potential magnitude) of the junction barrier (measured in *electronvolts*, see Appendix A). Optical losses, such as reflections from the cell's surface, can be reduced by proper optical design, whereas the conversion efficiency of carriers across the junction barriers presents a fundamental limitation.

These limitations on PV conversion efficiency are due to an inability of the semiconductor junction to provide for the conversion of the entire *optical spectrum* of incoming sunlight. An optimum photovoltaic conversion takes place when the energy of an incident photon is *just equal* to the energy difference of the barrier junction, thus absorbing *all* of the photon's energy in taking the charge carrier over the barrier. If the photon energy is less than the barrier energy, then no conversion takes place. If it is greater, then part of the photon energy is not converted and therefore is wasted.

The entire spectrum of sunlight photon energies spans a 10-to-1 range. It is known that PV conversion will only be optimal for one narrow range within that spectral spread. Mass-produced, *single-junction* silicon PV cells – the cheapest and easiest to manufacture – generally convert no more than half of the 30 percent, theoretically possible, incident solar energy into electricity. R&D efforts sponsored both by government and industry have been directed toward improving the efficiency of silicon PV devices, since this is one of the most important factors determining the cost of solar-converted electricity. An important direction for this research, also, is to find other semiconductors, such as gallium arsenide, that have higher theoretical conversion efficiencies, and work on reducing their cost (Razykov et al. 2011).

Technical progress on conversion efficiencies has been reported in recent years, on Si, CdTe, and $Cu(In,Ga)Se_2$ (copper indium gallium diselenide) PV cells, in either mono- or polycrystalline, thin-film forms, reaching up to 20 percent efficiency (Goswami 2015). Important cost reductions have also been projected for both mono- and polycrystalline Si in wafer form, reaching down near $1.50 per watt (peak power) and $0.76 per watt for Cd cells.

Multi-cell collections may be arranged in a module or large array (see Fig. 11.4). The cells can be interconnected into series and/or parallel circuits to yield the design objectives of voltage and current outputs. Each cell in an array develops its own small voltage and supplies current. The cells are interconnected in the array to supply an overall voltage and current for direct use or for storage in batteries. An improved form of solar cell called the *tandem cell* (Green 2006) has been developed and gives much improved performance. In this version several junctions are fabricated together with varying barrier potentials. These tandem junctions are able to cover more of the solar spectrum, since each cell in the tandem cell absorbs a particular portion of the spectrum.

Solar Economics[11]

To properly discuss the economics of solar, one must divide the market into two parts. First, there is small-scale solar, in which solar panels are affixed to the roof of a home. These can power some of home use and on occasion produce more electricity than is needed and so the power is sold back to the grid. Second, there are large PV arrays, as in Figure 11.4, requiring fields of panels, in some cases at peak power producing more than 500 MW.

Present-day applications of PV technology are limited by capital costs and solar availability. Northern latitudes simply do not receive enough sunlight – termed *insolation* – to justify the cost, even though those costs have come down dramatically.[12] The costs to manufacture PV cells have fallen steadily over the past 35 years. In the early 1980s, the cost was about $16 per watt of capacity. As of 2016, the cost is less than $1/W, generally about $0.75/W. Some developers claim they will soon be producing cells for as little as $0.55/W.[13]

But the cell cost or even the combined cells of a panel do not constitute the total (or even the major) cost of installing a rooftop system. Far more important are the grid connection and all the metering and wiring that that entails. One major cost is the inverter, which turns the direct current (d.c.) that comes from the panels into the alternating (a.c.) current that is standard in all homes and the grid. Once these costs are factored in the cost rises to an average of over $4/W. Of course, the cost will vary with roof size, number of panels, and desired electrical output, as well as connection

Figure 11.4   A solar photovoltaic array at the Indianapolis International Airport. This 76 acre array has a capacity of 17.5 MW.

costs. On average the price of a rooftop PV system will be somewhere between $15,000 and $30,000, but just about every state in the United States offers incentives – tax rebates, high guaranteed prices for any excess electricity generated, and so on – to encourage solar. In some states, the average out-of-pocket cost to the homeowner will be less than $10,000. Nonetheless, depending on the insolation it can take up to a decade (or more) to recoup one's investment through lower net electricity costs.

It should be noted that one effect of rooftop solar today is something like a regressive tax (Brown & Bunyan 2014). Because utilities have to pay homeowners at the retail (or some high fixed) rate for the electricity they generate above their needs, and because the utilities also must receive a "fair" return on their capital, they need to raise electric rates. But higher rates fall most heavily on people who cannot afford a solar PV system or, indeed, people who cannot afford a house. That is, the brunt of this kind of incentive system disadvantages the poor.

Utility-scale projects are also falling in price and they have benefited from federal government programs. Though there are reports from time to time in the press that solar PV has become cheaper than coal or even natural gas,[14] these claims ignore the cost of intermittency, which requires maintenance of backup, and reports will also downplay the cost of connecting to the grid. Many large-scale solar sites are – understandably – in lightly inhabited desert areas that are far from main transmission facilities, thus adding substantially to the cost of a large PV system. Still, some large-scale PV projects do make economic sense – where sunlight is intense and the site is too remote – e.g. an island – for any larger grid connection. With development of low-cost storage, more of these kinds of projects would become cost effective.

In addition to cost reductions from cheaper storage, costs could also fall further depending on improvements in semiconductor technology and manufacture. New developments are being sought to permit the manufacture of large quantities of PV cells cheaply in thin sheets at efficiencies close to the limit. These objectives may be reached only through development of new semiconductor materials and techniques of manufacture.

It should be noted that there is a technique that raises efficiencies significantly: *concentrating PV*. The idea is to concentrate the sunlight to anywhere from 100 to 10,000 times its incident intensity onto the solar cell. By such condensing of the solar energy, it is argued, the utilization of the cell per unit of area can be much higher, because the amount of energy converted will be greater. But the per-unit cost of fabrication of such cells will have to fall significantly for commercial viability.

Solar Thermal Power

Most of recent interest in solar electric power has focused on PV, which has come down in price sufficiently so that, as noted earlier, in some parts of the world it has come close to grid parity. There is an alternative means of using the Sun's energy to generate electricity: focusing the thermal energy onto a point to boil water, to generate steam, and thus turn a generator. There are other ideas of how the thermal energy from the Sun might be utilized to generate electric power, but the basic idea is to concentrate the sunlight onto a point source and let the heat energy generate electricity. This approach seems farther than PV from actual commercialization. As Table 11.1 showed, solar thermal is expected to cost almost double (per kWh) that of PV as of 2020.

Still, some large demonstration projects have been undertaken. The most notable – and largest – project is the 377 MW BrightSource Corporation facility in the Ivanpah Valley of California's Mojave Desert. The solar plant, known as Ivanpah, covers 5.6

Figure 11.5   The array of heliostatic mirrors and one of the receiver towers at the Ivanpah solar thermal generating facility.
Source: Steve Proehl/Getty Images.

square miles of desert and uses 352,000 mirrors (see Fig. 11.5) to concentrate light onto three receivers, in this case three boilers placed atop 459-foot towers, which in turn power electric generators. The mirrors are *heliostats*, that is, each is oriented in elevation and azimuth to reflect the Sun's rays into a receiver at the top of the tower. These mirrors track the Sun through its daily trajectory in the sky, reflecting the solar rays into the receiver.

The project has been controversial. Some habitats of endangered species have been disturbed; birds are killed in flight from the heat. But the plant has also performed poorly, according to analysts. It was expected to produce 940,000 MWh of electricity, but during the year from July 2014 to June 2015 it produced only a little above 550,000 MWh.

Moreover, and most unexpectedly, it emitted 46,000 metric tons of CO_2. The reason for this is that it takes a long time for the sunlight to be sufficiently strong to get cold boilers to produce steam. Consequently, every day natural gas is burned to pre-heat the water as well as to continue heating on cloudy days (which do occur even in the desert).

Also in the spring of 2016, Ivanpah revealed the danger of misaligned heliostats. Instead of boiling water, misalignment led to a fire that shut down one of the generators. This came about just as reports were released that indicated that the electrical output from Ivanpah was improving. But with so many issues to be resolved we should consider such power systems to be in need of considerably more development.

Wind Power

Wind Technology

Wind energy has been used throughout recorded history. Windmills, of one sort or another, are known to have provided pumping and milling energy in the earliest Mediterranean and Eastern civilizations, and wind energy was of crucial importance in Europe even into the Industrial Revolution. Ocean-going ships still used wind (sails) predominantly through the middle of the nineteenth century. With the development of technologies that used fossil fuels in the eighteenth century and electricity in the nineteenth century, wind power was employed less and less often. While the multi-bladed windmill (used as a water pump) was still a familiar sight in rural areas of America, by the mid-twentieth century wind power played little role in the economies of any of the world's industrial countries. Beginning with the energy crises of the 1970s, however, interest in wind power was revived, and by the 2000s it was clear that it could begin to make a valuable contribution to the production of electricity worldwide.

Today, modern wind-power generators have been developed to supply electricity to utility grids and remote-site users around the world. Use of wind-generated electric power, though expanding, does face the limitations of the wind's intermittency and its lack of power density. Wind-generated electric power is said to be *available* annually in many regions worldwide, meaning that the time durations during which useful electricity output can be produced over a year's time in that region may, given the incentive

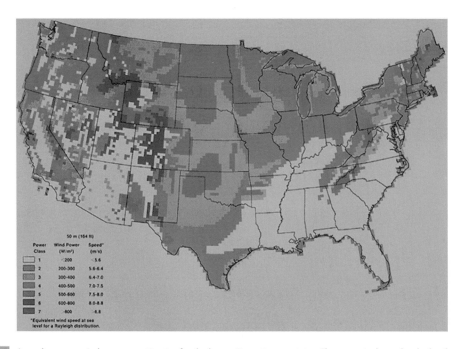

Figure 11.6 Annual average wind resource estimates for the lower 48 contiguous states. The range is shown by shade of gray–black, from lightest (lowest wind speeds) to darkest (highest speeds). The range of speed is from less than 12.5 mph (5.6 m/s) to greater than 19.7 mph.
Source: Wind Energy Resource Atlas of the United States, National Renewable Energy Laboratory (NREL), m&p 2–6.

system, be cost effective for investment. The wind is deemed available if its speed is consistently sufficient to produce useful power output, but not so very strong that it is destructive of the machine itself. "Good" wind sites are those where the annual average wind speeds are at least 13 mph, as indicated in Figure 2.12 for the United States.

This overview for the United States is shown in more detail (equivalently for comparable annual average speeds 5 meters/second [m/s] – which is about 11 mph – or more) in Figure 11.6. Similarly, Figure 11.7 shows annual average wind speeds in Europe, both *onshore* and *open sea*. Favorable regions in the United States are along the coasts, in mountainous terrains, and in the Great Plains. Similarly, good average wind speeds are found in Europe along the coasts and in surrounding areas of open seas.

"Wind farms," consisting of many wind turbines, have been constructed in North America and Northern Europe. The concept behind wind farms not only includes the summing of individual turbine outputs but also includes the *diversity* of outputs from individual turbines, each from its own location in the farm area, giving a summed output which may be smoothed out in time. The largest wind farm in the United States is the Alta Wind Center in Kern Country, California, in the Mojave Desert (Fig. 11.8). It can, at peak power, generate 1547 MW, the output of a large nuclear power plant. But its annual output is on average 2680.6 GWh, which is only 20 percent of what its output would be if it ran constantly at full power for a year. Of course, not even a nuclear plant manages that degree of reliability, but it still will on average produce about three-quarters more electricity in a year.

Wind Power Basics

The modern wind turbine has a set of blades, which look much like aircraft propellers, which drive the shaft of an electric generator. These blades are designed on the aerodynamic principle of *lift* similar to that of aircraft wings or sail boats (tacking up wind). Lift forces are created on a tapered aerofoil (tear drop in cross section) when its leading edge is oriented at a small angle to the direction of the incoming wind. Lift may be interpreted as due to a difference in air pressure between the two sides resulting

(a)

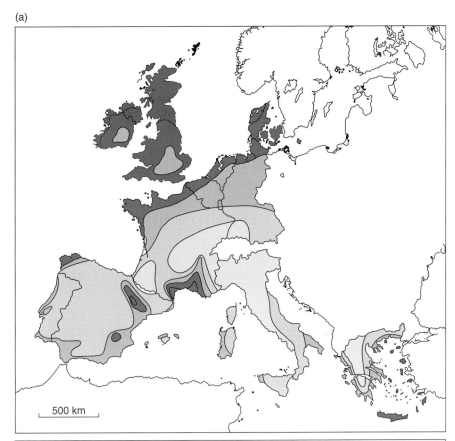

Wind resources at 50 metres above ground level for five different topographic conditions									
	Sheltered terrain[2]		Open plain[3]		At a sea coast[4]		Open sea[5]		Hills and ridges[6]	
	ms^{-1}	Wm^{-2}	ms^{-1}	Wm^{-2}	ms^{-1}	Wm^{-2}	ms^{-1}	Wm^{-2}	ms^{-1}	Wm^{-2}
	>6.0	>250	>7.5	>500	>8.5	>700	>9.0	>800	>11.5	>1800
	5.0–6.0	150–250	6.5–7.5	300–500	7.0–8.5	400–700	8.0–9.0	600–800	10.2–11.5	1200–1800
	4.5–5.0	100–150	5.5–6.5	200–300	6.0–7.0	250–400	7.0–8.0	400–600	8.5–10.0	700–1200
	3.5–4.5	50–100	4.5–5.5	100–200	5.0–6.0	150–250	5.5–7.0	200–400	7.0–8.5	400–700
	<3.5	<50	<4.5	<100	<5.0	<150	<5.5	<200	<7.0	<400

Figure 11.7 European Wind Atlas: (a) Europe Section, October 13, 2005 update; (b) European wind resources over open sea, December 29, 2009 update.
Source: The World of Wind Atlases.

(b)

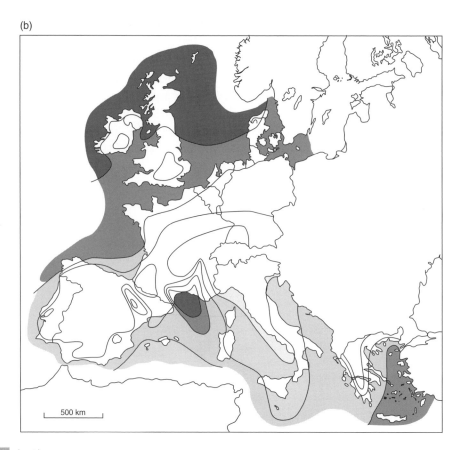

500 km

Figure 11.7   (*cont.*)

from the differing path lengths the air must take in flowing around the foil (see Fig. 11.9). The lift force turns the windmill blades for useful power output.

With lift forces on the blades turning the shaft, the mechanical power gets coupled to the generator. This is done through a geared system of shafts, as indicated on Figure 11.10. This system contains controls to correct for speed variations and *yaw* (back and forth) fluctuations resulting from the variations in speed and direction of the wind source (Fig. 11.10).

The power output of a wind generator (Gipe 2004, Manwell et al. 2009) varies in direct proportion to the area of interception of the windmill, which is approximately the projected area swept out by the rotor blades. The maximum power output of a windmill, as limited by its swept area, ranges up to about 3 MW for a 90-meter (nearly 300-foot) diameter rotor. This corresponds to converting the energy in a column of air moving at 15 m/s or 33 mph by an intercepted cross section of over 6000 m^2 (about 65,000 square feet).

As we might expect, it is not possible to convert all of the energy that exists in a given moving column of air. Aerodynamic theory, for example, shows that the maximum power that can be extracted from such a wind stream is 59 percent of the power flowing through that cross section (Gipe 2004).

Figure 11.8   Alta Wind Center in Kern Country, California.
Source: www.energy.ca.gov.

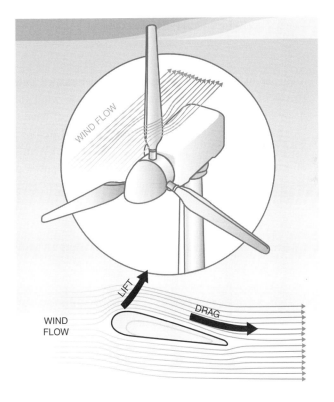

Figure 11.9   Wind power: turbine dynamics, showing *lift*.
Source: science.howstuffworks.com

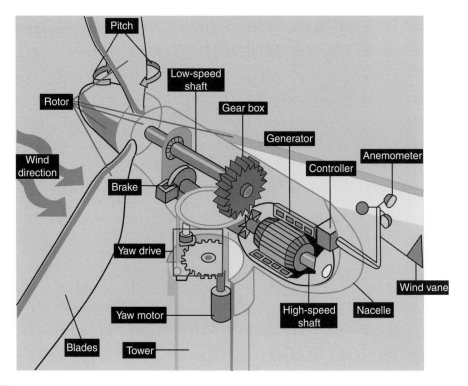

Figure 11.10 Wind turbine: internal view, showing gears and controls.
Source: Wikipedia Commons File: EERE illust turbin.gif.

So far in our discussion we have either related output power to specific wind speeds or to a rated power capacity. However, wind speeds are variable, indeed in many locations highly variable. The measures of wind machine energy production, like measures of solar energy systems, must therefore be in terms of an expected average rate of energy production (expressed in kW or MW). As noted earlier, this information is typically presented as the average power per unit of area, measured in average watts per square meter (W/m^2) for a given location.

Another way to measure the expected output of windmills, however, is to indicate the number of hours per year they are available to deliver a certain power output. This measure is essential because of the variability of the wind – not only are there periods when the wind does not blow, there are times when it blows *too* hard. In such conditions, all machines must be shut down, e.g. during hurricane or gale-force winds. This is done by *feathering* the blades into the wind, so that no lift forces are exerted on the blades. Therefore, when either the wind is calm or is blowing too hard, the machine is *not* available to deliver power. Conversely, when the wind is blowing in the usable range of velocities, the machine is said to be *available* and the number of hours per year that the wind machine can be expected to deliver electricity output is its *availability*.

Figure 11.11 shows the worldwide regions where a wind machine will deliver its upper-limit, rated power output. The numbers shown in the key, starting at values

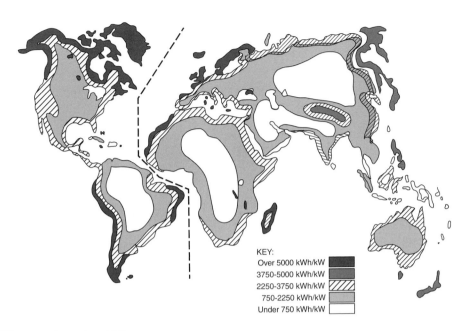

Figure 11.11 Worldwide wind availability.
Source: Mitre Corporation (1975), for the National Science Foundation, Wind Machines, NSF-RA-N-75-051. US Government Printing Office, Washington, DC.

under 750 kWh/kW and going to over 5000 kWh/kW, are equivalent to hours per year in which the machine is available to produce its rated output power. Thus, the best performance shown is over 5000 hours (57 percent of the 8760 hours per year) and the worst shown is under 750 hours (8.6 percent). No matter what measure is used, however, it is clear that only certain regions in the world are promising for wind generation.

The consequences of this variability of the wind, limiting the annual power output of a wind turbine, are clearly *operational*, in that the output power is not steady. The other consideration is *economic*, in that the return on investment for the turbine is limited to income from power less than its full output throughout the year. The fraction of the actual energy output over the year, divided by the full rated power output for the year is the capacity factor. This factor is not only an operational measure, but, as discussed in Chapter 9, is used as a determinant of the revenue returned on the investment. Capacity factors for typical wind farms run in the range 20–40 percent (with about 32 percent as the US national average), which could be compared with 70–75 percent for baseload coal-fired plants and close to 90 percent for nuclear plants in the United States.

There is another consideration. The wind data given above are on a broad *regional* basis. Actually, wind speeds depend on the local conditions (Gipe 2004, Manwell et al. 2009), such as terrain, buildings, and vegetation, and thus the performance of a wind machine is said to be very *site specific*. Siting guidelines, accounting for wind flow over hills, buildings, tall trees, etc. can be specified in terms of the obstacles' relative heights and spacing or simply in dimensions (in feet or meters). In general, they allow for spacing and heights, and leave ample room for free flow of the wind, down-wind as well as up-wind.

Operating Experience

A number of operational problems became evident when early wind turbine demonstrations were carried out in the late 1970s. Blades failed under the stresses of high wind speeds. There were fatigue failures in the turbine machinery due to the repeated stresses of fluctuating wind speeds. Many of these difficulties resulted from attempts to build very large machines, some with blade diameters up to 300 feet (90 m), mounted on towers nearly 350 feet (110 m) high. Some of these support structures failed because of the yaw and *teetering* motions that occur incessantly in turbulent winds. Wind machines are generally designed to operate with velocities up to about 35 mph, but must also be able to survive gales of 100 mph (in "stowed" or feathered positions).

In response to these mechanical failures, designers fashioned much lighter rotors, gimbaled mountings and more flexible supporting structures. A second generation of smaller wind generators, incorporating the new design features, was developed in the 1980s. These turbines had blade diameters ranging from about 15 to 75 feet, with output powers ranging correspondingly from about 10 kW to around 500 kW. Operating availabilities over 95 percent and annual capacity factors approaching 40 percent were attained using these intermediate-capacity machines. Typically, they are operated in wind farms, which benefit from centralized monitoring and control of arrays of wind machines.

This second generation of wind machines was installed and operated privately, with the aid of government financial incentives, rather than as a wholly sponsored federal demonstration effort. At wind farms built in recent years, machines have operated with capacities running up to (and sometimes more than) 3 MW, using light-weight, high-durability, fiber-glass construction, with much improved performance over previous designs. The promotional incentives were largely in the form of tax credits, as well as operating guarantees, under rules established under PURPA (see Chapter 9), the 1992 Energy Policy Act, the Energy Act of 2005, and programs that came out of the spending package meant to help revive the US economy after the financial crisis and "Great" recession of 2008–9 (discussed below). Still, it is noteworthy that wind farms in the United States are privately owned and maintained by American companies, such as NextEra Energy Resources, the largest wind farm operator in the country. Incentives for wind and other forms of renewable generation have been introduced throughout the world. China actually has the largest wind power capacity of any country – about 100 GW. However, in 2013 wind accounted for less than 3 percent of China's total electric power production.

Wind Farm Siting

The Altamont Pass in California was the location of the first large-scale wind farm in the United States; it opened in 1981. Winds there are consistent and strong due to the "sea breeze," caused by differential solar heating of the land versus the water near the coast. The breeze is further enhanced by being channeled through the pass, which is in the hills east of San Francisco Bay. As Figure 11.11 shows, the highest wind availability is typically in coastal regions (because of the land–sea effect) or (as shown on Figure 11.6) in high mountains. But in the United States there is also significant wind availability in an area stretching from northern Texas to the Canadian border.

Promising siting areas are found especially in the flat or gently rolling prairie crop lands of Kansas, and both the crop lands and semi-desert regions of Oklahoma and of north and west Texas. The wind generating capacity of the region as of 2015 was almost 27,000 MW. In most of these areas, dual use of wind farms with agriculture – successfully pioneered in California – has been possible. With such dual use, land owners will be compensated directly with income from land rents for every acre occupied by wind generators, in addition to their income from agriculture.

The northerly section of the Great Plains has also had major development of wind resources. As of 2015, 8000 wind turbines rated at 12,000 MW had been installed in six states: the Dakotas, Minnesota, Montana, Iowa, and Nebraska. There have been expectations that the capacity in this region would at least double by 2030. The wind velocities there are not very high, only moderately high, but they are consistent, and this has encouraged wind farm development. The Upper Great Plains is not only technically feasible for wind farm development, it is more socially acceptable than in more heavily populated regions of the United States. *Social acceptability* is an important element for alternative technologies. Wind farm development includes considerations of land use, visual impacts, noise, communication interference, and bird (and bat) kills. A rural siting mitigates some of these concerns, where joint use of the land with agriculture tends to compensate for impacts to those exposed.

Counting only those rural areas that qualify as socially acceptable, the total area usable for wind-power generation in the Upper Midwest is over 300,000 square miles. If this land area were used for wind generation, it could produce electrical energy over ten times the annual demand for the region. It has been estimated that 90 percent of the nation's (cattle) range land, 70 percent of the agricultural land, and 50 percent of forest ridges could be available for wind-energy production, without hindering their basic uses.

A concern remains about bird and bat kills, however. Creatures do fly into the revolving turbine blades and kills have aroused controversy in some of the existing California wind farms. As noted in Chapter 10, studies indicate that hundreds of thousands of bats and birds (including 83,000 raptors, such as the legally protected golden and bald eagles) are killed annually by wind turbines. The number is probably higher given that wind turbines have been added regularly during the decade of the 2010s, in fact, increasing in terms of megawatt capacity 16-fold since 2001. By 2030 mortality is expected to be in the tens of millions of birds and bats killed each year. In theory, raptor kills – especially of eagles – should lead to hefty fines for wind farm operators, but the federal government has waived the rules for renewable energy companies.[15] One should note that feral cats and building collisions kill far more birds annually, the latter more than 100 million each year. Still, wildlife groups fear that wind turbines will "ensure extinction certainty"[16] for many endangered avian creatures, especially bats. Bat kills have important ecological implications. Bats are mostly insect-ivores and the elimination of the bat population means an increase in agricultural pests, which in turn means increased use of pesticides or poor harvests (Boyles et al. 2011).

Another consideration for wind farm siting is the relationship of the locations of the turbines to the intended user or distribution network, usually a utility grid. Not only is proximity important (for *transmission* which is costlier the farther away the wind turbine is from the grid connection), but there is the additional problem that the wind-generated electricity may not be available when it is most needed – at times of peak demand. In the case of Altamont Pass and the other California wind farms, the diurnal rise of sea-breeze winds, as the Sun rises, creates an excellent match with daily electric demand of the utilities along the California coast. A similar coincidence has been forecast for at least one utility in the Upper Midwest. Whenever such a situation exists, more of the installed wind (intermittent) capacity can be counted as equivalent to installed conventional capacity.

We must reiterate that if wind generation at a particular site cannot match the high load demand it is to serve (which it cannot do reliably year-round), then energy storage will be required to make wind energy available when it is needed. Until utility-scale energy storage technologies are available at affordable costs, however, wind generation will be limited to supplying energy on a random basis. Wind, being an intermittent source, cannot be expected to supply much more than 15 percent of the total electric (kWh energy) demand in any region of the country. Its role will be confined, in such a case, to displacing a portion of higher-emissions fossil fuels in electric generation.

Environmental Impacts

Wind energy does lower CO_2 emissions compared with coal or natural gas generation, and does not emit SO_2 or NO_x or particulates. But it is not entirely benign with respect to the environment. Of course, bird kills are widely noted. But there are other problems less well established. For example, many people have complained of noise and low-frequency vibrations from wind, and doctors have seen cases of people living near large wind facilities who experience "symptoms that include decreased quality of life, annoyance, stress, sleep disturbance, headache, anxiety, depression, and cognitive dysfunction" (Jeffrey et al. 2013). How widespread such problems are is not clear, but it has been argued by doctors that "there is sufficient evidence to support the conclusion that noise from audible [wind turbines] is a potential cause of health effects" (Jeffrey et al. 2014).

There are also pollution problems from the mining of rare earth minerals – which are important in wind turbine construction. The rare earth neodymium is used in magnets that allow windmill operation at slower wind speeds than more traditional magnetic materials. A two-megawatt wind turbine contains about 800 pounds of neodymium, as well as a significant quantity of another rare earth, dysprosium (Stover 2011). Although rare earths are not actually all that rare, much of the output comes from China, where there are serious concerns about local pollution from rare earth mining and preparation; reportedly, people living nearby have had high rates of cancer and other diseases.

It is also worth considering that wind probably does not reduce CO_2 emissions by as much as one might at first think. If backup is fossil-fuel generation, then each time backup is required there will be CO_2 emissions. Also the platforms and towers for the turbines are made from large quantities of steel and concrete, two energy-intensive industries. If one looks at the amount of this material relative to the amount of power produced, wind is much more steel and concrete intensive than nuclear power – using four times as much concrete and five times as much steel per gigawatt hour per year versus a nuclear power plant (Nicholson 2013).

The bottom line is really this: no energy technology is completely benign with respect to the environment. But it remains true that on balance wind (as well as solar) will reduce the amount of CO_2 that is emitted each year if they are employed in place of fossil fuels.

Economics and Market Penetration

In the 1992 Energy Policy Act, wind-generated electricity was given a direct subsidy. For each kilowatt hour generated, the owner of the wind turbine would receive 1.5¢ (subsequently adjusted for inflation to 2.3¢/kWh), what was called the Production Tax Credit (PTC), which remains in effect. Investment costs by the 2010s will depend on the type of unit, the terrain on which it is set, and the distance from the site to a grid connection. In general, though, a commercial turbine installed could cost anywhere from about $1500/kW to $3000/kW. Though these prices are comparable to that of conventional electric generating technologies, it must be emphasized that the lower *capacity factors* of wind generation mean that these unit investment-cost figures are not yet comparable with conventional power plant costs. But since wind generation has no fuel costs and receives a direct subsidy, wind has had a competitive cost advantage of electrical energy delivered (see Chapter 9). Indeed, given the PTC, it has often been the case that on a given day, wind will be the lowest cost power available to the grid operators.[17] The cost goals of government and private promoters has been to achieve capacity (investment) costs well under $1000/kW and generating costs under 5¢/kWh for good wind resource sites. Despite the limitations apparent at this time, the US DOE has been optimistic about the future market penetration of wind power, forecasting an expansion of wind power sufficient to displace 3–4 quads of primary energy (see Chapter 4) by 2040. Aggregate US wind capacity in 2015 was over 74 GW, about 7 percent of total US generating capacity.

Despite its low power density and other concerns, it appears that wind generation technology will be in the forefront of alternative energy development in the near- to medium-term future, more important than solar-electric generation. It is especially important in the development of renewable electricity in Germany – as we describe in the next section.

Energiewende: Performance

There is considerable controversy over the evaluation of the German *energiewende*. Costs have been significant, and are expected to exceed €1 trillion to reach the stated goal of 80 percent renewables by the 2050s. The costs have been passed along to consumers who are paying about three times the average US rate for electricity – causing some distress, especially among Germany's relatively poor, many of whom have fallen into energy poverty. Energy intensive companies are moving some production to places where energy is cheap – such as the United States – although the German government has provided subsidized electricity rates to large firms to keep them in Germany.

The high costs that the program imposes stem from the system of feed-in tariffs (FIT) whereby grid operators are (1) required to buy all power that is produced by renewable sources; and (2) to pay a fixed price (though the price is generally reduced over time) usually set by the government for a set number of years – typically 20. The cost of the FITs overall (first introduced in 1990) has been great and will total an estimated $750 billion by 2022 (Poser et al. 2014). The FITs are set according to the technology used, and do not in any way correspond to the wholesale price of electricity. The most extreme case has been the FIT for solar, which has been set at more than 50 euro cents per kWh, about four times the retail price of electricity in the United States, five times the FIT for Germany's onshore wind (Hake et al. 2015), and seven times the wholesale price of German electricity generally. Because of the high FITs, German power companies have had to raise electricity prices, since they are required to pay above market rates to renewable generators, including a large number of households with rooftop PV. It may be argued that whatever one thinks of the *energiewende* in principle, solar is a poor choice for Germany given its high latitude. In fact, the capacity factor for German solar is less than 10 percent (Wilson 2013).

Nevertheless, many experts argue that the *energiewende* is a success and proof that a sophisticated industrial electric system can run largely on renewables. In 2013, Germany had derived 25 percent of its electricity from renewables (though this included hydro) but, as noted, CO_2 emissions, which had been falling steadily, flattened and even started rising again because of the closure of several of Germany's nuclear power plants – replaced by coal-burning power facilities mostly.

There are fears that Germany will suffer major power outages the more it relies on intermittent sources of electricity, but thus far it has not happened, largely because German grid operators can buy hydroelectric power from Scandinavia and Austria and nuclear power from France. Experts generally believe that the German system is not imminently in danger

Proponents of the *energiewende* also claim that over time renewables will be less costly than fossil-fuel resources. This argument is predicated on including costs to the environment in the latter, including an estimate of future damages from climate change as well as the costs of nuclear waste disposal and nuclear accidents. When these are included, it is argued that the *energiewende* is not only technologically possible, it is "socially efficient" (Hirschhausen 2014).

Questions linger, however, especially about the cost as well as the wisdom of closing nuclear power plants. Nuclear plants do have costs in terms of waste disposal and plant decommissioning, costs that are hard to quantify and potentially are large (see Chapter 7). But the fear of accidents, probably the main reason for popular demands for plant closures, stem from two cases: Chernobyl and Fukushima. Yet none of the German plants were of the Chernobyl design and the Fukushima accident came about from a tsunami, which is not likely to hit Germany. The *Financial Times* editorialized that the closure was "a huge mistake" and in an interview with the

German magazine, *Spiegel*, climate scientist James Hansen said that Germany was following a "false energy policy" by closing its nuclear plants and relying on intermittent solar and wind.

It is also questionable whether such a program is relevant to the United States. The United States has been cutting emissions through the conversion of coal-burning plants to ones that use inexpensive and plentiful (likely to remain so for at least several decades) natural gas. Germany does not have the same option. If it turned to gas, it would have to buy it primarily from one unreliable source, Russia. But natural gas seems likely to be the major source of electric generation in the United States for the next decades. The EIA projects that by 2040 solar and wind generation will expand considerably, but will produce only a third as much electricity as natural gas.

Energy Transitions: Government or Market (or Both)?

We return here to the main question raised at the beginning of the chapter. A transition to a lower fossil-fuel-intensive, CO_2 emitting economy seems likely and perhaps necessary in the decades ahead. But how are we to get there? Should the US government command a gradual (or immediate) cessation of fossil fuels? Or perhaps government should embark on an American *energiewende* leading to sky high prices for electricity? Alternatively, should we wait for prices of fossil fuels to rise and those of solar, wind, and accompanying storage to fall far enough for them to be competitive in the marketplace and await the kind of processes of adoption and diffusion we discussed in Chapter 10? This scenario might include government research funding but would allow gradual forces of diffusion to work.

As Vaclav Smil (2010) points out, the energy system of today is vast. It includes thousands of miles of pipelines, millions of miles of transmission lines, storage facilities capable of holding billions of barrels of oil and trillions of cubic feet of gas, and transportation facilities in the form of tankers, railroads, trucks – an investment in total of more than $5 trillion. It is difficult to imagine, he argues, it being cast aside in a decade or two – especially now when the supply of fossil fuels is great and the cost of them, low. "[T]he *scale* and the different specific engineering challenges of the decarbonization project," argues M.J. Kelly, "are without precedent in human history." [Emphasis in the original.] Kelly claims that an attempt to reduce CO_2 emissions by 80 percent by 2050 will, first, not actually succeed and may in fact "make matters worse" (Kelly 2016).

Of course, others would maintain that we must act immediately to rearrange our energy system to prevent dangerous climate change. As the Worldwatch Institute argues, "Humanity can prevent catastrophic climate change if we act now and adopt policies that reduce energy usage by unleashing the full potential of energy efficiency in concert with renewable energy resources... What is needed is a transformation of the entire global energy system."

An important question is: how could such a project be undertaken? It seems it could only be accomplished coercively or if the vast majority of people endorsed it. In Germany, popular support for the *energiewende* has been essential to its growth and to the successes that are claimed for it. But there is no popular consensus in the United

States where climate change is not considered an immediate threat. There would be resistance if government actions were costly and restrictive of freedom. Public opinion could change, but barring some dramatic evidence of impending climate catastrophe, it is doubtful that any such energy transition would be sustained as power shifted in Congress and the presidency from election to election. Still, continued research (and incentives for it) would be advisable in the hope that a transition would take place within the rules and benefits of the marketplace where the government is most importantly the referee of fairness. True, it would mean some risk that action will be too late, but as we noted in Chapter 2, catastrophic predictions related to energy have been wrong for the past few centuries. Moreover, some scientists have retreated from the prediction of near-term catastrophe, arguing that the sensitivity of the climate to a doubling of CO_2 is lower than previously thought. In the most recent (2014) IPCC report there was no consensus as to sensitivity and no best guess for the expected increase in temperatures should doubling occur. Of course, most scientists continue to see human-caused climate change as a threat, but one that seems farther in the future and diminished from the absolute catastrophe scenarios of recent years.

A noncatastrophist viewpoint does not mean that governments should do nothing with respect to climate change. Publicly (as well as privately) funded research should continue in several areas, among them: adaptation strategies for a warming world; improvement in climate sensitivity measurements; development of nonfossil energy technologies, especially grid-level storage capabilities for intermittent electricity sources; and enlargement of research efforts into new nuclear power, nuclear fusion, and different means of powering transportation.

Still, if we do not take vigorous action to curb emissions, it means we are taking a risk. We will be saying in effect that we think catastrophe is unlikely in the next generation or two. We have time, we can transition slowly guided largely by markets, we can adapt and develop and the future can take care of itself. That is the policy much of the world seems tacitly to be following.

But can we confidently say that that is all the world needs?

References

Boyles, J.G., P.M. Cryan, G.F. McCracken, and T.H. Kunz. 2011. "Economic Importance of Bats in Agriculture." *Science*, 332 (6025) 41–42.

Brown, A. and J. Bunyan. 2014, "Valuation of Distributed Solar: A Qualitative View." *The Electricity Journal*, 27 (10), 27–48.

Bryce, R. 2015. *Smaller Faster Lighter Denser Cheaper: How Innovation Keeps Proving Catastrophists Wrong*. Public Affairs, New York.

Giberson, M. 2013. *"Assessing Wind Power Cost Estimates."* Report, Institute for Energy Research, Washington, DC.

Gipe, P. 2004. *Wind Power: Renewable Energy for Home, Farm and Business*, Revised edition. Chelsea Green Publishing, White River Junction, VT.

Goswami, D.Y. (ed.). 2015. *Advances in Solar Energy: Annual Review of Research & Development*, Volume 17, Routledge, London.

Green, M.A. 2006 *Third Generation Photovoltaics: Advanced Solar Energy Conversion*, Springer-Verlag, Berlin, Heidelberg.

Hake, J.-F., W. Fischer, S. Venghaus, and C. Weckenbrock. 2015. "The German Energiewende: History and Status Quo." *Energy*, 92 (3), 532–546.

Hirschhausen, C. v. 2014. "The German "Energiewende: An Introduction." *Economics of Energy & Environmental Policy*, 3 (2), 1–12.

Hosenuzzaman, M., N.A. Rahim, J. Selvaraj, M. Hasanuzzaman, A.B.M.A. Malek, and A. Nahar. 2015. "Global Prospects, Progress, Policies, and Environmental Impact of Solar Photovoltaic Power Generation." *Renewable and Sustainable Energy Reviews*, 41, 284–297.

Jacobson, M.Z. and M.A. Delluchi. 2009. "A Path to Sustainable Energy by 2030." *Scientific American*, 301 (5), 58–65.

Jeffrey, R.D., C. Krough, and B. Horner. 2013. "Adverse Health Effects of Industrial Wind Turbines." *Canadian Family Physician*, 59, 473–475.

Jeffrey, R.D., C. Krough, and B. Horner. 2014. "Industrial Wind Turbines and Adverse Health Effects." *Canadian Journal of Rural Medicine*, 19 (1), 21–26.

Kelly, M.J. 2016. "Lessons from Technology Development for Energy and Sustainability." *MRS Energy & Sustainability: A Review Journal*, 3 (3), online at www.mrs.org/energy-sustainability-journal/.

Manwell, J.F. et al. 2009. *Wind Energy Explained: Theory, Design and Application*, 2nd edition. John Wiley & Sons, Chichester, UK.

Nicholson, M. 2013. "Nuclear Has One of the Smallest Footprints." *The Breakthrough*, at http://thebreakthrough.org/index.php/programs/energy-and-climate/nuclear-has-one-of-the-smallest-footprints.

Poser, H., J. Altman, F. ab Egg, A. Granata, and R. Board. 2014. "*Development And Integration of Renewable Energy: Lessons Learned From Germany,*" Finadvice, report for the Edison Electric Institute (EEI) and Finadvice's European clients, Switzerland.

Razykov, T.M., C.S. Ferekides, D. Morel, E. Stefanakos, H.S. Ullal, and H.M. Upadhyaya. 2011. "Solar Photovoltaic Electricity: Current Status and Future Prospects." *Solar Energy*, 85, 1580–1608.

Simmons, R.T., R.M. Yonk, and M.E. Hansen. 2015. "*The True Cost of Energy: Wind,*" Report, Strata Institute of Political Economy, Utah State University.

Smil, V. 2008. *Energy in Nature and Society: General Energetics of Complex Systems*. MIT Press, Cambridge, MA.

Smil, V. 2010. *Energy Myths and Realities: Bringing Science to the Energy Policy Debate*. AEI Press, Washington, DC.

Smil, V. 2015 *Power Density: A Key to Understanding Energy Sources and Uses*. MIT Press, Cambridge, MA.

Stover, D. 2011. "The Myth of Renewable Energy." *Bulletin of the Atomic Scientists*, online at http://thebulletin.org/myth-renewable-energy.

Wenham, S.R., M.A. Green, M.E. Watt, and R. Corkish. 2007. *Applied Photovoltaics*, 2nd edition. Earthscan, London.

Wilson, R. 2013. "Low Capacity Factors: Challenges for a Low Carbon Energy Transition – The Energy Transition," at www.theenergycollective.com/robertwilson190/288846/low-capacity-factors-challenge-low-carbon-energy-transition.

NOTES

1 This is due to the Merit Order Effect, which refers to the order in which electric generating facilities are given access to the grid, based on their marginal cost of production. Because renewables have very low marginal costs and because they are typically subsidized, they can offer power at very low prices, indeed even negative prices. This can force conventional power producers off the grid temporarily and also make conventional power production a money-losing business.

2 NPR interview 2010, transcript at www.npr.org/templates/story/story.php?storyId= 129253742.

3 In September 2016 there was a blackout of the entire province of South Australia. It seems it was caused at least in part by the heavy reliance on wind-generated electricity in the province. See Uhlmann, C., "South Australia blackout: When the lights go out it's a sign the electricity grid isn't working well." At www.abc.net.au/news/2016-10-06/uhlmann-on-power-blackout-in-south-australia/7906844.

4 www.autoblog.com/2015/10/08/gm-li-ion-battery-cost-per-kwh-already-down-to-145/

5 Nuclear power is sometimes used as a backup but it too must be kept hot and in readiness.

6 It is worth noting that geothermal is already a cost-effective source of electric generation, but of course it is limited to a few regions of the world (see Appendix C).

7 This section relies heavily on Smil (2008, 2015). Peter Grossman thanks Robert Bryce for comments on this section.

8 More technically: ". . . the universal measure of energy flux [watts/square meter or] W/m^2 of horizontal area of land or water surface rather than . . . the working surface of the [energy] converter" (Smil 2010).

9 In 2015, the DOE issued its "Wind Vision Report," which put the plant boundary at 1 MW per 85.24 acres. Industry puts the boundary at about 60 acres/MW.

10 An acre is 4047 m^2 and so 85 acres equals 343,938 m^2. 1 MW = 1 million watts; dividing watts by 343,938 equals 2.9 W/m^2 but given the low capacity factor (about 0.32) the actual rate of output, the power density, is 0.92 W/m^2. Bryce (2015) compiled a list of 16 large-scale wind projects from 40 MW to 2500 MW and found that the average capacity density was 2.3 W/m^2 but after accounting for capacity factors the power densities were less than 1.0 W/m^2.

11 For a general discussion of PV economics and global prospects see Hosenuzzaman et al. (2015).

12 Germany has subsidized a great deal of solar energy development despite its location above latitude 45° N. As discussed later in this chapter, this has led to capacity factors of German solar facilities that are in many instances below 10 percent.

13 A Chinese company has sold solar panels with a cost per watt of $0.60. But the company that was selling at this price went bankrupt. Other manufacturers in China with prices below $1/W are thought to be heavily subsidized by the Chinese government.

14 For example, see "Solar Power Crosses Threshold, Gets Cheaper Than Natural Gas," August 2015 at www.eenews.net/stories/1060023749.

15 At solar thermal generators, which use concentrated solar energy to heat a boiler, birds flying over the reflecting mirrors that concentrate the sunlight have caught fire in flight.

16 This phrase is taken from an article by Mark Duchamp at the website of Save the Eagles International: http://savetheeaglesinternational.org/new/us-windfarms-kill-10–20-times-more-than-previously-thought.html. One distinction between turbine and other bird killers (buildings and feral cats) is that feral cats and buildings do not generally kill raptors whereas, as we noted in the text, studies put the number of raptors killed by windmills in the thousands.

17 Essentially the PTC transfers costs from rate payers to tax payers – which may be more progressive than the methods used in rooftop solar PV.

Scientific Principles

Energy Conversion

The purpose of all energy technologies is to provide energy in useful forms, such as mechanical work, electricity, and useful heat. Indeed, when we discussed society's demand for energy in Chapter 4, we had in mind energy as the equivalent of a material commodity for a consumer or as an input factor in production. That is, we viewed energy as a quantity purchased or allocated to provide a means of accomplishing useful ends.

In order to be a useful commodity, energy in general must be converted to its end-use form. For example, the potential energy in water stored behind a high dam is converted to a useful form (mechanical work or electricity) once the water has fallen and turned the shaft of a water wheel or turbine (see Chapter 3). The energy tied up in the hydrocarbon compounds of fossil fuels becomes useful when the fuel has been burned and converted into heat. The fossil-fuel heat so derived may be useful for space heating or a manufacturing process, or the heat may itself be converted into mechanical work or electricity.

The final conversion process is of prime interest to us in this appendix. The process of converting heat into useful work energy is said to be the technological underpinning of modern industrial society. The *heat engine* is the historical name for all such machines. Industrial societies today are powered by heat engines of all sorts (steam turbines, gasoline engines, and so on) that are fired almost exclusively by fossil fuels (coal, oil, and natural gas). Such engines universally convert the heat of combustion into a useful form, such as mechanical work or electricity. Engines converting the heat of combustion of biomass alcohols would be in the same category. Even a nuclear reactor is merely a heat source to drive a steam turbine.

The scientific basis for understanding the workings and fundamental limitations of heat engines resides in *thermodynamics.* Thermodynamics involves a general set of principles that apply to energy conversion, heat transfer, or any one of many processes that involve changes in the form of energy. Thermodynamics applies regardless of the medium bearing the energy, the type of physical reaction involved, or the scale of the system in which the process is taking place. It is based, in the best tradition of modern science, upon repeated observations and experiments that fit into a consistent theoretical scheme.

Also of great importance to modern society is the generation of electricity. Electricity provides the means of transmitting and distributing energy that has already been converted to useful form by a thermal process or other means. In Chapter 3, we described a conventional power plant including the electric generator.

Here, following our discussion of thermodynamic principles, we will also review the principles underlying electricity generation.

Units of Measure

There are several different systems of units in use today for measuring physical quantities such as *energy* and *power*. Each system has a *fundamental* set of units that measure length, mass, and time. Thus, the modern metric system is called the *MKS system*, standing for *meters*, *kilograms*, and *seconds* – specifying the three fundamental units. (Also known as the SI system for *Système international d'unités*.) If we are making electrical measurements, we must add one additional fundamental unit, either for *charge* or *current*.[1]

All units of measurement that are not fundamental are called *derived* units, meaning that they can be expressed equivalently in terms of the basic units. Force, for example, is measured in newtons (N) in the MKS system, but can be expressed equivalently as kilograms divided by seconds per second or:

$$[N] = \frac{[kg]}{[s^2]}$$

This formula comes from Newton's law for the force required for acceleration of a mass. In this way, the MKS system is made self-consistent and calculations can be carried out with the assurance that the numerical answers will always be in units of the system.

Conversion of units becomes necessary when switching from one system of measurement to another. In the energy field, units in the British imperial system (similar to the US customary system), such as British thermal units (BTU), horsepower, and gallons, are still encountered, due to their common usage and familiarity in the United States and some English-speaking countries.

Here is a listing of MKS units as used in our discussions of energy throughout this text. Abbreviations are shown in brackets.

joule [J]:	energy
watt [W]:	power, time rate of energy delivery or usage
meter [m]:	length (basic)
kilogram [kg]:	mass (basic)
second [s]:	time (basic)
newton [N]:	force
degree Celsius [°C]:	temperature, centigrade scale
degree Kelvin [K]:	absolute temperature, Kelvin scale (absolute zero is −273 °C)
kilocalorie [kcal]:	heat energy[2]
liter [l]:	volume
ampere [A]:	electric current (see note 1)
volt [V]:	electric voltage[3]

electronvolt [eV]:   the energy (1.602×10^{-19} J) required to move one electron
                          over an electric potential difference (see next section) of
                          1 volt.

A comparable listing of imperial units, many of which are used in this text, follows
next. The British system has been called the *foot-pounds-seconds* (*FPS*) *system*, with
reference to its fundamental units.

horsepower [HP]:   a measure of power, the time rate of energy delivery or usage
foot: a familiar measure of length
slug: mass, not generally familiar in usage[4]
second: time, same as MKS
pound [lb]:   force, a familiar measure of weight – the force of gravity
degree Fahrenheit [°F]:   temperature
degree Rankine [°R]:   absolute temperature on the Rankine scale (absolute zero is
                          –460 °F)
BTU: heat energy[5]
gallon: a measure of volume

The following are a number of useful conversions between metric and imperial
units.

Energy
  1 J = 0.949×10^{-3} BTU
  1 kWh = 3412 BTU[6]

Power
  1 HP = 0.746 kW

Mass
  1 kg = 2.2 lb
  1 metric ton (10^3 kg) = 1.1 ton (2200 lb)

Temperature
  [°F] = 9/5[°C] + 32°

Temperature/energy equivalent[7]
  1 keV $\simeq 10^7$ °C

Heat energy
  1 J = 0.24×10^{-3} kcal
      = 0.940×10^{-3} BTU

Volume
  1 gallon (US) = 3.78 liters
  1 barrel (42 US gallons) = 160 liters[8]

Forms of Energy

The laws of physics tell us that energy exists in the classical forms of *work*, *kinetic energy*, *potential energy*, and *heat*, and in the hidden (nuclear) form of mass-equivalent energy.

Work

Work is force acting over distance, as expressed in:

$$W = Fd \, [\text{J}]$$

where F is force [N] and d is distance [m] for translational displacements along straight or curved paths, including purely rotational displacements.

Kinetic Energy

Kinetic energy is the energy of motion, as expressed in:

$$K = \frac{1}{2}mv^2 \, [\text{J}]$$

where m is the mass of an object in motion [kg] and v is the velocity of the mass [m/s] for translational motion. Kinetic energy exists also for rotational motion.

Potential Energy

Potential energy is energy that can be, but has not yet been, delivered to other forms. The gravitational potential to deliver mechanical energy is expressed as:

$$P = mgh \, [\text{J}]$$

where m is the mass of an object [kg], g is the gravitational constant [N/s^2], and h is the height of the object above some reference plane. Potential energy exists in systems such as electrical and chemical systems.

Heat

Heat energy is expressed colloquially as *heat content*, that is, the amount of heat energy (such as BTU) inside a solid body, liquid, or gas.[9] In engineering usage, however, *heat* is usually defined as the energy transferred from one mass to another when a temperature difference exists between the two; thus, the term might better be called *transferred heat,* denoted by the letter Q.

Insolation

Insolation – a measure of solar flux. The measure of incident solar-thermal power per unit of area. In MKS units – W/m^2; in British units – BTU/ft^2 hr.

Mass-Equivalent Energy

This amazing discovery of modern physics shows that mass and energy are equivalent, leading to the conversion of mass to usable energy in the nuclear reactor.

Thermodynamic Definitions

In order to conduct an orderly review of the laws of thermodynamics, it is first necessary to define certain terms and concepts used throughout.

A *system* is the specific combination of *working substance* and identified space in which a thermodynamic process takes place or in which certain thermodynamic conditions obtain. The working substance may be gaseous, liquid, or solid.

A *boundary* is the closed surface, real or imagined, enveloping the system and separating it from the rest of the universe.

The *surroundings* include all of the matter and space outside the system boundary.

In a *closed system*, there is no exchange of matter (mass) with the surroundings, but an exchange (transfer) of heat may occur with the surroundings.

In an *open system*, there is an exchange of matter, often a flow of matter, to and from the surroundings that is usually accompanied by an exchange of heat with the surroundings.

An *isolated system* is presumably not influenced by its surroundings in any way thermodynamically; that is, it has no transfer of energy or exchange of mass with its surroundings.

We can use these terms with reference to a modern steam power plant. We can define the thermodynamic system as the entire plant, or we can say that each of its internal parts, such as the boiler, the steam turbine, the electric generator, or the condenser, constitutes a separate system (see Chapter 3). If we elect to define the entire plant as the system, then the system boundary may be defined as the *building envelope* itself (Fig. A.1). The entire plant is certainly not an isolated system and it is not even a closed system, if we consider the mass of fuel, air, and cooling water flowing in and the mass discharge of ash, stack gases, and cooling water necessary for its operation. However, it is possible, if the system boundary is suitably defined, to imagine a closed system for this plant having exchanges of energy with the surroundings but no mass flowing in and out.

Such an imaginary boundary passes through the combustion chamber (furnace), so as to exclude the combustion gases and the mass of the coal and ash. The boundary must also pass through the condenser, so as to exclude the flowing mass of cooling water (see Chapter 3). Through such a boundary, albeit abstract, flows *only* energy: the heat transferred in the boiler from the furnace into the working substance (water/steam), the condensing heat transferred from the working substance to the cooling water and, of course, the electrical energy transmitted out of the generator toward its load.

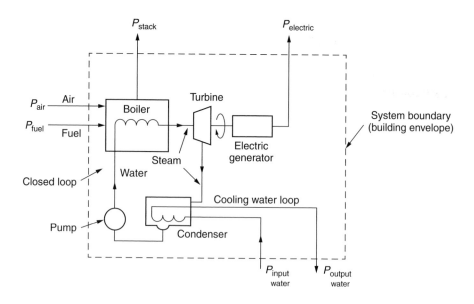

Figure A.1 Schematic diagram of a thermal power plant. $P_{input} = P_{fuel} + P_{air} + P_{input\ water}$; $P_{output} = P_{electric} + P_{stack} + P_{output\ water}$.

Source: Adapted from R.C. Dorf, 1978, *Energy, Resources and Policy*, Ballinger, Cambridge, MA.

The working substance of such a steam generating plant flows literally around a *closed loop* (see Fig. A.1). It starts as water that is heated into steam in the boiler, steam passes through the turbine, it is condensed back down to water, and then starts circulating all over again. This circulation could be contrasted with the old-fashioned steam engine where the steam (working substance) is discharged from the piston after each power stroke, thus making the engine an *open* system in thermodynamics.

Finally, isolated thermodynamic systems are idealizations, only approximated in the laboratory or when thought of as models of practical operation. The truly isolated system would have no mechanical or electrical means of receiving work energy from the surroundings and would be totally insulated thermally (that is, no heat transfer could occur through conduction, radiation, or other means) from the surroundings. Its only value is that it can provide a conceptual model to illustrate the workings of thermodynamic laws.

Conversion of Energy: The First Law of Thermodynamics

The First Law of Thermodynamics, most simply stated, is:

Energy can be neither created nor destroyed.

It is a law because it has been tested repeatedly in experiments and has never been found violated. It also serves as part of the basis of an entirely self-consistent scheme of physical theory and underlies a cosmic hypothesis that states: "The total amount of energy in the universe is constant."

It is not obvious, at the outset, how such a broadly stated law can be applied to practical energy technologies. One approach is to consider various thermodynamic systems. The simplest case is the *isolated* system, where the interpretation is easy – the total energy contained within the isolated system cannot change. A more practical case is the closed system, where transfers of energy with the surroundings can occur, but mass transfers do not. Here, the First Law can be stated in the form of an energy balance for the system, as follows:[10]

Energy in = Energy out [J]

The application of this energy balance to a conventional power plant is shown schematically in Figure A.1. This plant is a closed cycle that recirculates steam and water (see also Figure 3.3). As such, the system boundary excludes the mass flow of fuel (in), cooling water (in and out), and combustion gases (air in and stack gases out). But it includes the energy flows in and out of these masses. In the case of a power plant with a constant rate of energy exchange, the balance equation may be written in terms of power (time rate of energy flow – see Units of Measure section) as follows:

$$
\begin{aligned}
P_{\text{in}} \;\; &= P_{\text{out}} \;[\text{W}], \;\; \text{where(referring to Fig. A.1)} \\
&= P_{\text{fuel}} + P_{\text{air}} + P_{\text{input water}} \\
&= \text{total rate of energy flow in } [\text{W}] \\
P_{\text{out}} \;\; &= P_{\text{elect}} + P_{\text{stack}} + P_{\text{output water}} \\
&= \text{total rate of energy flow out } [\text{W}]
\end{aligned}
$$

Thus we see the energy transfers into and out of the closed steam cycle of the power plant. We know from the First Law that as much energy *goes in* as must *come out*, or for conditions of a steady rate of energy production, the *total power in* must equal the *total power out.*

However, not all of the power out of the plant is useful. For an electric power plant generally, the electricity is the only useful output. A measure of the fraction of useful energy out is the *plant efficiency*:

$$
\begin{aligned}
\eta &= \frac{E_{\text{useful}}}{E_{\text{fuel}}} \\
&= \frac{P_{\text{elect}}}{P_{\text{fuel}}} \text{(for the power plant with steady power out)}
\end{aligned}
$$

The plant efficiency is sometimes called the *First Law* efficiency, because it is based on that principle. It is also a practical measurement, however, because it only rates the useful output against the energy input from the fuel, and fuel is the only energy input that has a direct operating expense. It might also be noted that the energy content of the incoming air (P_{in}) is negligible compared to the fuel input energy and thus the energy balance equation can be written approximately as:

$$
P_{\text{fuel}} + P_{\text{input water}} = P_{\text{elect}} + P_{\text{output water}}
$$

Which relation can be solved for the output in the form:

$$
P_{\text{elect}} = P_{\text{fuel}} - P_{\text{stack}} - P_{\text{net output water}}
$$

where the last term has been collected together into the single term:

$$P_{\text{net output water}} = P_{\text{input water}} - P_{\text{output water}}$$

representing the net output of *waste heat* energy carried away by the cooling water. The importance of this waste energy will be considered during our discussion of the Second Law of Thermodynamics.

For an open system, according to our thermodynamic definitions, we must consider whatever mass flows take place through the system boundaries. For example, consider the steam turbine alone; that is, define system boundaries as solely enclosing this single component of the steam cycle. Here the general statement (energy in = energy out) must be written specifically in terms of the various forms of energy (work, kinetic energy, potential energy, and heat) in order to account for conversions between energy forms in the open process. In the steam turbine, for example, work is extracted as the mass of steam expands and speeds up, thus producing major changes in pressure, volume, kinetic energy, and internal heat content. These changes are all accounted for in the thermodynamic calculations, which require a conservation of energy – and also of mass – from the input to the output.

Limits on the Conversion of Energy: The Second Law of Thermodynamics

No thermal energy conversion process can be entirely complete. The steam of the power plant, for example, cannot be converted entirely into work or into electrical energy in any real-life system. Complete conversion would require not only that there be zero losses and zero friction, but also that surroundings be maintained at absolute zero temperature. These requirements are embodied in the Second Law of Thermodynamics. In order to better understand this law, it is first necessary to extend our set of thermodynamic definitions.

First, a *property* is composed of the observable, macroscopic quantities that characterize the working substance within a system. For example, temperature, pressure, and mass are properties of steam in a boiler. Properties are said to be *extensive* if they depend on the size of the system; for example, total volume and total mass are extensive properties of a system of a particular size. On the other hand, properties are often given on an *intensive* basis, reduced to a per unit quantity of volume or mass. Finally, the properties of a working substance are said to be well defined thermodynamically when the system is in an *equilibrium state*.

A *state* occurs when there is sufficient equilibrium so that the properties of the substance within are uniform and can therefore characterize the system as a whole. A minimum number of such properties for a system are required to define a state. For example, for a system containing an ideal gas, three properties – *pressure, temperature*, and *volume* – are necessary to define the *system state*. The state in this case will be defined only when the temperature and pressure of the gas are uniform throughout the volume enclosed by the system boundary and are not changing in time, that is, they are in *equilibrium*.

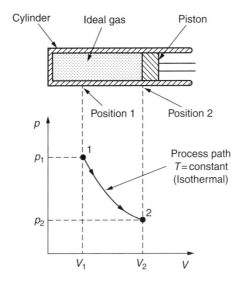

Figure A.2 Pressure–volume diagram for an ideal-gas process. The system's boundaries are the cylinder walls and the face of the piston, thus defining a cylindrical volume (V) filled with an ideal gas as the working substance. The process is controlled by having the piston move from position 1 to position 2, while maintaining the gas at constant temperature (T). (Note that this system is *closed*, but *not isolated*.) The process is portrayed thermodynamically as starting at state 1, characterized by properties p_1, V_1, and T, and proceeding smoothly to state 2, which is characterized by p_2, V_2, and T.

In a *process*, one equilibrium state changes to another. A process is usually depicted as a path on a diagram such as Figure A.2, which shows a process of isothermal (constant temperature) expansion of a gas.

In a *reversible process*, the system can return to the original state by return along the same path. This can only take place under certain strict conditions. Ideally, the reversible process takes place slowly enough so that transitional effects smooth out over the system, thereby maintaining a condition of *quasi-equilibrium* while changes in state are taking place. Thus, for example, the process depicted in Figure A.2 would be in *quasi-equilibrium* if the expansion from state 1 to state 2 were slow enough so that the temperature remained the same throughout the volume and the pressure remained evenly distributed while dropping in value. *Reversibility* then permits another slow, quasi-equilibrium process to take place, returning the system exactly to state 1.

An *irreversible process* does not maintain quasi-equilibrium and therefore cannot return to the original state without input of additional energy. Other sources of irreversibility are friction and energy losses that are not recovered on the reverse path. Perhaps the best example of a nonequilibrium path is the *throttling* process depicted in Figure A.3, where the turbulence is an excellent example of such a nonequilibrium condition.

Work applies the basic definition of work ($W = Fd$) to the substance in a thermo-dynamic system. Work can be done *on* the substance (*negative* work) or it can be

Orifice   Turbulence

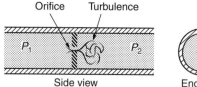

Side view

End view

Figure A.3 The throttling process. Gas in a high-pressure (p_1) chamber flows through a small orifice into a chamber at lower pressure (p_2).

delivered by the substance (*positive* work). An increment of work occurs with an incremental change in volume at a constant pressure as in:

$$dW = p\,dv$$

Thus, an expansion is an increase in volume, with a positive quantity of work delivered by the substance, and a compression is a decrease in volume, with a negative quantity of work done on the substance. In Figure A.2 the total work delivered by the ideal gas over the path 1–2 is:

$$W = \int_{V_1}^{V_2} p\,dV > 0$$

An integral, taken between two limits as shown here (Figure A.2), can be interpreted as the area under the curve of the function being integrated (here p).

Internal energy is the energy associated with microscopic thermal motion on a molecular or atomic scale in the working substance of a system. It thus includes the kinetic energy of molecules in thermal motion in a hot gas, thermal motion of electrons in liquids and in solids, or crystal lattice vibrations in solids. Internal energy, as denoted by U, is measured in units of heat energy and is distinct from the quantity of transferred heat (Q). It is a so-called *extensive* property, because it depends on the total mass of the system substance. It can be reduced to an *intensive* property if we measure it on a per unit of mass basis.

An *adiabatic process* is a process in which *no transfer of heat* takes place into or out of the system. A transfer of heat (Q) must involve changes in the internal energy and/or the work of the system, as expressed by:

$$dQ = dU + dW$$

for an incremental change in heat. An adiabatic process has *no transfer* of heat and hence $dQ = 0$ everywhere along its path from one state to another. Therefore, increments of work (dW) must be exactly compensated by incremental changes in internal energy (dU), with accompanying changes in system temperature. The p–V path for an adiabatic compression is illustrated in Figure A.4.

In an *isothermal process*, the temperature remains constant over the path from one state to another. Such a process can take place by design or occur naturally, as we will see for saturated steam. In the illustration of the piston shown in Figure A.2, a means of maintaining the temperature constant in the gas is assumed, but is not shown,

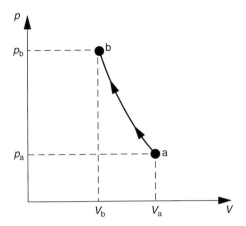

Figure A.4   The adiabatic process. The gas is compressed from volume V_a to V_b, with system boundaries that are perfect thermal insulators, thereby allowing no transfer of heat outside. Note that the path of the adiabatic process in the p–V plane is steeper than that of the isothermal process (Fig. A.2). The temperature in this adiabatic compression rises.

whether it is from the *transfer of heat* into the system or from the *internal energy*, during the expansion indicated.

Other processes involve a constant property. They include constant volume and constant pressure (*isobaric*) processes. Each has a characteristic path, dependent on the other properties.

A *cycle* is a sequence of processes for which the end state is the same as the initial state. A hypothetical example of a four-path cycle is shown in Figure A.5, with each path resulting from a different process. In the case of a closed system, the cycle represents all of the changes in property that the same mass of working substance goes through over and over again in continuous energy conversion.

Limits on energy conversion are best discussed in terms of the *Carnot cycle* (after N.L.S. Carnot, 1796–1832, a French engineer), an idealization of heat–energy conversion from which is derived the theoretical limit to thermal efficiency. This cycle can be visualized as the physical process shown in Figure A.6, which illustrates a reciprocating engine much like the steam engine (Chapter 3). Note in particular the connecting rod from the piston to the rotating flywheel, as in the steam engine, which requires that reciprocation of the piston must accompany rotation of the flywheel. A stroke in each direction is seen to involve two processes; thus, four processes are included in a complete reciprocation up and down. The four processes of the Carnot cycle shown as paths in the figure are:

Process 1–2: isothermal heat input
Process 2–3: adiabatic expansion
Process 3–4: isothermal heat rejection
Process 4–1: adiabatic compression

The ideal means of achieving the conditions required for each process are indicated in the figure at the bottom of the piston. It should be noted in particular that all four

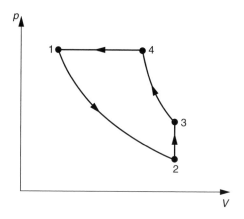

Figure A.5 A thermodynamic cycle. The cycle is composed of four paths: path 1–2: an isothermal expansion process; path 2–3: a constant volume process; path 3–4: an adiabatic compression process; and path 4–1: a constant pressure process. This particular hypothetical cycle would require work into the system substance, that is, it would represent a system for conversion of work to heat, rather than heat to work.

processes of the Carnot cycle are *reversible processes*. The Carnot cycle operation is called an engine because the system delivers net work when operated this way. Net work for a complete cycle can be represented using our mathematical definition of work *on* or *by* a substance as:

$$W_{\text{net}} = \int_{V_1}^{V_2} p\mathrm{d}V + \int_{V_2}^{V_3} p\mathrm{d}V$$
$$+ \int_{V_3}^{V_4} p\mathrm{d}V + \int_{V_4}^{V_1} p\mathrm{d}V$$

and can be understood graphically to be the area enclosed by the four paths on the p–V diagram. The energy required for the system to give up net work is supplied in process 1–2 of each cycle. The action of the Carnot system simply converts heat energy into work.

One of the most useful features of the Carnot conceptualization is that it depicts an *ideal* (reversible) engine working between two thermal reservoirs – one at a high temperature (T_H) and the other low (T_L). The high-temperature reservoir is the *source* of heat transfer into the engine and the low-temperature reservoir is the *sink* for rejected heat. This is depicted in Figure A.7a, where heat (Q_H) is transferred from the hot reservoir to the engine, which in turn converts part of that input energy into work (W) and rejects the remainder (Q_L) into the cool reservoir. The Carnot engine is used as an ideal standard to compare any other heat engine that operates between a high-temperature source and a low-temperature sink. Comparisons of thermodynamic cycles are commonly done using the property called *entropy* (S), which is most clearly defined in terms of transferred heat and temperature. Thus, the entropy increment is:

$$\mathrm{d}s = \frac{\mathrm{d}Q}{T} \quad [\text{J/K or BTU/}^{\circ}\text{R}]$$

where $\mathrm{d}Q$ is the increment in heat transfer (in J or BTU) and T is the absolute temperature (K or $^{\circ}$R).

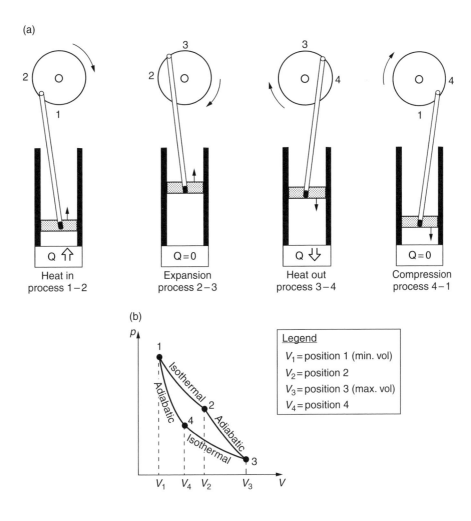

(a)

Heat in
process 1–2

Expansion
process 2–3

Heat out
process 3–4

Compression
process 4–1

(b)

Legend

V_1 = position 1 (min. vol)
V_2 = position 2
V_3 = position 3 (max. vol)
V_4 = position 4

Figure A.6   The ideal Carnot cycle. (a) The cycle in operation. (b) The cycle p–V diagram.

Absolute temperature is used because it measures total thermal energy from its reference at 0 K or 0 °R. Using this definition, the entropy change for a process going from one state (a) to another (b) would be:

$$S_b - S_a = \int_a^b \frac{dQ}{T} \quad [\text{J/K or BTU/}^\circ\text{R}]$$

The Carnot cycle provides particularly simple interpretations of entropy changes. For example, the two isothermal processes in Figure A.7b are simply reduced to summed changes in Q:

$$S_2 - S_1 = \int_{Q_1}^{Q_2} \frac{dQ}{T_\text{H}} = \frac{(Q_2 - Q_1)}{T_\text{H}}$$
$$\text{with } T_\text{H} = T_1 = T_2 \text{ and}$$
$$S_3 - S_4 = \int_{Q_4}^{Q_3} \frac{dQ}{T_\text{L}} = \frac{(Q_3 - Q4)}{T_\text{L}}$$
$$\text{with } T_\text{L} = T_4 = T_3$$

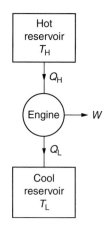

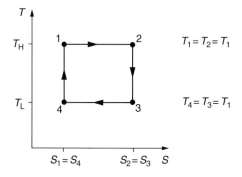

Figure A.7 The Carnot heat engine. (a) Schematic diagram. (b) T–S diagram.

The other two processes are *adiabatic* processes, and according to definition:

$$\mathrm{d}Q = 0$$

meaning no incremental transfer of energy, and hence:

$$S_2 - S_3 = 0$$

and

$$S_4 - S_1 = 0$$

These two paths of constant entropy characterize *isentropic* processes, which are both adiabatic *and* reversible.

Entropy and absolute temperature provide a useful alternative to displaying the paths of processes and cycles. The T–S diagram for the Carnot cycle is shown in Figure A.7b, using the same state numbers as Figure A.6. On a T–S diagram, area generally represents heat transferred in a process from any state (a) to any other (b) as the quantity:

$$Q_b - Q_a = \int_{S_a}^{S_b} T \mathrm{d}s$$

which is the area in the T–S plane under the path from a to b.

In the Carnot cycle, this T–S area representation is particularly simple when area is the rectangle as shown in Figure A.7b. The transferred heat in this case is:

$$Q_2 - Q_1 = \int_{S_1}^{S_2} T\mathrm{d}S = T_\mathrm{H}(S_2 - S_1)$$
$$= \text{rectangular area under path } 1 - 2$$

and

$$Q_3 - Q_4 = \int_{S_4}^{S_3} T\mathrm{d}S = T_\mathrm{L}(S_3 - S_4)$$
$$= \text{rectangular area under path } 4 - 3$$

The net area of the closed cycle path on the T–S diagram represents that portion of the heat transferred that is *available* for work in the engine. Thus, in the Carnot cycle:

$$W = \int_{S_1}^{S_2} T\mathrm{d}S + \int_{S_2}^{S_3} T\mathrm{d}S$$
$$+ \int_{S_3}^{S_4} T\mathrm{d}S + \int_{S_4}^{S_1} T\mathrm{d}S$$

but, since[11]

$$W = \int_{S_1}^{S_2} T\mathrm{d}S + \int_{S_3}^{S_4} T\mathrm{d}S$$

we therefore have:

$$\begin{aligned} W &= T_\mathrm{H}(S_2 - S_1) - T_\mathrm{L}(S_3 - S_4) \\ &= T - S \text{ area in rectangle } 1 - 2 - 3 - 4. \end{aligned}$$

Our earlier identification of the heat input process $(1-2)$ and the heat rejection process $(3-4)$ then enable us to write for the heat transfer process:

$$(\text{heat in})Q_\mathrm{H} = T_\mathrm{H}(S_2 - S_1)$$
$$(\text{heat out})Q_\mathrm{L} = T_\mathrm{L}(S_3 - S_4) = T_\mathrm{L}(S_2 - S_1)$$

and therefore, the *ideal available work* is:

$$W = Q_\mathrm{H} - Q_\mathrm{L}$$

This is to say, theoretically, in the Carnot engine all of the available energy goes into work (W) and thus we have an equality sign in this equation. It should be noted, moreover, that the heat (Q_L) that is rejected into the low-temperature reservoir is all *unavailable* for work, even in the ideal Carnot engine. These results enable us to write down a particularly simple expression for the ideal Carnot efficiency of a heat engine. The Carnot efficiency is defined as:

$$\eta_\mathrm{C} = \frac{\text{work out}}{\text{energy in}} = \frac{W}{Q_\mathrm{H}}$$

And therefore, in the case of the Carnot cycle, the simple result is:

$$\begin{aligned} \eta_\mathrm{C} &= \frac{(Q_\mathrm{H} - Q_\mathrm{L})}{Q_\mathrm{H}} = \frac{T_\mathrm{H}(S_2 - S_1) - T_\mathrm{L}(S_2 - S_1)}{T_\mathrm{H}(S_2 - S_1)} \\ &= \frac{(T_\mathrm{H} - T_\mathrm{L})}{T_\mathrm{H}} \end{aligned}$$

which represents the theoretical *upper limit* for any heat engine working between the temperatures T_H and T_L.

There are several reasons why no practical heat engine can achieve the Carnot limit. Heat losses from the system, such as radiant heat loss, would be one obvious way in which the available energy would be reduced. Also, friction causes efficiency loss, and it is readily understood as an irreversible process. Irreversible processes in general, whether from friction, turbulence, or other nonequilibrium conditions, represent the essential ways in which practical engines differ from the ideal. The expansion of steam in a practical steam engine is another way. It is not perfectly isentropic, that is, it is neither completely adiabatic nor entirely reversible in compression. Later in this appendix, we will compare the efficiencies of practical engines with the ideal.

An appreciation of the nature of irreversible processes, and how they turn energy into forms unavailable for useful work, gives us a better understanding of why a machine of perpetual motion could never exist in reality. Such a machine would be capable, for example, of utilizing energy converted by friction into heat. The Second Law tells us that such energy is *un*available for the work necessary to keep the machine moving. This machine might instead be able to derive net work from a *cool reservoir* (Fig. A.7a) by transferring heat out of the reservoir and then returning it. But here again the energy is unavailable, in this case because it resides in the *lower* temperature reservoir.

The unavailability of energy in a low-temperature reservoir is a form of the Second Law (attributed to Lord Kelvin, 1824–1907, British physicist). The modern heat pump (or refrigerator) illustrates that energy can be transferred from a lower temperature to a higher one only with a net expenditure of work.

The Rankine Cycle: Heat Engines Using Steam

Let us look again at the thermodynamic cycle of the reciprocating steam engine (see Fig. 3.1). As we noted in Chapter 3, this engine operates with a piston and therefore a description of it can parallel that of the Carnot cycle. Now, however, we will be considering a real-life operation, using a real physical working substance – *steam* – rather than an idealized cycle and substance.

The *p–V* diagram of the steam engine was given in Figure 3.2. The thermodynamic cycle that the paths in this diagram follow is the *Rankine cycle.* In the Rankine cycle, none of the paths correspond exactly to the ideal processes of the Carnot cycle and two of the paths have mixed processes. Paths 2–3 and 4–1 are, respectively, expansion and compression, but neither can be termed truly adiabatic because of the presence of irreversibilities such as turbulence. Path 1–2 is part compression and part expansion, with the expansion process having a slowly dropping pressure as the cylinder volume starts to increase at the end of the steam input operation. Finally, the exhaust path (3–4), after a short portion in expansion (at the end of the piston displacement), undergoes an almost purely constant pressure process in ejecting the steam into the atmosphere.

Figure A.8 displays, for the steam engine, the various states of the water/vapor working substance by means of the steam dome lines on a *T–S* diagram. On the left

(a)

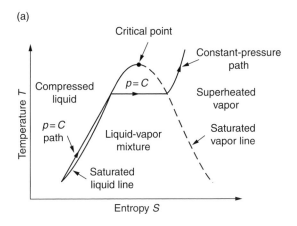

(b)

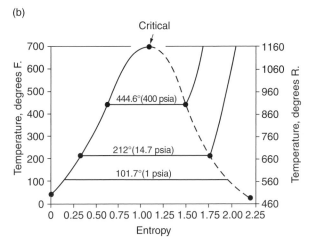

Figure A.8   The *T–S* diagram for water and steam. (a) A single constant-pressure path.
Source: Adapted from Burghardt (1978).
(b) The steam dome.
Source: Adapted from E.B. Norris and E. Therkelson, 1939, *Heat Power*, McGraw-Hill, NY.

in Figure A.8a is the *saturated liquid* line, marking the division between pure liquid and a liquid–vapor mixture. On the right is the *saturated-vapor* line, marking the division between the liquid–vapor mixture and the region of pure vapor, called the *superheated* region. These two lines of division meet on the peak of the steam dome at the *critical point* (Fig. A.8a,b).

If water is heated under pressure, its temperature and entropy will increase along a path just above the saturated liquid line on the *T–S* diagram as shown on Figure A.8a. When the boiling temperature of the compressed water is reached, the temperature then remains constant as more heat is added and the *T–S* path becomes a horizontal line. This constant temperature path continues, and only entropy increases as heat is added, until the saturated vapor line is reached. On Figure A.8b, we can see the heat input paths of the water–steam mixture at different pressures (measured in pounds per square inch absolute, psia): (a) for $P = 400$ psia, $T = 444.6$ °F; (b) for $P = 14.7$ psia (atmospheric pressure) $T = 212$ °F; and (c) for $P = 1$ psia, $T = 101.7$ °F.

If the saturated vapor were to be cooled at a specified pressure, it would trace a return horizontal path at a constant temperature across the steam dome back to the liquid line on the left.

If the steam is heated beyond the saturated vapor state, the temperature would again begin to rise, as shown for the path in Figure A.8a, into the superheated vapor region. Superheated path segments for the pressures 400 and 14.7 psia are also shown in Figure A.8b. We will see that taking the steam into the superheated range can improve the thermal efficiency of a steam plant.

Modern steam power plants (see Chapter 3) use a closed version of the Rankine cycle. Both the $p–V$ and $T–S$ paths for the closed Rankine cycle (shown in Figure A.9) may be compared with those of the Carnot cycle. The heat input path for the Carnot cycle, for example, is entirely isothermal (see Fig. A.7), whereas for the Rankine cycle, this path (1–2) starts by rising in the $T–S$ plane just above the saturated liquid line (see Fig. A.9). The remainder of the heat input process becomes isothermal only after the boiling temperature is reached. As we have noted, this constant temperature path is determined by the vaporization properties of steam for the pressure existing in the boiler.

Point 2 on the saturated vapor line in Figure A.9 marks the end of the $T–S$ heat input path and the beginning of the Rankine expansion process. This expansion process is nearly isentropic (constant S) and therefore drops in the $T–S$ diagram in a nearly vertical path, similar to the ideal Carnot expansion, to point 3. Path 3–4 is the steam exhaust process, which we can now see is a constant pressure *and* constant temperature path, due to the properties of steam condensation. This isothermal path therefore appears similar to the Carnot heat rejection path in the $T–S$ plane. Finally, the Rankine closed-cycle compression process shows as a short *isentropic* (vertical) path (4–1) on the $T–S$ diagram, spanning only a small part of the temperature change achieved in the boiler. Unlike the Carnot compression path, the temperature does not rise over the full range. This compression path for the Rankine cycle is best performed when the working substance is a liquid rather than a vapor or a vapor–liquid mixture.[12]

The Rankine cycle can be, and usually is, extended into the superheated region by continuing heat input, at the same pressure beyond the saturated vapor state (point 2 on Fig. A.9a). When this is done in superheater coils in the boiler, the heat input path is extended to a state in the superheated steam region. On the $p–V$ plane of Figure A.9a, of course, the path extension 2–2′ is merely a horizontal line, since the superheated extension is still at constant pressure. In the $T–S$ plane, the path 2–2′ rises to higher temperatures.

The expansion process from the superheated state is nearly isentropic (path 2′–3′). The chief value of the superheated extension (path 2–2′) is an incremental increase of *available energy* for the cycle, which is greater per unit of heat input than the increments leading to the saturated vapor state (point 2). This increase in available energy is represented on the $T–S$ diagram (Fig. A.9b) by the area bounded by the points 2, 2′, 3′, and 3. According to the principles set down earlier in this appendix, this will increase the thermal efficiency of the cycle, thereby delivering more useful work per unit of input heat and fuel.

(a)

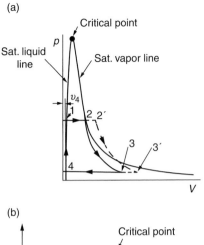

(b)

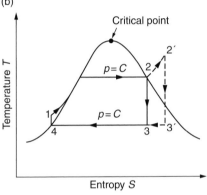

Figure A.9 Rankine cycle diagrams. (a) p–V diagram.
Source: Adapted from Faires (1948).
(b) T–S diagram.
Source: Adapted from Burghardt (1978).

Another significant contribution to thermal efficiency improvement is obtained in the Rankine cycle through the employment of the condenser. The cooling effect of the condenser on the cycle efficiency can be understood through the use of the T–S diagram. On Figure A.9b, it is easy to see that lowering the exhaust temperature ($T_L = T_3 = T_4$) for a specified input temperature ($T_H = T_2$) will increase the available energy, because it increases the T–S area bounded by the paths of the cycle. Conversely, power plants operating *without* a condenser would discharge steam at a higher exhaust temperature (T_L) and therefore have less available energy in the cycle. The efficiency benefit of a condenser may be understood by considering pressure, as observed in Chapter 3. There it is noted than an increased pressure differential (reducing *back pressure*) is created across the steam-driven engine by the cooling action of the condenser.

There are several means of increasing the available energy in the Rankine cycle. One important method is regeneration, as shown schematically in Figure A.10a and b, and represents an attempt to approach Carnot cycle efficiency. It does this by recuperating the heat of vaporization of a fraction (labeled m in Figure A.10a) of the

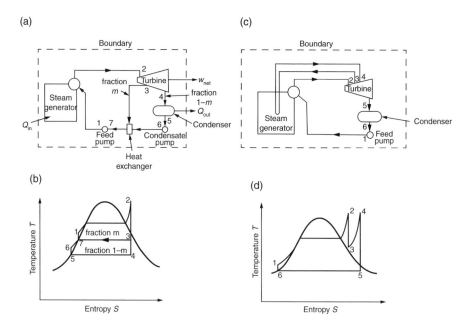

Figure A.10 Rankine cycle efficiency improvements: regeneration and reheating. (a) Schematic diagram for one stage of regenerative heating. (b) *T–S* diagram for one stage of regenerative heating. (c) Schematic diagram for one stage of reheating. (d) *T–S* diagram for one stage of reheating.
Source: Adapted from Burghardt (1978).

working substance to reheat the remainder $(1 - m)$ as feedwater back to the boiler. A single stage of regeneration (power plants usually have several) can improve the cycle efficiency by as much as 7 percent.

Reheating is still another means of improving thermal efficiency. Whereas reheating might be understood on a common sense basis, its function in terms of second-law efficiency can be seen explicitly on the *T–S* diagram in Figure A.10c and d. The addition of area in the superheated region (path 3–4–5) results in a gain in available energy (*T–S* area in Figure A.10d) and a net gain in efficiency.

Increasing Entropy: Further Ramifications of the Second Law

The property entropy (*S*) has another significant interpretation. This interpretation is that entropy is a measure of the *state of chaos* of a thermodynamic system. Systems of any size are considered in these theoretical speculations, ranging from the microscopic scale to the cosmos. On a cosmic scale, it is said that the state of chaos in the universe is continually growing and hence the entropy of the universe is always increasing.

Lest the scope of these generalizations becomes too great, let us consider a single system as depicted in Figure A.11. For a transfer of heat into the system from its

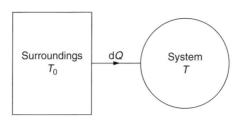

Entropy change for a single system.

surroundings, as indicated, $dQ > 0$ means heat *into* the system and $T_0 > T$ is required for this heat transfer. Therefore:

$$dS_{syst} = \frac{dQ}{T} > 0$$

that is, the change in entropy of the system is positive. The change in entropy of the surroundings for this heat transfer, however, is:

$$dS_{surr} = -\frac{dQ}{T}$$

a negative quantity, because the heat transfer $(-dQ)$ is out of the surroundings. The total (system and surroundings) entropy change is therefore:

$$dS_{syst} + dS_{surr} = \frac{dQ}{T} - \frac{dQ}{T_0} > 0$$

which is a positive quantity, since $1/T > 1/T_0$. If the heat transfer were the other way, *out* of the system, then $dQ < 0$ and $T > T_0$ would be required, resulting *again* in a positive total change in entropy. We therefore see that entropy increases no matter which way the heat transfer goes.

This result of *increasing entropy* has been generalized widely. *Irreversible* processes within a system, for example, lead only to increases in entropy. Even an isolated system with no transfers with its surroundings either has a constant entropy or an increasing entropy: it will *never* decrease. This is expressed simply in this relation for the entropy of an isolated system (dS_{isol}):

$$dS_{isol} \geq 0$$

which is called the Clausius relation (after R.J. Clausius, 1822–1888, German physicist). The inequality of this relation is said to apply to the universe as a whole, and embodies the statement that *the entropy of the universe is continually increasing.*

To better appreciate what this all means, let us look at thermal behavior on a detailed microscopic scale. On a scale of individual particles, or small groups of particles, the microscopic view of entropy entails a measure of the randomness of motion and the degree of mixing of particle types, velocities, and motions. Thus, in this microscopic view we would observe increasing randomness and mixing, with a trend toward a statistically uniform sea of chaotic motion. The universe, according to the cosmic interpretation of entropy, is becoming more and more such a sea of chaos, even though its total energy is constant (the first law of conservation of energy). The only way that energy can be made *available*, according to the theory, is for work to be done on a portion of the sea or for there to be a local release of concentrated energy (such as by heat of combustion).

In an isolated system, the microscopic theory of entropy stresses the increase in chaos or disorder that is to be expected if the theory of entropy increase holds true. This view carries over into another area, much more concrete in concept, namely, mineral resources. Economists and others have observed that the world's natural concentrations of minerals are being dispersed as a result of human activities.[13] Thus, concentrations of metal ores (such as copper), precious crystals (such as diamonds), and other useful elements are mined, fed into the manufacturing process, and sold to consumers all over the world. Many of the mineral-based products are discarded and end up in solid waste disposal sites, mixed with many other materials. Even with complete recycling, which requires an input of work, some mass is lost from any metal product due to ordinary processes of wear.

This picture of gradual dispersal of minerals fits the concept of increasing entropy in several ways. First, there is a random dispersal of the material. Next, connected with the dispersal is a mixing of the pure material with many other minerals (and organic materials too). Dispersal and mixing were among the irreversible processes discussed in connection with hot gases in engines in the previous section. In the case of the dispersal of minerals, as in the cases of throttling or mixing of gases, increasing entropy can be taken as a measure of increasing chaos. Therefore, the observation is that mineral use also represents a trend toward increasing entropy, along with all processes of energy conversion, and both appear to be inevitable as a result of human activity.

Indeed, it is the seeming inevitability of the entropy process that leads to another profound concept – the *Arrow of Time* (Eddington, 1958). This notion points out that since actions, reactions, and activity, in general, are continuously taking place in the universe, the ongoing increase of entropy is itself a measure of the passage of time and its irreversibility even carries with it the unidirectional aspect of the passage of time.

A conclusion sometimes drawn from the prediction of inevitable increases of entropy is that ultimately the universe will suffer a *heat death*. According to this extension of the theory, energy and heat will continue to become more diffuse, leading ultimately to a point when it is no longer feasible to concentrate heat, organize matter, or do useful work. By the time of this ultimate catastrophe, of course, life would long have ceased and the universe would be a cold, diffuse, disordered spread of energy and matter approaching the condition of *maximum entropy* (Georgescu-Roegen, 1971).

Conversion to Electricity

In a conventional power plant, the steam turbine drives the electric generator (see Chapter 3). Conventional electric generation takes place according to the principles of electromagnetism, which were formulated in the nineteenth century.[14]

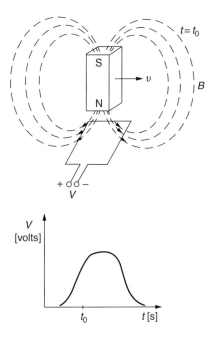

Figure A.12 Voltage induction in a coil by a changing magnetic field. (a) Magnet moving by a coil. (b) Induced voltage versus time.

The generation of electric voltage, for example, is understood through Faraday's law of induction. The understanding of the creation of magnetic fields by electric currents is attributed mainly to Oersted and Ampère, and the forces on current-carrying conductors in a magnetic field to Ampère and others. These effects are key to a basic understanding of present-day electricity generation, but the laws of electromagnetism are also the basis for other major areas of technology, including communications, broadcasting, computers, electromechanical devices, and instrumentation.

Faraday's induction law underlies the creation of electric voltage in the coil of a conventional electric generator. It is shown in its simplest form in Figure A.12, where a magnet is depicted being moved past a loop (a one-turn coil) of wire. If a voltmeter or oscilloscope were connected across the two open terminals of the coil, a voltage that varies in time would be measured (see graph in Fig. A.12). This voltage is *induced* by the changing magnetic field.

The discovery of induction is credited mainly to Faraday, who stated that the *electromotive force* (voltage) induced in a coil is proportional to the time-rate of change of the magnetic flux linking the coil. In the configuration of Figure A.12, we can understand Faraday's time-varying flux linkage as follows. As the magnet moves by the coil on a path parallel to the coil, the number of magnetic field lines that pass through the loop varies. The maximum number of field lines passing through the loop occurs, of course, when the magnet is directly opposite the coil. The sum of the field lines passing through the loop is the magnetic flux linking the coil. Therefore, in

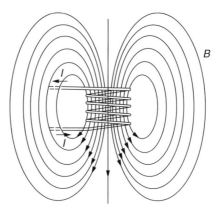

Figure A.13   An electromagnet. Current flows in the coil, thus creating a magnetic field surrounding the coil.

this configuration the flux linkage varies as the magnet passes by, starting from zero, rising to a maximum and then falling back to zero.

Faraday also stated that the strength of the induced voltage in a coil is proportional to the number of turns in the coil. Therefore, if our coil consisted of two turns around the same loop, instead of one, then the same experiment would yield twice the peak voltage in Faraday's law, as expressed in the formula:

$$V = N \frac{\mathrm{d}\Phi}{\mathrm{d}t}$$

where N is the number of turns in the coil, $\mathrm{d}\Phi/\mathrm{d}t$ is the time rate of change of the magnetic flux linkage in each turn, and V is the voltage induced across the open terminals of the coil.

Voltage induction in conventional electric generators takes place essentially as we have just described, but with several major modifications. One difference is that the motion between the magnetic field and the coil is rotary, not translational (see Fig. 3.14). Another important difference is that the magnetic fields are not supplied by permanent magnets but by electromagnets.

Electromagnets are based on Ampère's law, which states that the flow of electric current creates a magnetic field in a plane normal to the direction of the current flow. A simplified picture of an electromagnet is shown in Figure A.13, which depicts a current-carrying coil. The magnetic field lines created by the current flow in the coil suggest that this substitutes for those of the permanent magnet. Electromagnets in fact are a highly developed technology, and magnetic fields can be created that are far stronger than those that occur naturally with magnetic materials.

The operation of the conventional *alternating-current* (a.c.) generator takes place as the magnetic poles turn on the rotor past coils embedded in the stator (a simplified description of a conventional generator in Figure 3.14). The voltage induced in each stator coil alternates in polarity (positive and negative) as the passing magnetic poles alternate between north and south.[15] This alternation is consistent with Faraday's law, because the flux linkages reverse themselves as each pole face passes.

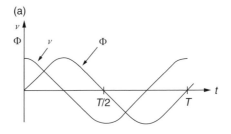

(a)

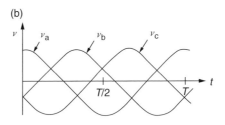

(b)

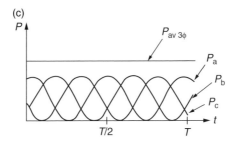

(c)

Figure A.14 Alternating-current generator waveforms in time. (a) Magnetic pole flux (Φ) versus time. (b) Three-phase voltage versus time. (c) Three-phase power versus time.

If the magnetic field is properly shaped at the pole faces, the flux linkage waveform is very nearly sinusoidal (that is, a sine function in time), as shown in Figure A.14a. Because the induced voltage depends on the time rate of change of the flux linking the coil, the voltage is also sinusoidal, but shifted in time. These sinusoidal variations of flux and voltage repeat themselves in a time period T, as indicated on the figure. This time period corresponds to a complete alternation through a pair of poles (for example, N–S–N). In American a.c. power systems, this period is commonly 1/60 second and the frequency of alternation is therefore 60 Hertz (cycles per second).

The basic theory of electricity tells us that the electric power of a generator at any instant is expressed as the product of the voltage times the current at that instant.[16] More precisely, the electric power output of a generator coil is:

$$P = VI \ [\text{watts}]$$

where: $P = P(t)$ is instantaneous power at time t delivered by the generator coil to an external circuit [watts], $V = V(t)$ is the instantaneous voltage across the terminals of

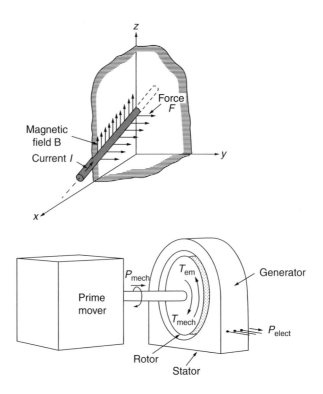

Figure A.15 Prime mover/generator reaction. (a) Magnetic force in a current-carrying conductor. (b) Torques and rotation.
Source: Adapted from Elgerd (1978).

the generator coil [volts], $I = I(t)$ is instantaneous current flowing through the terminals of the coil to an external circuit [amperes].

We can understand the basis of this simple formula if we recognize the nature of each of the two terms in the product $V \times I$. Voltage represents the potential energy to do work on an electric charge,[17] whereas current represents the rate at which charge flows. Therefore, the product VI is the rate at which work is supplied or delivered by a generator to an electrical load, such as an electric motor. The rate of work, however, is power, according to our starting definitions.

In considering the operation of the generator, we conclude that it only supplies energy to an external circuit when a current flows through the terminals of the coil, that is, no power (or energy) is supplied simply by inducing a voltage alone. Since the flow of a current marks the transmission of electrical energy out of the generator, we can now ask the elementary question: how is the current flow related to the conversion of energy in the generator, as driven by the prime mover?

The flow of current in a stator coil sets up its own magnetic field, as shown in Figure A.13. This stator magnetic field reacts against the electromagnets of the rotor. These reaction forces on current-carrying conductors are depicted in Figure A.15a. In the case of the a.c. generator, the magnetic field resulting from the stator current causes forces on the current-carrying conductors in the rotor (pole) windings. These forces on the rotor winding always tend to *oppose* the motion of the rotor pole and thereby set up a counter-torque to the mechanical torque supplied by the prime mover (see Fig. A.15b). Such a reaction force is observed in all sorts of electromagnetic systems and is generally predicted by Lenz's law.[18] The prime mover (steam turbine) must supply mechanical *work* against such a force (torque) of reaction in electric generation.

The Transformation of Electricity

After being generated, electricity must be transmitted and distributed. The conventional means of transmission is either through overhead lines (heavy wires) or underground cables. Either means of transmission results in electrical losses along the conductors.

Early in the evolution of electrical technology, it was recognized that electrical losses are reduced by operating transmission lines at high voltages. The reasons for this can be appreciated from two simple considerations: The first follows from the law of electric power as just described, namely, that power is equal to voltage multiplied by current. Thus, for any specified amount of power to be transmitted, the higher the voltage used, the lower will be the current required to transmit that amount of electric power. For example, if we elect to transmit power at twice the voltage (V) then we will find that half the current (I) is required to transmit that given number of watts (P). Second, the losses in conductors vary with the square of the current flowing in the conductor[19]; therefore, losses decrease *inversely* as the square of the operating voltage of an electrical transmission line. So if the voltage is doubled, the losses are reduced by a factor of four.

The means of achieving high voltages for low-loss transmission was the electrical transformer (Fig. A.16).[20] A transformer converts the voltage and current as generated to a higher voltage and lower current. Aside from small losses in the transformer, the electric power is transmitted *through* the transformer as indicated on Figure A.16.

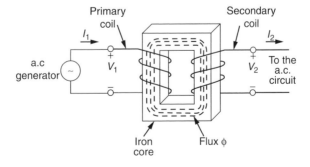

Figure A.16   Idealized electrical transformer. For a.c. power flow: $P_{in} = V_1 I_1$; $P_{out} = V_2 I_2$. For zero losses: $P_{out} = P_{in}$.
Source: Adapted from M. El-Hawary, 1983, *Electrical Power Systems: Design and Analysis*, Reston, VA.

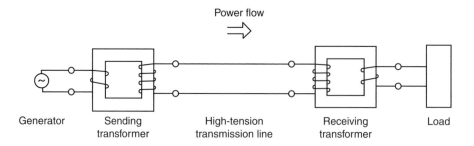

Figure A.17 A single-phase a.c. transmission line.

Transformers operate on the *principle of induction*, although applied in a different way than we have described for generators. The transformer, for example, has no moving parts. Instead it uses time-varying magnetic fields created by alternating currents to *induce* voltages. Therefore, in order to use transformers, it became necessary to use an a.c. electric system. The voltage magnitude induced across the secondary coil (V_2) varies directly with the a.c. primary voltage (V_1), and its magnitude depends on the ratio of the number of turns in the secondary coil to the turns in the primary, as expressed by:

$$V_2 = \frac{N_2}{N_1} V_1$$

where N_1 is the number of turns on the primary coil and N_2 is the number of turns on the secondary coil. This follows, of course, from the law of induction, so that the magnitude of an induced voltage in a coil is proportional to the number of turns in the coil. This formula expresses the *transformation of voltage*, which can be a *voltage step-up* if:

$$\frac{N_2}{N_1} > 1$$

or a *voltage step-down* if:

$$\frac{N_2}{N_1} < 1$$

In the idealized transformer shown in Figure A.16, we see two coils wound on a closed iron core. The primary coil on the left is connected to the generator that supplies the a.c. voltage (V_1) and current (I_1). The a.c. primary current (I_1) creates an alternating magnetic flux (Φ), which is concentrated in the iron core.[21] The alternating magnetic flux links the secondary coil and, according to the induction principle, induces a voltage (V_2). If the secondary coil is connected to a circuit, current (I_2) then flows out of the secondary coil and a.c. energy flows *through* the transformer. In practice most electrical transformers use separate coils that have no direct electrical contact with one another, but rather rely on magnetic induction.

High-tension (high-voltage) transmission circuits are illustrated in Figure A.17. The transformer at the sending end (which is fed by the generator) has a larger number of turns on the secondary coil than on the primary coil and therefore provides a step-up in voltage. At the receiving end, on the other hand, there is a step-down in voltage, because the primary side connected to the transmission line has more turns. If the step-up of voltage for this circuit were 10:1, the losses in the transmission lines would be 1/100 (1 percent) of those with no step-up.

Bibliography

Thermodynamic Principles

Krenz, J.H. 1984. *Energy, Conversion and Utilization*, 2nd edition. Allyn & Bacon, Boston, MA.

Severns, W.H. and H.E. Degler. 1948. *Steam, Air and Gas Power*. Wiley, New York.

Faires, V.M. 1948. *Theory and Practice of Heat Engines*. Macmillan, New York.

Burghardt, M.D. 1978. *Engineering Thermodynamics with Applications*. Harper & Row, New York.

Van Wylen, G.J. and R.E. Sonntag. 1976. *Fundamentals of Classical Thermodynamics*. Wiley, New York.

The Law of Entropy and Its Philosophical Implications

Georgescu-Roegen, N. 1971. *The Entropy Law and the Economic Process*. Harvard University Press, Cambridge, MA.

Eddington, A. 1958. *The Nature of the Physical World*. University of Michigan Press, Ann Arbor.

Weinberg, A.M. 1982. "Avoiding the Entropy Trap." *The Bulletin of the Atomic Scientists*, October.

Electromagnetism and Electricity

Singer, C. J., E.J. Holmyard, A.R. Hall, and T.I. Williams. 1958. *History of Technology: The Industrial Revolution*, Volume 4. Oxford University Press. Oxford, UK.

Elgerd, O.I. 1978. *Basic Electric Power Engineering*. Addison-Wesley, Reading, MA.

NOTES

1 The MKS unit for charge is the coulomb (the charge of about 2.56×10^{18} electrons) and for current is the ampere (1 ampere is a charge flow of 1 coulomb per second).

2 The kilocalorie is based on the heat required to raise the temperature of 1 kg of water by 1 °C. Heat energy can also be measured in (metric) joules along with mechanical energy or electrical energy, and so this is a duplication within the MKS system. The equivalence between the two is $1\ J = 0.24 \times 10^{-3}$ kcal.

3 Voltage is commonly thought of as a measure of the force pushing the electric current through conductors, much as water pressure pushes the flow of water through pipes.

Although this is not entirely inaccurate, a better conception of voltage as a measure of electric *potential energy* is given later on in this appendix (see note 17).

4 A mass of 1 slug has a gravitational force of 1 pound at the Earth's surface.

5 The BTU is based on the heat required to raise the temperature of 1 pound of water by 1 °F.

6 One kWh is the energy delivered or used at a power of 1 kW (10^3 watts) for 1 hour and is the familiar measure used in electric rates. The *heat rate* of a thermally driven electric generator is commonly measured in BTU/kWh; 3412 BTU/kWh represents a heat rate at 100 percent conversion efficiency.

7 This approximate conversion gives the average kinetic energy measured in keV (10^3 electronvolts, see Units of Measure section) of a particle in a hot gas of plasma (see Appendix C). The exact relation is (kinetic energy) $E = kT$, where $k = 0.862 \times 10^{-7}$ keV/K (Boltzmann's constant), and T is absolute temperature on the Kelvin scale.

8 The barrel (bbl) is the common measure of petroleum volume.

9 More precisely, heat content is subsumed into the scientific engineering term internal energy, which will be used in the discussion of thermodynamics that follows.

10 In the case of systems capable of storing energy, this equation becomes $E_{in} = E_{out} + E_{stored}$. We review practical systems with storage, including conventional hydroelectric stations, pumped hydroplants, and batteries in Chapter 3 and Appendix C.

11 In the closed path integrals on Figure A.7b:

$$\int_{S_2}^{S_3} = \int_{S_4}^{S_1} = 0$$

and

$$\int_{S_3}^{S_4} = -\int_{S_4}^{S_3}$$

12 It is possible, physically, to compress the steam–liquid mixture isentropically (vertical T–S path) to a higher temperature/pressure point at or near the saturated liquid line from which the heat input process in the boiler could proceed. Operation with the liquid–vapor mixture causes practical difficulties such as corrosion and pitting of mechanical parts.

13 We are discussing here only durable minerals that do not change in basic composition as a result of human use. Thus we are excluding minerals that are sources of chemicals (for example, calcite and sulfur), which undergo chemical reactions in manufacturing. We also exclude here uranium and thorium, which undergo the even more fundamental changes of nuclear reactions in their use as fissionable fuels, and fossil fuels, which undergo chemical change in combustion.

14 For a readable account of the development of the theory of electromagnetism see Singer, Holmyard, Hall, and Williams (1958). Major contributors to these theories were A.M. Ampère (French physicist, 1775–1836), C.A. de Coulomb (French physicist, 1736–1806), M. Faraday (British physicist and chemist, 1791–1867), K.K. Gauss (German mathematician, 1777–1855), J. Henry (US physicist, 1797–1878), J.C. Maxwell (British physicist, 1831–1879), and C. Oersted (Danish physicist, 1777–1851).

15 Magnetic fields have a direction associated with them. The field direction is customarily chosen to be positive as the lines emanate from a north pole.

16 Electric power is the time rate at which electrical energy is delivered by the generator at a particular instant.

17 The exact definition of a volt is the work done in moving an electric charge against the electric forces of repulsion by other charges. The MKS unit of 1 volt represents the change in electric potential when 1 joule of work is done against electrostatic forces on 1 coulomb of charge.

18 Lenz's law states that when a time-varying magnetic flux induces a voltage in a coil and the voltage is permitted to produce a current in the coil, the current will always flow so as to oppose the magnetic flux changes inducing the voltage (after H.F.E. Lenz, Russian physicist, 1804–1864).

19 Ohm's law tells us that the voltage *drop* (change) along a length of a conductor is

$$V_R = IR$$

where I is the current through the resistance and R is the resistance of a length of conductor. The law of electric power tells us that the power delivered (lost) to the length of conductor is:

$$P_R = V_R I = I^2 R$$

20 Faraday applied his theory of induction to both an electric generator and a form of the transformer in 1831. However, early in the electric age, most electric power came from direct-current (d.c.) generators. It was not until later in the century, after the alternating-current generator had been developed by W. Siemens (German engineer, 1816–1892) and others that the idea of using transformers in a.c. currents was developed. Credit for the introduction of a.c. systems to America goes mostly to George Westinghouse (US engineer and industrialist, 1846–1914).

21 Materials such as iron, which can be magnetized when put in the vicinity of a permanent magnet or a current-carrying coil, are highly *permeable* to magnetic flux. When magnetic field lines flow through a closed path that is composed of such material, magnetic flux becomes concentrated. In transformers, the iron core channels the magnetic lines so that a maximum flux linkage of the secondary coil is achieved. Iron is also used in generators to guide and shape the magnetic flux (see Chapter 3).

Establishing Criteria for Safety and Health Standards

A democratic society can choose which energy technologies to adopt. But to choose the benefit of any technology is also to assume certain risks. We have discussed some specific risks in connection with fossil-fired and nuclear generating technologies. But the issue of risk raises two more general, indeed philosophical, questions. First, how do we assess the risks? That is, how are risks to be determined and then factored into an analysis of costs and benefits? And from that point, how are they to be used to determine policy? What must be kept in mind is that not only is risk a potential cost, but that avoidance of risk has a cost that we are sometimes unwilling to pay. Second, what are the decision criteria and who – government officials, technologists, the general public, or some combination – decides what society pays?

These questions have produced a voluminous scholarly literature overlapping various fields – philosophy, economics, law, psychology, and so on. The responses to these questions may be crucial in energy-policy decision making. Without some conscious analysis of the method of risk determination, and some informed choice among alternative procedures, the exercise of choice on energy technologies may be less than rational or democratic.

Risk and Uncertainty

Before looking at the specifics of decision-making processes and analyses, we need to understand an important distinction: that is, the difference between risk and uncertainty. Most broadly, a risk is where a decision is to be made where there are two or more possible outcomes but the probability of each outcome is known or can be ascertained with some degree of assurance. Uncertainty is where the probability of each outcome is not known. In fact, one can think of a risk–uncertainty continuum, where on the far end of the "risk" side all states of the world (post-decision) are known as are the probabilities of attaining each. On the other end is complete uncertainty or ignorance (Tversky & Fox 1995), with contingencies that cannot be accounted for *ex ante*. Of course, one would likely be somewhere in the middle with respect to a decision. Some possible states of the world are known and some probabilities can be assigned either because of empirical evidence or simply as a subjective appraisal. Other states are less clear and "surprises" are possible.

There is a further issue of general importance with respect to assessment of energy and environmental policies: irreversibility. In other words, something like pollution from a coal-fired plant can be reversed; once the plant is shut down the pollution ends. Even a hydroelectric dam, which inundates fields and villages, can presumably be demolished, the villages rebuilt, the fields restored – albeit at very high cost. But climate change, on the other hand, may not be reversible. Indeed, as noted in Chapter 6 there are arguments that climate change could lead to extinction of human life on Earth. But this remains at the far end of ignorance; because probabilities are unknown, the costs and benefits of arriving at this state are nonquantifiable (Ackerman & Heinzerling 2002).

We will, however, largely confine ourselves to the risk side of the continuum – while noting that, in some cases, risk and uncertainty become very close and in the process calculating the cost of outcomes becomes increasingly difficult.

Assessing Risk

Clearly, some actions and activities, for example those that pose a danger to life and limb, such as handling nuclear waste, contain some element of risk – greater risk, for example, than working behind a desk. Acts that better people's lives – such as rural electrification – offer some benefit. But individuals will not always agree that one action is risky or that another action provides a benefit.[1] Indeed, any event may present a perception of risk to one and opportunity for benefit to another. And if it is difficult to determine what is a risk, it is harder still to determine how much a bad outcome of a risk will cost or how much a possible benefit through a risk reduction would be worth. Yet these kinds of evaluations are, and some could argue *must* be, made in the formulation of energy policy. Society needs to be able to analyze the potential consequences of its choices. As a result, policymakers have sought ways to assess and evaluate – and indeed, to quantify – the risks, costs, and benefits of decisions.

This effort at assessment and quantification originated in utilitarian concepts of the nineteenth century. Early utilitarian thinkers believed that happiness itself could be quantified by some objective measure. If it were possible to make such a determination, as some economists of the period believed, then decision makers would be able to count the happiness produced and pick actions that produced the most happiness. Decision making would be simple – a matter of counting. Of course, this was demonstrated to be simplistic, not simple, but the desire to find measures for quantification of risk remained for good reason: without them, policymaking is likely to be based on seemingly arbitrary criteria.

Whatever approach policymakers use, they must start from two realities of all cost–benefit decisions:

- Perfect safety is unattainable.
- Societal resources – money, labor, capital, raw materials, and so on – are finite.

The result is that society, through its officials or through direct referenda, must specify the level of safety to be achieved, and allocations, as we have noted, must be made between competing claims. In other words, society must choose to expend resources to promote safety in one area and not in another; there are always trade-offs.

In making decisions, US government officials have adopted an approach that can be outlined in the following fashion (from Hiskes & Hiskes 1986):

- Identify the problem and formulate policy objectives.
- Consider alternative courses of action.
- Determine consequences (positive and negative) of each alternative.
- Analyze the probability of each consequence.
- Assign costs and benefits to each consequence.
- Select the best alternative.

But what is the best way to analyze the costs and benefits?

Risk–Cost–Benefit Analysis

To evaluate competing claims in a systematic manner, policymakers have come to rely on a quantitative method called cost–benefit analysis (CBA) or risk–cost–benefit analysis (RCBA).[2] In CBA, engineers and/or economists weigh the costs of creating or controlling a particular technology against its benefits. As the degree of control increases the risks fall, but the costs of control rise. Thus, the objective is to arrive at the optimal (minimum) total cost, defined at the point where the cost of the last unit of safety is equal to the value of the benefit received from it.[3] In CBA, benefits and costs are figured on a market basis and are measured in standard units of currency. As we will discuss, the results of CBA (or RCBA) do not necessarily determine policy choices, but they do inform them.

In many cases, RCBA uses the same initial calculations as CBA, but by adding the element of risk, the approach becomes increasingly subject to controversy since riskiness of any activity is often subjective and contingent (Barnett & Breakwell 2001). RCBA weighs nonmarket variables, attempting typically to include the "costs" of hazards to human life, health, and safety. These risk variables are often quantified indirectly – by what people actually pay to reduce a similar risk or more directly by asking what they *would* pay for a particular risk reduction. Depending on the variation of RCBA methodology the analyst chooses, risk factors are typically considered on a probabilistic basis to arrive at an expected "likely" total cost.

Although, inevitably, any technological project involves a degree of uncertainty in its initial stages, CBA is straightforward when we consider, to use a classic example, the costs and benefits of building a new highway. There, analysts can estimate the construction cost and the value of total benefits, and they can compare those costs and benefits against other alternative uses for the financial resources. Dollar estimates make perfect sense. The costs of building the road, financing it, and maintaining it are estimated in a straightforward manner. It is also realistic to place a dollar value on the benefits, including the economic improvement possible through a better transportation system. Analysts use historical, empirical data; the effects of such projects have been observed before.

The difficulty in RCBA lies in the fact that the costs and the benefits of a technology do not always lend themselves easily to specific dollar values; indeed, some do not lend themselves easily to quantification at all. Consider the issue of pollution from fossil-fired electric power generation. The potential cost of property damage may be calculable on a dollar basis, although the amount an analyst will choose may vary widely, depending on whether controversial elements, such as the prevention and cleanup of acid rain, are added. Those who see acid rain as a direct consequence of fossil-fired generation may want to increase the potential value of property damage many orders of magnitude higher than someone less convinced of the cause of acid rain. The estimated cost is likely to vary as well, depending on the extent of control considered necessary.

But the issue of property damage is far less thorny than other aspects of air pollution. What, for example, is the cost if we feel that air pollution might destroy an outdoor stone sculpture? Perhaps we can make a tenuous valuation on the basis of comparative art auction prices, although that value cannot completely compensate for the loss of a unique aesthetic object.

More importantly, how do we account for the impact on our health and lives? To deal fully with the impacts of pollution, we have to add health and safety to the analysis. But what is the dollar value of human life and suffering? Some economists and engineers have tried to establish such values. Furthermore, the courts routinely make cash awards to those who have suffered from the impacts of technology – a practice we generally regard as acceptable, even just.[4] Yet such calculations are subject to enormous variation and disagreement by the very people who perform RCBAs (Rhoads 1980). Indeed, they use different units of measure. While some place a dollar value on human life, others measure the cost on its own terms, for instance, as excess deaths per 100,000 people. But since other costs and benefits have distinct monetary values, this leads to difficult comparisons where deaths are weighed against dollars.

To simplify the comparison, some analysts have attempted to arrive at an appropriate dollar value of life. A *human capital* approach to RCBA values a life as a stream of income through time. If the person becomes ill or dies then his or her income is disrupted, so that the analyst adds up instead a projected stream; this summed value, appropriately adjusted (discounted) for the time value of money, is then presented as the value of the individual life (Robinson 1993).[5] But, of course, to do so raises serious ethical questions. Is a person's value strictly economic? Is he or she to be measured only as the sum of expected earnings? Can we argue that a project that could kill 100 people and injure 1000 has a human cost that is only the sum of lost work days, hospital bills, and other financial losses? Few would say this is the case; in fact, any such measurement of human life, at first glance, might seem contrary to ethical norms (Johansson-Stenman 1998).[6]

The human cost inherent in any energy technology is so problematic that it has sometimes been simply left out by those decision analysts who would be most expected to quantify it. Some have argued that giving any value for nonmarket, subjective costs is wrong in principle, and their solution is to perform RCBAs that only include readily quantifiable elements. They maintain that only by avoiding human costs can an unbiased, objective, value-neutral study be achieved.[7] But is this really a better, fairer approach than a human capital valuation?

There are crucial questions raised by studies that claim to be value-neutral. Implicit in the concept of such a study is the idea that facts and values can be clearly separated and also that an RCBA can and should deal only with objective facts.

Many philosophers have strongly questioned whether a fact–value dichotomy can be truly said to exist in RCBA or even in science (Michalos 1980, Shrader-Frechette 1985). Science does deal in facts, but these *facts* are subject to change and *evaluation* as the data change. If indeed science dealt only in immutable facts, then how could there be new theories that attempt to evaluate data and explain causality (Shrader-Frechette 1985)?

The "facts" of RCBA are even more problematic. First, the very act of choosing details deemed objective over those deemed subjective is itself evaluative and seems to contradict the premise of a value-neutral study. Second, the "objective" facts usually included in RCBA involve market valuations, which may be observable and factual, but are at the same time extremely variable and subject to fashion, taste, and many other highly subjective factors.

Even the prices of the necessities of modern life such as energy resources – presumably less affected by taste – truly reflect value only at the instant they are given. Consider the price of a commodity such as oil. On a market basis in the 1970s, it went from a few dollars per barrel to over $30 in less than a decade, and then within a couple of years slipped to less than half that peak amount. Since RCBA is intended to evaluate projects at the planning stage, if the price of oil is to be included, it must be given at some estimated future price – that is, at a time when the plans will have been put into effect. What is the objective valuation of the future price of oil? Clearly, it may depend on when the study is performed, and the views and values of the person performing the study. Indeed, there are probabilistic estimates for decisions under uncertain futures, called *Bayesian estimates,* that are acknowledged to be subjective (Raiffa 1970).

Since even the most objective facts may be inseparable from the decision analyst's values, it seems especially difficult to justify ignoring nonmarket elements because they are too value-laden. But even if facts could be separated from values, it would still be questionable to leave out human costs, where they exist. Doing so might produce a study that is in some sense more objective, but far from complete. Such an approach appears if anything more biased than the one that gives an arbitrary value of human life. At least the latter acknowledges human costs. To avoid confronting them may buy objectivity at the cost of irrelevance.

Some theorists who argue for the elimination of nonmarket variables from RCBA do so not to ignore them, but rather to avoid quantifying factors they believe cannot be quantified. These theorists have suggested that to quantify nonmarket variables gives inordinate power to the decision analysts, since their cost–benefit numbers would then seem to be all-inclusive (Mishan 1972). According to such a viewpoint, RCBA should focus on what is quantifiable and analysts should simply note those factors that cannot be quantified. Then it is left to the policymakers (not the decision analyst) to decide which factors – market or nonmarket – should weigh more heavily. In such a policymaking process, it could then be decided that even though an RCBA demonstrated cost effectiveness, a project simply would not be viable for ethical, political, or social reasons. This approach keeps RCBA in its place as a tool for policymaking, rather than having it determine policy.

But an equally good case can be made for quantifying all the important variables, however difficult such a task may be. The issue may hinge on the answer to this question: if costs and benefits are not quantified, exactly how are they to be weighed in the political process against costs and benefits that *are* quantified? In other words, although we can say that pollution is bad for our health, how do we weigh that statement in order to determine the required amount of pollution control, which may range in cost from a few dollars to over $100 million for a single power plant?

Philosopher K.S. Shrader-Frechette has noted that quantifying nonmarket variables at least provides a basis for discussion and argument. She argues, for example, that by giving explicit numerical parameters, the Rasmussen study on nuclear plant safety (Chapter 7) allowed for explicit challenge; vague generalities would have been harder to confront (Shrader-Frechette 1985). The issue for those who believe in complete quantification is how best to do it. Clearly, just assigning arbitrary values for human life and other nonmarket variables is not adequate. But, as we will see, various ways have been developed to improve the valuations of RCBA.

United States Government Decision Making

Policymakers in the United States rely widely on CBA, which has been mandated for all federal technological projects.[8] Many of the analyses have proceeded with the assumption of value neutrality. There are numerous examples of government-mandated cost–benefit studies that rigorously avoid quantification of nonmarket elements. For instance, in 1975 an RCBA on oil tanker safety by the Congressional Office of Technology Assessment ignored the costs of oil spills. In particular it omitted such costs as aesthetic damage to beaches, the cost to other vessels of avoiding contaminated waters, and the potential damage to the fishing industry. All of these costs were difficult to quantify and were thus left out (critiqued in Shrader-Frechette 1985).

It should be noted that value neutrality seems to contradict the basic spirit of government policymaking. In the very act of setting policy objectives, the government begins with a set of values, not facts. Policies are adopted (we can assume at least in part) because they are beneficial and will improve the general welfare of society. In making such a choice, policymakers use their judgment of how a government, an economy, and a social structure should function. Often, value neutrality is simply impossible. Government policymakers have had no choice at times but to confront nonmarket issues, in particular the potential human costs of technology. However, at such times, rather than giving exact directions for the establishment of standards for health and safety, they have tended to create vague criteria. In each of the electric-generating technologies we have covered, regulatory criteria avoid explicit calculation of dollar costs and benefits to human life and suffering. Officials have mandated that the projects must have an *acceptable level of risk*, or, equivalently, they have used the principle of a *threshold level* below which no measurable risk presumably exists.[9]

A clear example of how this basic approach works in practice is the Nuclear Regulatory Commission's policy on nuclear power plant safety. Following a review of plant safety in the aftermath of the Three Mile Island accident (Chapter 7), the

NRC issued a document entitled *Safety Goals for Nuclear Power Plants*. These goals are stated qualitatively and in nondollar quantitative terms (paraphrased here):

- Members of the public should bear no significant additional risk to life and health, beyond an established baseline level, because of the operation of a nuclear plant.
- The risks of generating electricity by nuclear plants should be no greater than those due to viable competing technologies (mainly coal-fired generation).[10]

Phrases such as "acceptable level of risk" or "significant additional risk" are vague and unhelpful as part of the analytical input to policymaking. But in this case, does the quantifiable standard – comparing risks – really provide a better measure? Only if it is clear just what the probability of an accident is. The NRC established a standard of what that probability *should be*. It stated that each nuclear reactor should be built so that there is only one chance in 10,000 that it will have a core meltdown.[11] This formulation stems from the Rasmussen study, which estimated the probability of a meltdown and of a worst possible accident. As related in Chapter 7, the results showed that such probabilities were very low and were intended to demonstrate that the hazards could be ignored in everyday life by the average citizen.[12] As we noted, however, there has been considerable question about the accuracy of the Rasmussen probability estimates and the standard has evidently not been reassuring to the population at large.

Alternative Methods of Establishing RCBA

Although government regulators have in the past tended to use vague or untestable standards for levels of public risk, there are alternatives that have been utilized often in recent years especially with respect to energy and the environment.

Revealed Preference

One method that has had a number of proponents, in the technical community especially, is called the revealed-preference (RP) approach. This method says that there is a level of acceptable risk, and it offers a statistical procedure on how to determine it. The idea is that we accept certain risks in our daily lives. We drive cars, walk the streets, use electrical appliances, ride in elevators, take medications, and so on; all of these can, and have, caused injuries and deaths. Of course, we also spend money to avoid risks: we buy insurance, we add security equipment and fire detection equipment, and so on.

In essence, RP is saying that if a technology can be shown statistically to have little risk, indeed less risk than most of the above, and if it has substantial benefits as well, then we should have no hesitation in adopting it. This argument was employed by Rasmussen, as well as by Nobel Prize-winning physicist Hans Bethe, in supporting nuclear power development in the United States.[13] However reasonable such a position might seem, the risks being compared are qualitatively different – some are voluntary, others involuntary. We may indeed be willing to assume a risk in order to drive a car, recognizing that it benefits us directly to have this method of transportation at our disposal. But the decision to build a power plant is often governmental and/or corporate, and those potentially at risk may have little or no influence on the decision. In other words, driving a car would be a voluntary risk, while the power plant would be distinctly involuntary. The two risks are, thus, qualitatively quite different. To accept one voluntarily is not a sufficient condition for allowing the second to be imposed on you. To argue that a person ought to accept one risk because he or she accepts the other is an example of the naturalistic fallacy in ethics.[14] Statistics about relative dangers, even if they are absolutely correct assessments, are far less important to the argument.

Moreover, such a general observation of behavior does not get the kind of quantitative measure most analysts are looking for. There are, however, two basic ways of quantifying revealed preference: the travel–cost method (TC) and the hedonic price method (HP) (Atkinson and Mourato 2008). The former is often used in analysis to value natural spaces. In other words, if a power plant or a dam is to be constructed in an area thought of as a natural preserve or recreational area, cost–benefit analysts can calculate the number of people who travel to this area; how far they come to get there; and whether they typically spend money in restaurants or stay overnight in local hotels. Through these factors, an analysis can establish a baseline for the locale's *value* as a tourist location and thus the opportunity cost if the plant, dam, etc. is constructed.

Hedonic pricing (HP) is a process of decomposing the various attributes of a good or service and establishing values for those attributes. So, for example, an HP analysis of housing prices would calculate how air quality, distance from shopping, and other attributes affect the market price of the house. With respect to health and safety, one datum has been the differential in wages between jobs with different levels of physical risk and an analyst may assume that risk accounts for the differences (Atkinson and Mourato 2008).

Of course, with respect to a new power plant, there is still the question of voluntary versus involuntary risk, and some advocates of the revealed-preference method have attempted to introduce a compensating factor into their calculations. For example, involuntary risks may be assigned a higher value and therefore require a greater amount of benefit to offset them (Starr 1976). However, attempts at finding an appropriate level of compensation have produced widely disparate results.

But such calculations tend to obscure larger issues, not of reasoning, but of equity in a democratic society. It can be considered from this standpoint: who bears the costs and are they the same people who reap the benefits (Sunstein 2005)? Thus, when governmental authorities decide ultimately whether to build a power plant, we find that their decision requires some individuals to bear special risks. Although the authorities may authorize building the plant, presumably in order to provide benefits for the many, the decision makes the burden of risk not only involuntary but also unequal. So, whenever a power plant is built, some are put at greater risk than others by mere proximity to the plant, if nothing else. In fact, it may well be that some are imposing risks on others that they themselves would be unwilling to assume – a proposition that seems incompatible with our notions of equity and justice.

Expressed/Stated Preference

One creative solution to the ethical and evaluative dilemmas of RCBA methodology was proposed by E.J. Mishan, an economist specializing in RCBA theory. Mishan (1976, 1981) proposed an approach that takes the cost–benefit evaluation, in principle, away from the remote expert analyst (an engineer or an economist) and places it with the population at risk. In other words, instead of analysts attempting to calculate the value of health and well-being to a community affected by a change in risk, the analyst needed to determine the community's own valuation of that change in risk. That is, what is the valuation that each individual in the population at risk places on an improvement (or degradation) of his or her own level of risk?

This has become a common form of cost–benefit study and is termed the expressed or stated preference (SP) method. Actually, SP can take two forms: willingness to pay (WTP) – that is, how much would one pay to avoid a risk; or willingness to accept (WTA) – how much would one want to receive to tolerate an increase in risk. WTP and WTA fall under the category of *contingent valuation*.

As in revealed preference, the emphasis is on the cost of *risk*; *it does not try to put a value on human health or lives*. The cost of reducing a risk is far more easily and ethically quantified than the value of a life. But this method improves on revealed preference by referring to those affected. It calculates what a change in risk means using the underlying principle that each person knows best his or her own interests. It also gathers data on the preferences of those affected by the specific risks.[15]

Stated preference is usually conducted by asking a representative sample of those at risk (directly or by questionnaire) how much a risk meant to them and how much they would be willing to pay to reduce or eliminate it. In other words, if a project meant near-certain death or injury to some members of the community, they probably would be unwilling to accept it at any cost or be willing to pay a substantial amount to reduce that risk. If, on the other hand, they were persuaded that the risks were negligible, they might not be willing to pay anything. The statistical facts, although important, would not be as important as the meaning or perception of risks, including the consequences of the hazard in question.

Although this method provides data on public perceptions and feelings, it is designed only to create a better means of quantifying risk and is not intended as a referendum of public opinion on a technological project. Nevertheless, because it would focus on the choices of the population instead of the experts, it does at least appear more democratic than either the human capital or revealed-preference approaches.[16]

Still, there can be problems with SP or any system if the problem is large enough. For example, what is the "value" of the ecosystem? One might think that an ecosystem could be decomposed into various parts and people asked about each of those separately and then the numbers are added up to give a value of the ecosystem. But, as Arrow et al. (2000) note, such a *system* is inevitably more valuable than the sum of its parts.

Stated preference also may provide a somewhat inaccurate picture. For example, the way questions are phrased can bias survey results. Also, since the people at risk generally lack technical expertise, there is a worry that researchers are apt to so oversimplify the issues that they no longer reflect the reality of the problem. There is a drawback to the method, too, when we consider energy projects that are meant to last (or whose effects will last) over many generations. In such cases, we will have some people judging the acceptability of risks today for people in the future whose values and belief systems may well be different, and whose choices about risk are impossible to predict (Grossman & Cassedy 1985). Indeed, one generation acting for its self-interest could irreversibly damage the world for the next, especially if we consider that people resemble the rationally self-interested human being of economics. As Gowdy (2004) argues, "People's individual preferences may be incompatible with long-term human survival."

Ethically Weighted RCBA

An alternative approach to overcoming the problems of RCBA is to adopt an ethical weighting in calculating costs and benefits. This seems a natural element in the process. Johansson-Stenman (1998) observes, "[T]he issues dealt with in environmental economics are often of an ethical *nature*. Hence, the choice is not really whether we should impose ethical values or not, but whether we should deal with ethics explicitly or implicitly." (Emphasis in the original.) Operationalized, decision analysts would have to use ethical values as the overriding criteria in their RCBA. In other words, any technological proposal that would lead to a violation of fundamental ethical values would require significant revision or would be rejected outright – regardless of its market cost effectiveness.

One such concept would require the adoption of a single precept of distributive justice to provide the framework for analysis (Kneese, Ben-David, & Schulze in MacLean & Brown 1983). For example, one could use the utilitarian concept of the greatest good for the greatest number or a simplified egalitarian idea that actions must improve the lot of those who are worst off.[17] Under such constraints, decision analysts would have to determine (if they are using the second concept) the group worst off and see whether it would benefit from the project under study. If those who were most disadvantaged became still worse off, either they would require very large compensation or the project would be eliminated even if there would have been a general improvement of society's welfare.

Through such a system of ethical weight, we might also be better able to make judgments about intergenerational issues. For example, using the utilitarian formulation, we could say that we must include all those in the future who are likely to be affected. If they are likely to be put at great risk or to be burdened with greater costs while we in the present generation reap the benefits, then the project would be unacceptable. Of course, using any given rule might leave unclear the ethical validity of one or another course of action. In that case, policy could be set by the usual factors of standard cost–benefit calculations. However, at least the problem then would have been scrutinized from an ethical perspective and could be cleared of violations of ethical norms.

Of course, there is an obvious problem with this approach. It is apt to reduce all ethics to one easily phrased precept. In a pluralistic society such as ours, there is not likely to be a single norm of distributive justice that all would endorse. One way to overcome that problem might be to specify two or three, or even a dozen, different precepts that are widely held and then perform, in essence, several different RCBAs. This, however, could lead to confusion for policymakers. A.V. Kneese and his colleagues have demonstrated that as few as four different rules of distributive justice can lead to four different conclusions – sometimes diametrically opposite ones.

There are other important questions with respect to SP. What, for example, are the "motives" behind the willingness to pay (Johansson-Stenman 1998)? And how much do people care about the "means" used to achieve whatever ends a project intends (Gowdy 2004)?

Yet it might be useful for policymakers to have several different positions to consider.[18] One problem with a conventional RCBA is that by providing one set of figures, it tends to disallow ambiguity. Several ethically weighted analyses, in contrast, might point up ambiguities and thereby stimulate a fuller discussion of the social and moral implications of a given technology than might otherwise have taken place. In this way, too, major variables are quantified without any one set of numbers being deterministic. Policy cannot then be decided solely on the basis of a single RCBA. Decision analysts would only provide a range of alternatives from which the policymakers could choose. What is left unclear is just how many RCBAs would be enough to represent adequately the span of values in society. There is no simple answer to this problem.

Cost–benefit analysis might indeed be improved, but what seems evident is that no one method is likely to give us definitive answers as to how best to decide criteria for health and safety. And if RCBA cannot provide definitive answers, then we must consider who should decide. Who should decide the form of RCBA, and the way safety standards are formulated? Indeed, since standards are inseparable from basic technology, the question must be broadened: who should decide policy over technology?

Who Should Decide?

Some have argued that only engineers and scientists have the expertise to understand new technology and that, therefore, only they are competent to make decisions about it. But this position necessarily makes assumptions that are highly debatable. It assumes either that the issues are strictly technological, or that technical experts have the widest insights into all the ramifications of a technology, or both. Of course, this has been disputed for many years (see Laski 1931). And none of these assumptions are likely to be the case as long as people's health, wealth, and value systems are to be considered. In fact, as we have seen, experts have often adopted dubious assumptions in their analyses of technological issues.

Another reason experts do not make ideal arbiters of public policy is that they often disagree, and society is left with the dilemma of which expert to believe. The ordinary citizen cannot trust any given expert simply because he or she appears to use the facts; all sides will employ scientific facts in their arguments. As a result, the public and government officials are often left confused and uncertain by debates on technical matters. This can be seen, for example, in the debate over federal government air quality standards.

The Clean Air Act of 1970 (and its amendments in 1977) established a primary ambient air quality standard that, as we saw in Table 6.3, was closely correlated with the best estimate judgments for an effects threshold. In other words, the standards adopted were related to an estimate of the point below which no measurable risk presumably exists (Cassedy 1992–3). The standards were apparently based on widely held beliefs of the scientific community.

Despite the assumption of a scientific consensus, the threshold concept in relation to air pollution was soon challenged. First, it was established that certain groups within the population have markedly lower thresholds for health distress due to air pollution. Infants and the elderly are known to be especially vulnerable. But laboratory tests also began to show distinct physiological effects on the general public at pollutant concentrations considerably below the established thresholds. These examples have led some experts to the conclusions that (a) there may be no single threshold standard for air quality that is appropriate for the entire population; (b) the present standard is too lax; or (c) the methods that regulatory bodies use to establish standards are inadequate (Wilson et al. 1980). The debate, however, did not change the views of all the experts and had little impact on the federal regulators; they left the standard unchanged.

Not only do experts disagree, they sometimes do not even see problems the same way. For example, engineers and geologists were asked to estimate the risk of an earthquake at a nuclear reactor site in California. Although we would generally view both engineers and geologists as members of the scientific community, they produced very different results. The latter said that the risk could not be quantified; the former reported that the risk was "so low that it need not be included in the design bases" (Hund 1986). The disagreement lay not in the facts but in the way each group saw the world. The geologists tried to understand the nature of the site. While they could identify a fault structure, they could not determine whether or not it would produce a significant quake during the lifetime of the nuclear facility. The engineers, on the other hand, used standard risk assessment methodology. They could not predict whether or not there would be an earthquake either, but the method allowed them to set odds that were extremely low (Hund 1986).

One idea has been advanced for resolving the disagreements among scientists or between groups of people within the technological community. It is the idea of the science court. This concept, as advanced particularly by Arthur Kantrowitz (1975, also Michalos 1980), calls for the establishment of a system of adversarial proceedings on major technological issues. In this court, disputants would have a chance to present evidence, cross-examine one another, and argue their case just as in a court of law. The decision would be rendered by a panel of scientists acting as a sort of jury. These individuals would not be in fields directly connected with the matter at hand. So, for example, no nuclear engineers or physicists would decide questions about nuclear facilities, since they might directly benefit from the outcome. Kantrowitz did not mean for this body to determine policy per se. The court would determine what could be done as a strictly technical matter, not what ought to be done finally; that decision would be left to policymakers. But at least the court would put to rest technical disputes and allow the policy process to proceed from there.

However, the neatness of this formulation belies several potential problems. First, the concept implies the fact–value dichotomy, which, as we have already seen, may be untenable. Second, there is a danger that court findings may be taken as truth rather than what, at best, would be an opinion about which side produced the best available evidence at the time of proceedings. Philosophers such as Shrader-Frechette have also raised the question of whether scientists in one field are the best judges of the claims of those in another. It can be asked, for example, whether scientists will tend to view the advance of technology as a good in itself, biasing them to favor a project and tending to require the onus of proof to be much greater on opponents of any given project.

Shrader-Frechette (1985) has argued instead for a technology tribunal. It would preserve the opportunity for an open hearing of conflicting views, but the tribunal would render a judgment that would be less final than the judgment of a court. In this concept, the jurors would include nonscientists who could view the project in a broader context. This would help avoid the bias of the expert who, as Harold Laski (1931) noted nearly a century ago, may tend to confuse the facts with "what he proposes to do about them." Of course, this idea is challenged by those who believe that nonscientists cannot understand the complexities of scientific concepts. But there is good reason to reject the notion that a sufficient understanding of a technological project requires years of study of science and mathematics and that technology is inevitably a mystery to all but an elite.

The issue of a science court or tribunal and beliefs about its composition are tied up with the larger question of who should make policy. The view we adopt probably will depend on what we see as the proper role of government officials, experts, and the people. Many would ascribe to government the right to make rules for the public good, even when that conflicts with the wishes of some citizens. The United States has laws against child labor and free use of some drugs and others that regulate the use of toxic substances – all of which are opposed by some, but are generally accepted. Although some government restrictions are legislated, others are determined by regulatory agencies; the Food and Drug Administration, for example, permits or denies the public access to drugs. The agency determines what is right for citizens. Government in these cases is considered to know best and to be acting paternalistically to protect its citizens. As part of the paternalistic view, expert opinion plays a key part in the decision-making process.

Along these lines, we can imagine how government might formulate energy policy decisions. Experts would provide information and analyses (including RCBAs), and then government officials would mandate binding rules about the use and development of energy technology and for the standards for health and safety – presumably for the benefit of society as a whole. Of course, few in our society would attribute to government an unbiased concern for the public interest or give it absolute authority. But in a paternalistic framework, the executive (or legislative) branch of the government could establish authorities to act where they perceived the need to do so. These authorities would not have to consult with the public or its representatives before taking any action (though they might be held accountable after the fact). Essentially, they would have a free hand to make and establish standards and implement policies.[19]

Others, however, see paternalism as authoritarian and argue along the lines espoused by John Stuart Mill. In this view, government has virtually no right to impose policies upon a citizen against his or her will. Mill (1947 [1859]) wrote:

> The only purpose for which power can be rightfully exercised over any member of a civilized community, against his will, is to prevent harm to others. He cannot rightfully be compelled to do or forebear because it will be better for him to do so, because it will make him happier, because in the opinion of others, to do so would be wise, even right.

Thus, the fact that an energy policy would benefit most people in a society – unless it could be shown to be a matter of life or death – is not, in this view, sufficient reason for imposing it on others, particularly if it placed anyone at risk. All those affected would have to consent to a policy before it could be adopted.

Although allowing individuals to restrict technological development of society as a whole may seem too extreme to some people, many would still argue for a greater democratization of the process. They would argue against government-imposed policy, especially where there is risk to life and health. In this view, policies would be arrived at through democratic means including local or even national referenda. It could be required that voters be supplied, perhaps through the internet, with all the relevant information, including technological information. But the idea of referenda to decide technological development, while democratic, does require an enlightened public. In reality, it presents the possibility of questions being decided by better advertising campaigns rather than on their merits.

Also, if we use democratic methods, we pay a price in efficiency. Even with our current, somewhat limited, right to challenge policy in the United States, projects are held up for years, whereas in countries with more authoritarian forms of government projects are planned and implemented much more quickly. Speed may even mean greater social equity. For example, completion of a nuclear power plant may mean less hardship for the poor as a result of greater economic development. However, the more democratic we make the process, the slower the wheels grind and the more this may lead to hardship.

Indeed, the undeniable complexities, philosophical as well as technical, surrounding technological decision making can leave one cautious about adopting a set policymaking system. It perhaps explains why we have not set up a definitive system to date. Instead, we continue to try in a somewhat haphazard fashion to balance efficiency and democracy, government authority and individual rights, and expert opinion and public perception.

References

Ackerman, F. and L. Heinzerling. 2002. "Pricing the Priceless: Cost–Benefit Analysis of Environmental Protection." *University of Pennsylvania Law Review*, 150 (5), 1553–1584.

Arrow, K., G.C. Daily, P. Dasgupta, S. Levin, and K.-G. Maler. 2000. "Managing Ecosystem Resources." *Environmental Science and Technology*, 34, 1401–1406.

Atkinson, G. and S. Mourato. 2008. "Environmental Cost–Benefit Analysis." *Annual Review of Environment and Resources*, 13, 317–344.

Barnett, J. and G.M. Breakwell. 2001. "Risk Perception and Experience: Hazard Personality Profiles and Individual Differences." *Risk Analysis*, 21 (1), 171–177.

Cameron, T.A. 1992. "Combining Contingent Valuation and Travel Cost Data for the Valuation of Nonmarket Goods." *Land Economics*, 68 (3), 302–317.

Cassedy, E.S. 1992–3. "Health Risks Valuations Based on Public Consent." *IEEE Technology and Society Magazine*, Winter, 7–16.

Cohen, B.I. and I. Lee, 1979. "A Catalog of Risks," *Health Physics*, 36, 708–721.

Gowdy, J.M. 2004. "The Revolution in Welfare Economics and Its Implications for Environmental Valuation and Policy." *Land Economics*, 80 (2), 239–257.

Groom, B., C. Hepburn, P. Koundouri, and D. Pearce. 2005. "Declining Discount Rates: The Long and the Short of it." *Environmental and Resource Economics*, 32, 445–493.

Grossman, P.Z. and E.S. Cassedy. 1985. "Cost–Benefit Analysis of Nuclear Waste Disposal: Accounting for Safeguards." *Science, Technology and Human Values*, 10 (4), 47–54.

Hiskes, A.L. and R.P. Hiskes. 1986. *Science, Technology, and Policy Decisions*. Westview Press, Boulder, CO.

Hund, G.E. 1986. "The Fault of Uncertainty." *Science, Technology & Human Values*, 11 (4), 45–54.

Johansson-Stenman, O. 1998. "The Importance of Ethics in Environmental Economics with a Focus on Existence Values." *Environmental and Resource Economics*, 11 (3–4), 429–442.

Kantrowitz, A. 1975. "Controlling Technology Democratically." *American Scientist*, 63, 505–509.

Laski, H. 1931. *The Limitations of the Expert*. Fabian Tract No. 235, The Fabian Society, London.

MacLean, D. and P.G. Brown (eds.). 1983. *Energy and the Future*. Rowan and Littlefield, Totowa, NJ.

Michalos, A.C. 1980. "A Reconsideration of the Idea of a Science Court." In P. Durbin (ed.), *Research in the Philosophy of Technology*, 3, pp. 10–28. JAI Press, Greenwich, CT.

Mill, J.S. 1947. *On Liberty*. Bobbs-Merrill, Indianapolis, IN.

Mishan, E.J. 1972. *Economics for Social Decisions*. Praeger, New York.

Mishan, E J. 1976. *Cost–Benefit Analysis*. Praeger, New York.

Mishan, E.J. 1981. *Introduction to Normative Economics*. Oxford University Press, New York.

Raiffa, H. 1970. *Decision Analysis: Introductory Lectures on Choice Under Uncertainty*. Addison-Wesley, Reading, MA.

Rawls, J. 1971. *A Theory of Justice*. Belknap Press, Cambridge, MA.

Rhoads, S.E. (ed.). 1980. *Valuing Life: Public Policy Dilemmas*. Westview Press, Boulder, CO.

Ricci, P.F. and L.S. Molton. 1981. "Risk Benefit in Environmental Law." *Science*, 214, 1096–1100, December 4.

Robinson, R. 1993. "Cost–Benefit Analysis." *BMJ*, 307, 924–926.

Shrader-Frechette, K.S. 1983. *Nuclear Power and Public Policy*. D. Reidel, Dordrecht, Netherlands.

Shrader-Frechette, K.S. 1985. *Science Policy, Ethics, and Economic Methodology*. D. Reidel, Dordrecht, Netherlands.

Starr, C. 1976. "General Philosophy of Risk–Benefit Analysis." In *Energy and the Environment*, Pergamon, New York.

Sunstein, C.R. 2005. "Cost-Benefit Analysis and the Environment." *Ethics*, 115 (2) 351–385.

Tversky, A. and C.R. Fox. 1995. "Weighing Risk and Uncertainty." *Psychological Review*, 102 (2), 269–283.

Wilson, R.S., S.D. Colome, J.D. Spengler, and D.G. Wilson. 1980. *Health Effects of Fossil Fuel Burning; Assessment of Mitigation*, Ballinger, Cambridge, MA.

Whittington, D. and W.N. Grubb. 1984. "Economic Analyses in Regulatory Decisions: The Implications of Executive Order 12291." *Science, Technology & Human Values*, 9 (1), 63–71.

NOTES

1 In fact, economics uses a metric of risk aversion, and there are examples of aggregated "societal" risk levels. But the measure of risk covers a gamut from those who are highly risk averse to those who welcome risk and are referred to as "risk loving."

2 Some scholars and practitioners prefer to refer to benefit–cost analysis or BCA.

3 Or, more formally, the marginal cost must be equal to the marginal benefit.

4. The courts and insurance companies award money for injuries and death, but that is different from the matter we are discussing here. Court awards represent compensation *after the fact*; human capital RCBA is an attempt to place some market or intrinsic value on human life and suffering *ex ante*, or before it can occur.

5 A major area of contention in CBA (and RCBA) is the idea of "discounting the future." That is a dollar today is presumed to be worth more (has more buying power) today than a dollar next year. Inflation would reduce its purchasing power. A dollar next year will be worth more than a dollar the year after that and so on. Thus the cost of environmental damage – say estimated at $1 million, 10 years from now is less than $1 million today. Put another way, one could put less than $1 million in a bond today and have $1 million in 10 years. The problem with this procedure is that if we put an explicit value on a human life today of $1 million, and then assume a discount rate of 1.5 percent, it would then be more valuable to save one person today than about 2 billion people 500 years in the future. Many have argued that the costs of future catastrophes should not be subject to discount rates since it is easy to trivialize massive future losses (for example, from climate change). See, for example, Groom et al. (2005).

6 For a general critique of all valuations of human life and suffering, see Ackerman and Heinzerling (2002).

7 For a critique of the value-neutral study see especially Shrader-Frechette (1985), Chapters 3 to 6.

8 In 1981, President Reagan signed Executive Order 12291, effectively requiring all government agencies to perform cost–benefit analyses for all their major regulatory decisions. For discussion of the implications see Whittington and Grubb (1984).

9 US courts have rules that risks may be too slight (*de minimus*) to be considered. See Ricci and Molton (1981).

10 In adopting these goals, the NRC was criticized by its own Advisory Committee, which charged that a more stringent principle had been abandoned. This principle stated that risks should be *as low as reasonably achievable* (ALARA) and was based on the idea of utilizing the best available technology.

11 This refers to chances per reactor year of operation (see Chapter 7).

12 Such a notion has an interesting intellectual history. Around the turn of the century, the mathematician Edouard Emile Borel (1871–1956) speculated that there were probability levels sufficiently low that the average person would conclude he could not observe the event in his lifetime.

13 Other notable proponents have included Chauncy Starr, a founder of the Electric Power Research Institute and I. Bernard Cohen. Cohen and I. Lee (1979) developed an extensive table of risk probabilities for decision analysts to use.

14 The *naturalistic fallacy* can be any one of three different errors: (1) the confusion of ethical and factual statements; (2) the derivation of ethical statements from nonethical statements – or the confusion of *is* and *ought;* and (3) the definition of ethical concepts in nonethical terms. For a discussion of the naturalistic fallacy with regard to nuclear power see Shrader-Frechette (1983), Chapter 6.

15 Specificity of risk points up another problem with the revealed-preference method. The assumption in compensation efforts for involuntary risks is that the public regards one involuntary risk in the same way that it regards another. However, there is no reason to believe that this will be the case. The expressed-preference approach, by focusing on specific risks, avoids that problem.

16 Cameron (1992) argues for a hybrid system combining SP and RP which, she argues, should produce "a more comprehensive picture of preferences..."

17 The formulation used by Kneese and his colleagues for the egalitarian view is that we must try to "maximize the utility of the individual with the minimum utility ... until he catches up with the next worse off individual." Or more to the point, until "all utilities are identical." This is taken from the theories of John Rawls (1971) seen in earlier chapters of this book. Inevitably, to create an easily phrased precept of distributive justice, Kneese and his colleagues have had to simplify Rawls's ideas greatly.

18 Shrader-Frechette (1985) has proposed a different system of ethically weighted RCBA based on the idea of lexicographically ordered ethical claims. Although her system is too complex to describe here, it is worth noting that it requires a weighting scheme for claims that may also lead to several alternative RCBAs. She defends the idea of alternative RCBAs, noting that this would mean that "citizens could select the *values* by which they would choose social policy as well as the policy itself."

19 The Obama administration's Clean Power Plan, which was to have gone into effect in 2016, is an example of how the US executive branch can in fact impose rules affecting energy and the environment without explicit consent of Congress or the general public. These rules could presumably be reversed by the courts, Congress, or, in this case, by the incoming Republican Trump administration.

Synopses of New Technologies

Technical Efficiency

Energy conservation, as we noted in Chapter 4, can be understood variously to mean resource substitutions, doing without, or *technical efficiency*, the last of which we will focus on here.

Technical efficiency is best discussed with reference to the laws of thermodynamics (Appendix A). The First Law of Thermodynamics states that energy can *be neither created nor destroyed*; it can only change form or be stored temporarily. According to the Second Law, energy tends to be transformed from forms available for useful work and *transfer of heat* to forms that are *less available* or *unavailable* to do work.

The efficiency of a thermal plant is usually defined according to the First Law. A thermal plant supplies useful heat either for a stage of industrial operation or for a *heat engine*, such as the modern turbine. Plant efficiency merely compares the *useful output* (useful heat or electricity) to the *fuel input* as a measure of the best use of the fuel.

Energy inputs and outputs for an *industrial process heat* (IPH) plant are shown in Figure C.1. This figure may be compared with Figure A.1, which is a similar diagram for an electric power plant. For both plants, we can identify a fuel input, useful energy output, and losses. Both thermal plants can be made to reduce the nonuseful heat output and thereby make better use of the fuel input. This is done with *heat-recovery* technology, as indicated in Figure C.2. A *recuperator* is used on the process water/steam output, and an *economizer* is used on the stack (hot flue gas) output. Each of these devices is a *heat exchanger*, which transfers part of the outgoing flow of heat back to preheat the air or water flowing in. (Such devices are not new, but were neglected until the advent of high fuel prices in the 1970s.) Using heat recovery, the useful output energy can be increased for the same fuel input or, alternatively, the same useful output can be obtained with less fuel input. Either type of recovery leads to improved plant efficiency.

There are limits to efficiency, however, as a consequence of the Second Law of Thermodynamics. As nonoutput steam has more and more heat exchanged out of it, we would expect the temperature of the effluent to decrease. But the output temperature could not go *below* the temperature of the surrounding air or water of the plant, because we could not then transfer any more heat into it. We therefore realize that the outside, surrounding temperature places a theoretical

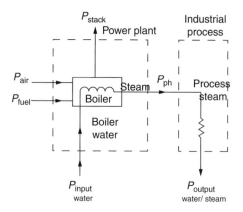

Figure C.1 Industrial process heat plant (no heat recovery). p_{ph} is process heat, rate of heat transfer; $p_{input} = p_{fuel} + p_{in} + p_{input\ water} \simeq p_{fuel}$, $p_{output} = p_{ph} + p_{stack} + p_{output\ water/steam}$; plant efficiency: $\eta = p_{ph}/p_{fuel}$.

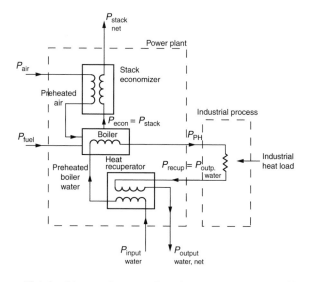

Figure C.2 Technical efficiency with industrial process heat. p_{econ} is stack economizer input, rate of heat transfer; p_{recup} is heat recuperation input, rate of heat transfer.

limit on the amount of low-temperature energy that can be recovered. The efficiency of a basic IPH system can usually be raised from about 87 percent to close to 95 percent through the use of a *heat recuperator*, but to recover *all* of the heat is not possible.

These thermodynamic considerations also give us insight into the conversion of thermal energy into mechanical work and electricity (Fig. A.1; Appendix A). The mechanical work is, of course, the action of high-pressure steam on the turbine blades, which in turn spin the shaft of the electric generator. The efficiency with which the thermal plant delivers such useful work is likewise limited by the upper and lower temperatures of the plant, as expressed in the Carnot limit. Correspondingly, the purpose of the cooling-water loop is to reduce the lower temperature (T_L)

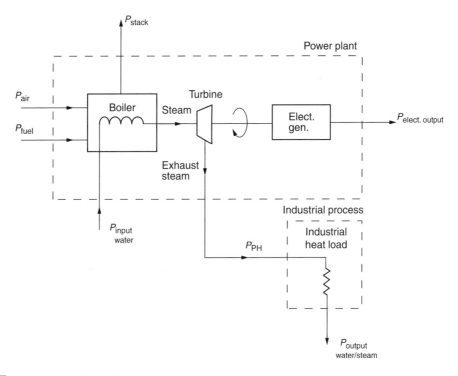

Figure C.3   Cogeneration: technical efficiency in electricity generation. $p_{output} = p_{elect\ out} + p_{ph} + p_{stack} + p_{output\ water}$.

of the power plant cycle. Remember from Appendix A that the Carnot efficiency (Figure A.7) is given by $(T_H - T_L)/T_H$; thus, if T_L falls, efficiency rises. Similarly, the higher the starting temperature is, the higher the ideal (Carnot) limiting efficiency of the heat engine will be. The practical upper limits on temperature are determined by the quality of the materials and their ability to withstand the high-temperature/high-pressure conditions and higher heat losses.

A major gain in useful energy output is achieved by the use of *cogeneration* (Fig. C.3), in which *both* electricity *and* IPH are outputs. In this particular version of cogeneration, the lower-temperature exhaust steam is used for IPH instead of being discharged as waste heat.

$$\eta = \frac{p_{electout} + p_{ph}}{p_{fuel}}$$

$$= \text{plant/efficiency}$$

A typical steam cogeneration system can result in a fuel saving of about 16 percent over the separate production of the same amount of IPH and electricity. This is depicted in the bar graph in Figure C.4, where the steam-turbine plant is compared with gas-turbine and diesel-cogeneration systems which can save 27 percent and 24 percent, respectively. Whether or not a cogeneration system will be a worthwhile investment will depend on the cost (including financing) of the equipment and fuel. It should be noted that in many instances – for example with large industrial energy users – cogeneration is already a cost-effective technology.

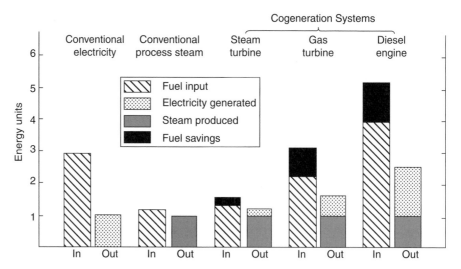

Figure C.4 Cogeneration systems: performance comparisons. Energy inputs and outputs shown for sample cogeneration technologies. The fuel savings (solid-black areas) represent the differences between fuel consumption by each cogeneration system and the fuel that would be required if separate equipment were used to produce the same amount of steam and electricity. Shown from left to right are the inputs and outputs of each of two conventional energy systems and three cogeneration systems. The input/output pair on the left is the conventional generation of electricity by means of steam, using a boiler and a steam turbine, with an output of 1 energy unit and an input of about 2.9 units ($\eta = 0.34$). Next is shown the input/output for conventional process steam using a boiler only with an output of 1 energy unit and an input of 1.15 units ($\eta = 0.87$). The next three pairs are for cogeneration systems, steam turbine, gas turbine, and diesel engine, respectively. In each cogeneration case, the output process steam is the same: 1 energy unit. Added to each cogeneration steam output is the electricity output, which varies according to the technology – steam turbines providing the least and diesel engines the most. The corresponding fuel input energy for each combined output is shown for each of the three cases in cross hatching and the fuel savings are shown in black for producing the same outputs separately and conventionally. The fuel savings shown are: from left to right, 16 percent, 27 percent, and 24 percent, for the three cogeneration technologies.

Source: Reproduced, with permission, from the *Annual Review of Energy*, Volume 3, © 1978 by Annual Reviews Inc., article by R.H. Williams, "Industrial Cogeneration."

Small Hydroelectric Generation

Small-scale hydroelectric plants are customarily those of output capacity 30 MW or under, typically in a range of 5 kW to 5 MW. As we noted in Chapter 3, most small hydroplants utilize *low heads* (30 feet or less), operating to a significant degree on the *flow* of the river as well as the pressure of the height of the *head* behind the dam.

A significant expansion in hydrocapacity is possible in the United States if the development of small hydrosites is considered.[1] The existing developed small hydrosites have aggregate capacity less than 500 MW, while the potential, developable capacity is over 20,000 MW. In regions where increased electric system capacity is needed on a relatively short-term basis, small hydroplants are particularly useful because they only take two or three years to construct.

A major factor in the adoption (or re-adoption) of small hydropower generation is, of course, its low operating cost. Small hydroplants have *unit capital costs* (e.g. $/kW) that are higher than thermal plants, but because of low operating/maintenance costs, long plant lives, and zero fuel costs, they often enjoy a cost advantage over thermal plants. If we consider the total generating cost of *small* hydropower (reduced to a per kWh basis) we find costs in the range of 5¢/kWh. This can be compared with about 9¢ for coal and 11¢ for nuclear plants, accounting for total generating costs (all in levelized, constant $2013).

Environmental degradation may be a barrier to hydroelectric development where construction means disruption of natural habitats or recreational areas. However, small hydroplants present fewer environmental concerns when compared to major hydroplants, because the impacts are on a small scale at each site and because many sites were (historically) already dammed. In some cases small hydroplant operations improve the environment by sprucing up facilities that had been abandoned in the early twentieth century, and left in disrepair.

Enhanced Oil Recovery

Enhanced oil recovery (EOR) is a means of extracting additional resources from already discovered and exploited oil fields. As discussed in Chapter 2, the amount of the resource that is *recoverable* depends on the technology of extraction as well as the amount in place.

Recovering oil is not as simple as drawing water from a well. Petroleum usually is embedded in porous rock formations. These formations are under pressure from overlying and surrounding geological formations and, often, from accompanying natural gas formations. When the oil-bearing formation is first drilled, a pressure differential is created at the point of drilling, where the pressure is atmospheric (whereas the surrounding formation has higher geological pressure). This pressure differential causes a migration of the oil through the rock formations toward the drilled well shaft. Once at the well, the oil comes to the surface under natural pressure. When the natural pressure is sufficiently high, the well erupts as a legendary *gusher.*

If the natural pressure is not sufficient to bring the oil to the surface through the shaft, as usually happens with older wells, then other methods must be employed to force the oil out. Petroleum extraction techniques are classified into *primary, secondary,* and *tertiary* practices. Primary extraction is where natural geological pressures are sufficient to cause the flow of oil to the well. It also applies to simple pumping, which becomes necessary when the pressure becomes too low.

As pressures fall further, secondary practices need to be initiated. The reservoir may be flooded with water to maintain pressures and to displace oil toward the well. Typically, about one-third of the oil in a given field can be recovered by primary and secondary techniques. In the earlier era of petroleum development, extraction

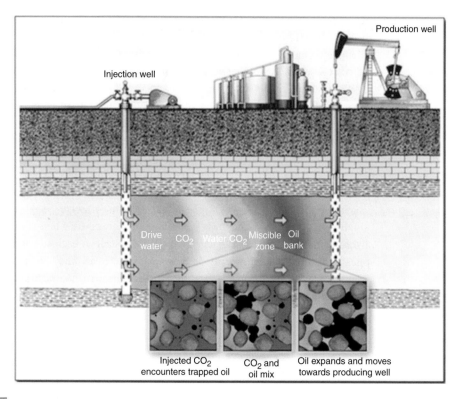

Injection well

Production well

Drive water CO_2 Water CO_2 Miscible zone   Oil bank

Injected CO_2 encounters trapped oil CO_2 and oil mix   Oil expands and moves towards producing well

Figure C.5 Tertiary enhanced oil recovery. Cross-section illustrating how carbon dioxide and water can be used to flush residual oil from a subsurface rock formation between wells.
Source: USDOE at http://energy.gov/fe/science-innovation/oil-gas-research/enhanced-oil-recovery.

was only rarely taken beyond the secondary stage. But when prices of imported oil have risen, tertiary methods – EOR – have become more attractive and indeed have been employed in fields such as the Permian Basin in Texas.

Application of EOR technologies increases the amount of the resource that is recoverable. Innovations in tertiary EOR could, it is estimated, produce as much as 280 BBL of oil, nine times current reserves.

The EOR technologies apply various forms of injected heat or chemicals. In recent years, a gas, especially carbon dioxide (CO_2), under pressure can be injected to extract additional oil. Figure C.5 shows a diagram of how water and CO_2 together can be utilized to force out oil trapped in rock formations. The CO_2 and water are injected into the petroleum formation through a perforated pipe. The gas mixes with the oil, thinning its consistency, making it less viscous, and driving it to the wellbore. After the oil reaches the well, it is pumped to the surface, processed to separate the oil from any water and then shipped or piped to a refinery. Carbon dioxide is especially attractive as an injection gas since there are also ongoing experimental efforts with respect to the capture of CO_2 during combustion at power plants. The idea then is to sequester the CO_2 to keep it from adding greenhouse gases to the atmosphere. This offers a productive use for captured CO_2, although some people might find it a bit problematic since it means that the combustion of fossil fuels is used to produce more fossil fuels.

The new generation of EOR technologies is designed especially around controlling the CO_2 after it is injected. Uncontrolled CO_2 in an oil well can migrate to any permeable area including places that are not likely to be productive. The new concepts involve embedding CO_2 in foams and gels along with (in some cases) "engineered nanoparticles." According to the DOE, seven experimental projects using these innovations have been ongoing since 2010.

Resource Recovery

In the 1970s, *municipal solid waste* (MSW) *resource recovery* was thought to be approaching commercialization. Pilot plants had been operating in more than 20 cities, providing energy and materials recovery from municipal wastes. Municipal waste is customarily defined as refuse from households and commercial buildings. The quantities of MSW generated in the United States are immense (over 250 million tons/year). The amount had been increasing due not only to population growth, but also to increasing waste generation per capita. But from about 2002–2012 the amount leveled off.

The material breakdown of typical municipal refuse is shown in Figure C.6. Approximately one-third of MSW is combustible, with a heating value of about 10 MBTU/ton. If we include the potential energy savings from materials in waste that

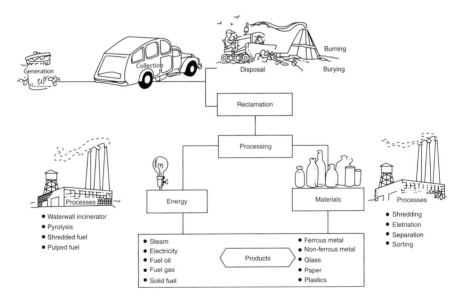

Figure C.6   Municipal solid waste collection and processing.
Source: S.L. Blum in P.H. Abelson and A.L. Hammond (eds.), *Materials: Renewable and Nonrenewable Resources*, p. 48. Reprinted with permission. © 1976 American Association for the Advancement of Science.

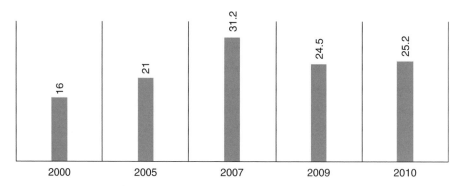

Figure C.7   Total revenue in the US from resource recovery (selected years), in $US billion.
Source: Data from the Statistical Abstract of the United States: 2012.

can be recycled[2] (about 30% metals, glass, etc.) then the MSW recovery can have an even greater impact on conservation on the national scale.

The process of MSW collection, processing, and either disposal or recovery, as it has been envisaged, is shown in Figure C.6. In resource recovery plants, wastes are divided into either recovered energy or recovered materials. The products of the materials separation process are:

1. Ferrous metals – mostly from metal cans
2. Nonferrous metals – mostly aluminum goods
3. Glass – from bottles
4. Paper fiber – from newsprint, if separated from combustible components.

These materials can be sold, where markets exist. Figure C.7 shows total revenue produced in selected years between 2000 and 2010.

Following materials separation, energy utilization can take place. This energy has, in the past, usually been derived by burning the reduced MSW directly. The refuse is fed as a fuel supplement for coal or another conventional fuel to generate steam for industrial process heat, district heating, or electricity generation. Such use of waste fuel would cut the cost of steam generation through reduction of conventional fuel purchases.[3]

Emission controls are especially important for resource recovery plants (or ordinary incinerator plants), because they are often in close proximity to residential areas. The major concerns in emissions are with particulates and the products of burning plastics (mainly hydrogen chloride from incineration of polyvinyl chlorides, or PVCs). In fact, this concern has caused sufficient public opposition to stop construction of resource recovery plants and to shut down plants already operating. Municipal waste, on the other hand, is lower in sulfur content than many conventional fuels, such as oil and coal, so the use of an MSW fuel supplement reduces sulfur-related emissions.

Resource recovery, like several of the other near-term alternative technologies discussed in Chapter10, has faced barriers to adoption. These barriers occur within

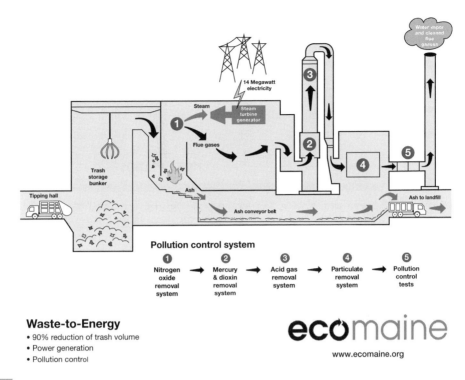

Pollution control system

①	→	②	→	③	→	④	→	⑤
Nitrogen oxide removal system		Mercury & dioxin removal system		Acid gas removal system		Particulate removal system		Pollution control tests

Waste-to-Energy
- 90% reduction of trash volume
- Power generation
- Pollution control

ecomaine

www.ecomaine.org

Figure C.8 Waste-to-energy system.
Source: ecomaine, at www.ecomaine.org/our-facility/waste-to-energy-plant/.

the institutional setting of municipalities. A municipality faces the responsibility to collect and dispose of MSW, come what may. This overriding requirement on municipalities, and hence on the practicality of any MSW technology, molds the way in which its feasibility and economics are evaluated. The most obvious factor in feasibility is the reliability of the technology. Will it function consistently? Will it operate in an environmentally acceptable way? Frequent or lengthy operational outages, without alternative means of disposal, can create intolerable buildups of waste, with attendant health and aesthetics problems. Thus, the *trialability* factor in the process of innovation (again, see Chapter 10) becomes critical, and many municipalities have been reluctant to take the risk of an unproven technology.

In fact, resource recovery plants, as defined here, ceased being built by the mid-1980s. The idea had foundered for a number of years for both technological and institutional reasons. On the technical side, many pilot plants experienced problems with processing or environmental controls. Even though solutions to these problems were matters of straightforward engineering, the municipalities and their contractors seemed incapable of overcoming them. In their place came *waste-to-energy* incinerators that, at most, produced usable heat or electricity, that is, heat to generate electric power using a conventional steam generator. As Figure C.8 shows, today's waste-to-energy plants basically are waste incinerators, but also contain pollution control systems not unlike those used in coal-burning power plants.

In recent years, some MSW facilities have been designed around a process in which the noncombustible waste is removed from the MSW and the remainder is processed into what is called "refuse-derived fuel (RDF)," or biomass fuel. An example is the Elk River Energy Recovery Station in Minnesota, which processes "up to 1500 tons" of MSW daily, and up to two-thirds of that is turned into RDF, which in turn provides fuel for three generators with a total rated capacity 29 MW.[4]

Heat Pumps

The conventional use of electricity for space heating is *resistance heating*, which is just the ohmic dissipation of electrical energy in a resistance (Appendix A). Heat pumps represent an instance of conservation through *technical efficiency*, as discussed in the first section of this appendix. A heat pump operates like an air conditioner in reverse – it takes heat from the outside and transfers it inside. The process provides electric space heating, but much more efficiently than conventional electric heating.

A heat pump, like a refrigerator or an air conditioner, is composed basically of a *compressor*, an *evaporator*, a *condenser*, and an *expansion valve* (see Fig. C.9). The compressor is a necessity – thermodynamically – because if we are to transfer heat from a low-temperature body to a higher-temperature body, mechanical work is required (see Appendix A). The compressor does the work by pumping the *refrigerant* (working fluid) from a low-pressure, vaporized state to a higher-pressure, liquid state.

The refrigerant is hot as a result of compression and heat exchange to the room or building then takes place from the condenser. After passing through the condenser, the refrigerant goes through an expansion valve into a low-pressure evaporator. In the process of expansion, the refrigerant drops in temperature and a cooling action takes place, drawing heat from the air volume surrounding the evaporator (even though the outside air is cold, it still contains latent heat that can be transferred). Work is again done on the refrigerant by the compressor and the cycle continues.

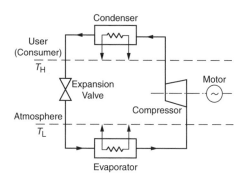

Figure C.9 Heat pump: schematic diagram.
Source: Adapted from Camatini (1976).

The effectiveness of a heat pump (or a refrigerator) is measured by its coefficient of performance (COP), defined as the ratio of the quantity of heat to be transferred to the quantity of work required for the transfer from lower (T_L) to higher temperature (T_H) shown in Figure C.9. The COPs of working heat pumps today approach a factor of two, meaning that it takes about one unit of energy expended as work to deliver two units of heat energy. Conventional electric heating has a COP of one, meaning that *twice* as much electrical energy is used for the same heating than through a heat pump.

Heat pumps have been available commercially in the United States and Europe since the 1950s. But the process of *commercialization* (Chapter 10) had not progressed very far prior to the 1970s. Part of the reason was that the capital investment requirement for a heat pump installation was significantly higher at that time than for a natural gas or oil furnace of comparable size. However, this technology has gained increasing market penetration in recent years, especially in European markets.[5] The introduction of *ground-source* or *geothermal* heat pumps has enhanced their utility in the temperate zones, where lower ambient temperatures of the air reduce heat transfer below economic levels.[6]

Synthetic Fuels

Synthetic fuels (or synfuels) are fuels, in gaseous or liquid form, that can substitute for conventional fossil fuels, such as natural gas or oil. The two principal resources for these alternative fuels are coal and biomass. Coal, as we learned in Chapter 2, is a huge fossil resource, whereas biomass is a renewable resource. Biomass resources include wood, grasses, and certain crops. We will discuss in this section both *coal-derived* and *biomass-derived* methods of creating synfuels.

The creation of synthetic fuels from their basic resources requires chemical reactions, of one sort or another, which produce new chemical compounds having the same (or at least similar) heat content and physical states (gaseous or liquid) as the conventional fuels they are to replace. As such, they are in a different category from *alternative fossil fuels*, such as oil shale, which require physical processing to create the fuel but not a new chemical form.

The question might logically be asked: why bother with the extra step of synthetic processing when coal or wood are usable as fuels – why not burn them directly? The answer is that, in their natural bulk form, they are not usable for such purposes as transportation fuels. In addition, they are difficult and costly to store and ship in their natural states. Liquid or gaseous forms are far easier and cheaper to transport and store in large quantities.

Consider automotive fuels. Conventional automotive fuels, gasoline and diesel, are of course derived from petroleum. Petroleum, in both crude and refined states, is currently transported very efficiently in liquid form through pipelines, in tankers and in trucks. Huge stocks (millions of barrels) of crude oil and refined products are routinely stored and transported daily, and even larger amounts of crude oil have been stored in strategic national reserves. The sheer bulk of coal or wood, for example, would make transportation and storage of fuels of equivalent energy value much more problematic. The energy in a million tons of coal or wood, for example, is equivalent to the oil transported (at a fraction of the cost) in one supertanker.

Liquid automobile fuels today are readily portable because of their energy density (see Chapter 11). A small tank holding 15 gallons of gasoline provides the energy to move the internal combustion vehicle for more than 300 miles. Such portability makes it very challenging for other automotive technologies to compete against. Any alternative fuel or energy source that can be substituted for petroleum in automotive applications must be equally *transportable*, *storable*, and *portable*.

Coal-Derived Synthetic Fuels

Chemists have long demonstrated that conversion of coal to liquid and gaseous fuels can be accomplished physically. But most synfuel development still has a long way to go before it can be produced at competitive cost to oil and natural gas. Indeed, what development engineers have shown so far is that there are a number of alternative means of producing liquid and gaseous synfuels. We should keep in mind, however, that as yet none of them has been conclusively demonstrated as having superior technology for commercial application.

Coal can be used as the basic ingredient for either synthetic petroleum or synthetic gas (Probstein & Hicks 2006), although for both it must be extensively treated and its chemistry altered. Coal is composed primarily of two elements: carbon and hydrogen. In general, it might be said that the main scientific goal in the development of all coal-derived synfuels is to create gaseous or liquid hydrocarbon compounds that have a higher ratio of hydrogen to carbon than exists naturally in coal itself. We will consider both gasification and liquefaction technologies.

Coal Gasification

The production of a synthetic substitute for natural gas is complex and costly. The start of the process, however, is comparatively simple, because it can be started by simply heating coal, in a process called *devolatilization*. This results in gasification, with the gas produced containing some methane (CH_4) and hydrogen (H_2), but the concentration is not usefully combustible. A more highly combustible gas is created by reacting the gas with steam and air, in addition to heat. These reactions will produce two gases primarily: *carbon monoxide* (CO) and *hydrogen* (H_2). Both gases are combustible and the combination can be used as a fuel, known as *town gas*, a term that dates back to the nineteenth century, when such gas was used for lighting and industrial heat.

But neither alone nor together is carbon monoxide or hydrogen comparable to natural gas in energy content. The simple gasification process using heat, water, and room air produces a *low-BTU* gas (LBG), about 150 BTU/SCF (BTUs per standard cubic foot). The energy content is limited because the CO + H_2 is diluted with noncombustible nitrogen (N_2) and carbon dioxide (CO_2). If pure oxygen is employed

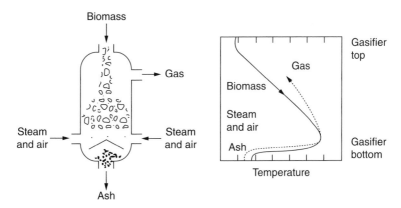

Figure C.10   Lurgi gasifier: a gasifier for either coal or biomass feedstocks.
Source: Adapted from Johansson et al. (1993).

instead of air, the coal gas output will be a *medium-BTU* product. Natural gas is a *high-BTU* product (HBG), on the order of 1000 BTU/SCF, and is composed mostly of methane (CH_4). Although some methane is created in simple gasification processes, it is likely to be only a small proportion of the output. In order to turn coal into a product comparable to natural gas, it must be subjected to further reactions and often requires chemical catalysts to raise the proportion of methane.

We can get a clearer sense of coal gasification by considering a Lurgi gasifier (see Fig. C.10). The Lurgi process, which has been known since the 1930s, is used commercially to produce gas, albeit low- or medium-BTU gas. However, Lurgi gasifiers have also been employed to create gases from coal that are then processed further into synthetic natural gas (SNG).[7]

Coal is fed into the gasifier, and as it is distributed around the top of the chamber, it is subjected to heat from below that dries the coal. But at temperatures of 1100–1500 °F, the coal is also *devolatilized*. As noted, this reaction leads to only a small amount of gasification. As we can see in Table C.1, the devolatilization reaction is a chemical process that produces $CH_4 + C + H_2$. But in the Lurgi system, the gasification process has just begun. As the coal falls through the chamber, it not only undergoes further heating, it reacts with both air and steam. The air and steam, along with continued application of heat from coal burning at the bottom of the gasifier, result in three different gasification reactions, which can occur simultaneously. The *carbon–oxide* reaction (see Table C.1) – from the effect of heat on carbon and oxygen – results in release of both noncombustible carbon dioxide as well as combustible carbon monoxide. Then, with the addition of steam, we also get *steam–carbon* and *water–gas* reactions. These produce carbon monoxide and hydrogen, and carbon dioxide and hydrogen, respectively. With the help of a catalyst, the last reaction can result in a high BTU product. Then, finally, the addition of a catalyst to the *shift conversion* yields a high BTU gas output, as does the input of methane (CH_4).

Despite the multiplicity of reactions, the Lurgi gasifier produces gas with a low percentage of methane. Greater amounts of methane can be generated even in variants of the Lurgi process through the application of both heat and high

Table C.1 Coal gasification reactions		
Chemical reaction	Reaction name	Component of grade
1100–1500 °F	Devolatilization	All
Coal → CH_4 + C + H_2		
C + O_2 → CO_2	Carbon–oxide	Medium BTU
2C + O_2 → 2CO		
C + $2H_2$ ↔ CH_4	Hydrogasification or methanation	Low or medium BTU
C + H_2O ↔ CO + H_2	Steam–carbon	Low or medium BTU
CO + H_2O ↔ CO_2 + H_2	Water–gas shift	Low or medium BTU
CO + H_2O ↔ CO_2 + H_2	Catalytic shift conversion	Feeds high BTU
$3H_2O$ + CO ↔ CH_4 + H_2O	Catalytic methanation	High BTU gas

Source: US Department of Energy (1981).

pressure. This leads to the reaction of carbon and hydrogen gas – called the *hydrogasification* (or *methanation*) process (Table C.1).

Coal Liquefaction

Coal can also be liquefied to produce substitutes for the products of crude oil. One way is to start with gasification, adjust the composition of the gas to achieve the desired ratio of hydrogen to carbon, purify it, and then react it in a separate process to turn it into a liquid. Since it starts as a gas, this method is called *indirect* liquefaction (Probstein & Hicks 2006).

There is a process of indirect liquefaction that is today producing significant quantities of synthetic liquid fuel in South Africa by a company called SASOL. The company uses an indirect liquefaction process that begins with a Lurgi gasification scheme (see Fig. C.10).[8] Once the gas is cleaned and purified, it is treated by a process called the Fischer–Tropsch synthesis. Developed in 1925, this process reacts the gas, under pressure, with an iron catalyst and releases heat. As the heat is removed by coolants, gaseous hydrocarbons are liquefied. The output from the Fischer–Tropsch synthesis includes most of the products of ordinary crude oil, from light to heavy hydrocarbons.

Though the SASOL plants have demonstrated fairly large-scale operations (they produce 150,000 bbl/day from 120,000 Mt of coal), the indirect method of liquefaction is relatively inefficient and thus, it is believed, will ultimately prove more costly than the alternative: *direct liquefaction* (Probstein & Hicks 2006). The basic chemical objective in direct liquefaction is the same as in the indirect processes: to obtain a higher ratio of hydrogen to carbon than exists naturally in coal. But in gasification, and thus in indirect liquefaction, the molecular structure of coal is broken down into smaller components. Direct liquefaction, by contrast, requires a coal input of a large molecular structure. This process, requiring high temperatures and pressures, is called a (chemical) *cracking process* (Probstein & Hicks 2006). Once the molecular structure is cracked, the coal can be hydrogenated directly, with pressurized hydrogen, which transfers the hydrogen into the output compounds in a solution.

There are basically three techniques for direct liquefaction: *pyrolysis*, *solvent extraction*, and *catalytic hydrogenation*. The pyrolysis process – known since the nineteenth century – is the simplest, since it merely causes a reaction in the coal at high temperatures in an oxygen-deficient atmosphere. Pyrolysis produces both liquid and gaseous outputs and a low-hydrogen solid waste called *char*. It is not considered promising for large-scale operation, largely because of the char, which needs to be separated from the liquid hydrocarbons and then disposed of.

Solvent extraction is more promising. A number of different solvent extraction processes have been developed. Essentially, in all of these processes, coal is mixed with and dissolved in a solvent. They differ though on exactly how hydrogenation is accomplished. Catalytic liquefaction, though far more experimental than the solvent process, was intended from the outset to give high yields of liquids.[9] In one example, H-coal, a slurry of coal and a solvent are again heated under high pressure and mixed with hydrogen gas.

Environmental Concerns

One of the most important questions affecting the ultimate feasibility of coal-based synfuels in general is the impact they would have on health and the environment.[10] Coal-derived synfuels could have profound effects on both, running from the mining process to utilization. In fact, a large synfuel industry would be expected to have the same environmental consequences as are caused by industries that presently use coal as an energy source. In addition, it would add problems commonly found in the refining and the burning of conventional liquid and gaseous fuels.

For example, in the production process of coal-derived synfuels, heat is applied either in the form of steam or directly as a requirement for the reactions of the process. The heat energy would be supplied by burning coal itself, and indeed more BTUs of coal would need to be expended than are created of gas or liquid. As a result, a synfuel plant would, like a coal-burning electric power plant, require controls on sulfur and nitrogen oxide emissions and on particulates. In fact, when synfuel development was being pushed around 1980, various government agencies, labor unions, and environmental groups expressed concern about the potential damage to air quality from synfuel plant operations.

As with conventional coal-burning power plants, environmental concerns do not end with air pollution. Coal ash must be disposed of as well. Even though the ash represents only 10 percent of the original coal mass, large-scale synfuel plants would use huge amounts of coal. If the United States met only one-third of its demand for natural gas with synthetic gas, the syngas industry would consume over 2 billion tons of low-sulfur western coal per year. This would be double the highest projected extraction rate of coal for conventional purposes. This would mean, first, that huge quantities of ash as well as scrubber sludge and desulfurization residues would be produced and would require safe disposal; if improperly handled, the waste could cause massive pollution of rivers and aquifers.

These spillover effects, and those typically found in conventional oil and gas refining, are at least predictable and technology exists to cope with them. Others are less predictable and may occur on such a scale that major engineering efforts will be required to solve them. In gasification and indirect liquefaction, there is concern about the environmental impacts of *acid gases*, primarily hydrogen sulfide, hydrogen cyanide, and naphtha. These gases are particularly hazardous: in high concentrations, they can be fatal; in low chronic exposures, they may be carcinogenic. Procedures are known for removal of these gases (Probstein & Hicks 2006), but they would add to the cost of production.

Finally, it should be kept in mind that carbon dioxide is the inevitable product of combustion of any sort and the conversion of coal hydrocarbons to other forms of fuel does not change the final product of combustion. Therefore, coal-derived synfuels do not provide any solution to the dilemma of the greenhouse effect. In fact, if their widespread use leads to increased and prolonged use of fossil fuels, the dilemma could be made even worse.

Biomass Synthetic Fuels

Biomass fuels are natural organic resources that can be made into direct substitutes for fossil fuels (Probstein & Hicks 2006, Gupta & Demirbas 2010). These organic resources include wood, crops, and organic wastes. If they were to be used on a mass scale in place of fossil fuels, they could in theory reduce air pollution as well as worldwide greenhouse gas accumulations. Plants, of course, through photosynthesis, convert atmospheric carbon dioxide to carbon and oxygen in the process of growth (see Fig. C.11). That means that, while the combustion of a biofuel releases carbon dioxide, the organic resource from which it was derived has compensated by removing it from the atmosphere while the energy crop or tree was growing.

The hope has been that the net accumulation of greenhouse gases would be much less than if fossil fuels are burned instead. But in recent years that idea has been challenged and some studies claim that in fact the current generation of biofuels, especially corn-derived ethanol, may well be worse for the environment than fossil fuels on several dimensions, including greenhouse gas emissions. For example, the Environmental Working Group claims that the original estimate of greenhouse gas emissions from corn-derived ethanol "... underestimated ... emissions by failing to account for changes in land use" (Cassidy 2013). Other forms of environmental damage include soil erosion and water pollution from fertilizer, herbicides, and pesticides as well as effluent from biofuel processing.

Widespread use of biomass would return humans to the energy resource used when we emerged on Earth as a species. Wood, of course, was historically the original biomass-energy resource and is still a major source in the developing world. Peat, agricultural wastes, and animal dung, to a lesser extent, fall into the same category. Since the start of the Industrial Revolution, utilization of wood and the other biomass fuels has fallen, although in the developing world crop residues continue to be the major source of fuel. In the industrialized world, even with some revival of the use of wood as a fuel, less than 3 percent of the primary energy output comes from biomass fuels of all types.

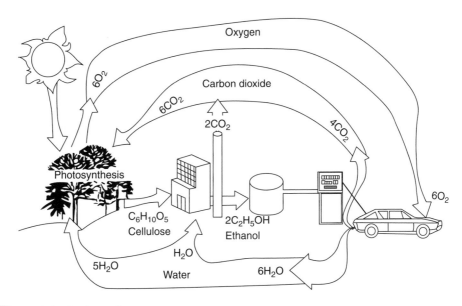

Figure C.11 The idealized carbon cycle.
Source: Courtesy of the US Department of Energy.

The potential base of biomass fuel sources, including managed forests, energy crops, and organic waste products, is large. Such resources can be converted into fuels, such as methane and ethyl alcohol fuels, that may substitute for natural gas or refined-oil products. Alcohol fuels are liquids, which have an ease of *transportability*, *storability*, and *portability*, like refined-petroleum products. They can replace oil for many of its current applications, most notably as a transportation fuel.

Biomass resources can serve as the feedstocks for the alternative fuels, just as crude oil serves as the feedstock for gasoline and heating oil. The first requirement for *biofuel* development, however, is efficient, low-cost means of producing and handling biomass feedstocks on a mass scale. Major innovations in agriculture are needed to grow energy crops and manage forests and to do so in ecologically acceptable ways.

Whether or not this can be accomplished remains a question, notwithstanding the fact that in the United States alone vast quantities of one kind of biofuel are being made. As of 2014, 13.5 billion gallons of ethanol were consumed, mainly blended into gasoline; another 10 billion gallons were (as of 2014) consumed in the rest of the world. At the same time, bio feedstocks are also used in a substitute for diesel, and about 7 billion gallons of biodiesel were consumed. Biodiesel is made primarily from vegetable fats – soybean oil for example – but can be produced from such things as the grease left over from cooking.

But whether this is environmentally sound is uncertain. For example, some deforestation in recent years has been due to tree clearing in developing countries to grow soybeans for biofuels. Since trees act as a great absorber of CO_2, on net biodiesel is often said to hasten climate change. Expansion of corn production in the Great Plains has utilized marginal lands, which have required considerable applications of fertilizers and herbicides, which have polluted nearby rivers and aquifers.

Overall, corn ethanol, which as we have noted consumes a large percentage of the corn crop, is said to reduce petroleum use in the United States by only 1.75 percent, and has had "a negligible net effect on CO_2 emissions" (Jaeger and Egelkraut 2011).

It should also be emphasized that this expansion of biomass fuels is due to government mandates, not because of independent commercial viability. Indeed, it has been demonstrated that on an energy-per-gallon basis ethanol (which has only about 70 percent as much energy as gasoline) is significantly more expensive than gasoline. But blending of large amounts of ethanol into gasoline is required by the 2007 energy bill, costing US consumers about $10 billion each year above what they would pay if gasoline did not have the blend requirement (Bryce 2015).

Biomass Resources

Special species of trees and crops have been chosen for biofuel study and prototype projects because of their characteristics of growth, manageability, and environmental impact. In order to convert these biomass resources into synthetic fuels, various biochemical processes are being considered. Two possibilities are fermentation and anaerobic digestion.

Some crops and crop management techniques might reduce the problem of CO_2 emissions. The use of fast-growing tree species, such as poplar, sycamore, and sweet gum, have been proposed for wood production in "short-rotation" tree farms. In addition to high productivity of wood (as much as 10 tons annually per acre), new growth of trees fixes carbon at about three times the rate of mature growth. This carbon fixing rate for new growth nearly matches the CO_2 production of the combustion of this wood. Woods are about 50 percent carbon composition by weight, which becomes CO_2 as a result of combustion.

Short-rotation, intensive-culture (SRIC) tree plantations have been shown to be highly productive (Cassedy 2000, Chapter 2). SRIC techniques use fast-growing tree species, planted at close spacing and harvested at short intervals, so as to maximize the productivity per acre of land used. As an example, a plantation of sycamores, with spacing averaging 2 meters (allowing about 3000 trees per hectare/1200 trees per acre) and harvested at about 5-year intervals, has produced an annual average of nearly 5 tons of wood per acre in temperate climates. The trees can either be replanted after harvesting or, with some species, the stumps will re-sprout after cutting, in a process called "coppicing." Even larger yields are possible in tropical climates, using species such as the eucalyptus.

At present, starchy food crops, particularly corn and sugar cane, are the biomass stocks primarily used in the production of biofuels. The fermentation of corn into ethanol is widespread. The United States uses almost 40 percent of the corn crop nationally to produce ethanol. As noted in Chapter 10, there is the hope of developing low-cost methods for conversion of cellulosic materials into ethanol, thus expanding the possibilities for biomass fuels. This not only increases the alcohol yield from the starchy crops, by allowing conversion of the stems and leaves, but would also permit the use of resources that are entirely cellulosic, such as wood and grasses, for production of fuel alcohol.

It has been estimated by the US Department of Energy that 200 million acres of land in the United States could be made available for energy crops (see Fig. 2.14). This acreage includes idled food-crop lands as well as pastures and forest lands that could support energy crops. In the United States, idled lands are located mostly in the eastern two-thirds of the country, with the greatest concentrations in the Midwest and South. The crop lands presently used in the United States for food production total more than 300 million acres. An additional 200 million acres are classified as crop lands but are not currently being used as such. Out of all these lands (500 million acres) half, or 250 million acres, are capable of growing energy crops without irrigation. The prospective energy-crop lands have the potential to supply over 25 quads per year of primary energy for the United States, nearly one-quarter of the nation's present total energy demand.

The new energy crops, such as wood and grasses, provide another benefit; they are grown as energy feedstocks only. Energy crops that are also food crops, such as corn and sugar, bring in the undesirable situation of coupling the food and fuels markets, thus distorting food prices. When prices in one market influence prices in another, it is called a "cross-market" impact. In Brazil, for example, significant cross-market impacts were experienced when the country began a large-scale program of ethanol production from sugar cane. The price of sugar shot up because of the large demand for cane in energy production (Monteiro et al. 2012). Such a cross-market dependency would not occur when nonfood feedstocks are used for biofuels.

The inclusion of cellulosic crops would not only widen the scope of biomass resources, but would also mean a lower requirement for fertilizer and agricultural energy inputs than required for crops such as corn. Perennial switch grass, for example, requires much less fertilizer for its (rapid) growth than grain crops because most soil nutrients remain after harvesting (cutting) the grass. Also, grass is harvested more efficiently than crops such as corn, thus requiring less expenditure of (fuel) energy in its agricultural production. Similarly, wood requires less than half of the energy input (fertilizer and fuel) of corn. An important measure of the effectiveness of any of these biomass feedstocks is the fraction of the output biofuel energy content required as input for its production.

Biofuel Processes

There are several technology pathways to convert biomass into liquid or gaseous fuels. These are either biochemical or thermochemical processes (Fig. C.12). Prime examples of the biochemical processes are fermentation for ethanol and anaerobic digestion for a methane-based *syngas*. The thermochemical processes include *pyrolysis*, in many cases followed by *catalytic* and *shift* processes of the type used in coal-based synfuels (this appendix). In the following sections we will consider the processes for the production of ethanol, a liquid fuel, and methane, a gaseous fuel.

Liquid Biofuels

Liquid biomass fuels hold a particular promise for replacement of fossil fuels for transportation – namely, gasoline and diesel fuel. If biomass liquid fuels such as

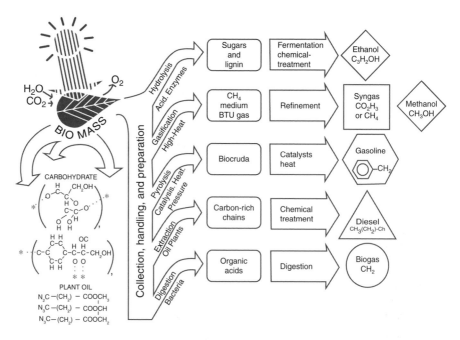

Figure C.12 Pathways for conversion of biomass feedstocks to liquid and gaseous fuels.
Source: Courtesy of the US Department of Energy, 1991.

alcohols can be mass produced competitively, then the technological and economic advantages of the internal combustion engine can be carried over into a new era of renewable transportation fuels.

Liquid biomass fuels would also preserve the portability of gasoline or diesel oil, meaning that a fuel tank of the biomass liquid would give approximately the same speed and a driving range of about 70 percent of the volume of gasoline because ethanol is less energy dense than gasoline. Refueling would be as fast and efficient as it is for gasoline. It should be noted that, while tanks can be built to accommodate both ethanol and gasoline, these "flex-fuel" engines are in the minority of the cars on the road in most parts of the world. At the same time, ethanol producers claim half of all new cars are "flex-fuel ready." In general, ethanol is always blended with gasoline, although flex-fuel cars are capable of running on blends of 85 percent ethanol and 15 percent gasoline (or E-85). Standard internal combustion gasoline engines run on blends of 10 percent ethanol (E-10) and can typically work well on blends of 15 percent (E-15). Above that, engines may be damaged. Indeed, some nonautomotive engines – lawn mowers, for example – may be damaged even by blends as low as E-10.

Alcohols are derivatives of water, wherein a hydrocarbon radical replaces one of the hydrogen atoms in water (H_2O) to combine with the leftover hydroxyl radical (OH). The simplest example is methanol, having the formula CH_3OH, in which the ethyl radical CH_3 has combined with the hydroxyl OH. Methanol is more popularly known as "wood alcohol," since historically it has been derived from wood. As a replacement fuel, however, methanol is not ideal since it is very toxic, dangerous if swallowed, inhaled, or even for repeated contact with the skin.

Ethanol (C_2H_5OH), known as "grain alcohol," is derived from biomass feedstock – historically, grains or sugars – by fermentation and distillation, the same processes used to manufacture alcoholic beverages. As noted, however, researchers have experimented with herbaceous crops, grasses, and even wood as feedstock for ethanol.

Fermentation has been used historically with either sugar or starchy feedstock. It takes place with sugar feedstocks without any pretreatment and on starchy feedstocks after a process of hydrolysis. Small-scale production of ethanol fuels has been operational, in the United States and elsewhere, throughout the twentieth century. Brazil had moved into mass production of sugarcane-derived ethanol by the latter part of the twentieth century. In the United States, mass production of corn ethanol has taken place in the twenty-first century.

Fermentation to ethanol results from the decomposition of glucose (sugar) by the action of enzymes (Gupta & Demirbas 2010). Molasses, for example, contains over 50 percent glucose, and with malt added to supply enzymes, molasses will convert directly to ethyl alcohol. When a carbohydrate (such as corn) is the feedstock, however, intermediate enzyme reactions must take place, first converting the starches to sucrose and then converting the sucrose to glucose. *Sucrose* ($C_{12}H_{22}O_{11}$) is common table sugar, usually derived from plants such as sugar cane or sugar beets. *Glucose* ($C_6H_{12}O_6$) is the sugar usually used in fermentation. The overall ethanol-fermentation process requires support processes that add to the costs of production. A major support requirement is boiler heat for distillation, alcohol dehydration, and silage drying.

Fermentation of starches requires extra steps and different enzymes (Fig. C.13). Starches must first be milled and cooked until they are gelatinized. Special strains of yeasts are then added to supply the enzymes that will create glucose. Then, other yeasts supply the enzymes that promote fermentation of the glucose to ethanol. Certain bacteria can also ferment glucose to ethanol efficiently, with a high fraction of conversion of the glucose.

Hydrolysis of *cellulosic* materials into fermentable sugars is even less straightforward than the hydrolysis of starches, but research and development efforts have yielded some encouraging results. Enzymatic hydrolysis appears to be a leading candidate for (hoped for) high-yield, low-cost cellulose conversion. The enzymes are produced by certain bacteria or genetically altered fungi.

A significant consideration in processing cellulosic matter for fermentation is the hydrolyzing of two distinct components – *cellulose* and *hemicellulose*. The cellulose polymer can be hydrolyzed directly into glucose. Hemicellulose, on the other hand, is hydrolyzed mainly into *xylose*, which has a similar polymer structure. Xylose then requires particular strains of yeast as catalysts to achieve fermentation. The additional breakdown of the hemicellulose fraction for fermentation is critical in expanding the productivity and lowering the production costs of alcohol from cellulosic crops and wood.

Research has found that cellulose feedstocks (e.g. wood chips) can be fed, following a pulverizing pre-treatment, into a process combining both hydrolysis and fermentation. This new combined process is called *simultaneous saccharification and fermentation* (SSF) (Fig. C.14). The cellulose is subjected to hydrolysis, aided by enzymes, in the presence of yeasts, which ferments the glucose as fast as it

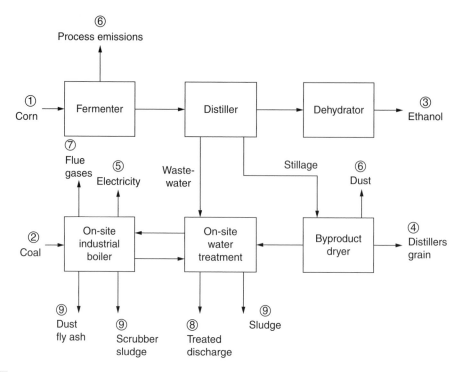

Figure C.13 Process flow diagram for a corn-to-ethanol fermentation plant.
Source: Energy Technologies and the Environment, June 1981, US Department of Energy, Report No DOE/EP0026. Available from National Technical Information Service, Springfield, VA22161.

is produced. Meanwhile, the xylose – also liberated by hydrolysis – is fermented, using special strains of yeast as catalysts with carefully controlled aeration.

Ethanol, from corn, continues to be the major liquid biofuel, but the reliance on the corn crop makes the 2007 law a controversial energy mandate. Moreover, it can never be more than a relatively minor component of the energy mix, since even at 15 billion gallons it replaces less than 10 percent of gasoline consumed. (In 2015, 140 billion gallons of gasoline was consumed but it would take 200 billion gallons of ethanol to provide the energy to replace it.) It will probably take a breakthrough in the production of cellulosic ethanol to make it a significant replacement for gasoline, but that breakthrough – forecast by some scientists and politicians for at least the last 25 years – still seems a long way off.

Gaseous Biofuels

Gaseous biofuels, mainly in the form of methane (CH_4), could replace natural gas to supply heat for residential and commercial needs and for industrial processing. Conceivably, a biogas could substitute for natural gas as the fuel for gas-fired electric power plants, also. In order to substitute directly for natural gas, however, a biomass methane would have to be of "pipeline quality", i.e. a heat content in the range of 1000 BTU per standard cubic foot.

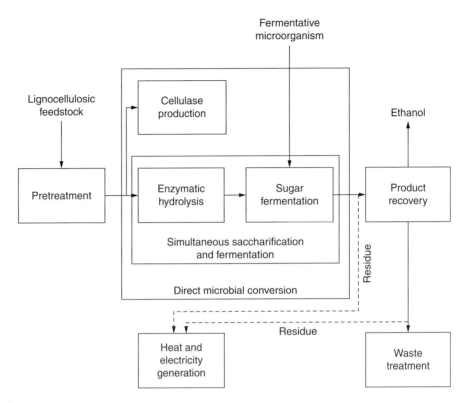

Figure C.14   Simultaneous saccharification and fermentation.
Source: Adapted from Johansson et al. (1993).
Note: Cellulase is an enzyme that assists in the hydrolysis of cellulose.

Gaseous biofuels may be produced by biochemical reactions, such as anaerobic decomposition, or by thermochemical means, such as *pyrolysis* or *gasification*.[11] The principal biochemical means for methane production is anaerobic digestion. Anaerobic digestion is the process whereby organic materials are broken down into methane, and other components, by the action of bacteria in an oxygen-deficient atmosphere. The process has been known and used since the late nineteenth century, but like other energy sources was abandoned in the industrialized world with the advent of cheap fossil fuels. With the oil crises of the 1970s, however, interest revived.

Methane digesters have gained wide use in the less-developed countries over the past few decades and millions of these devices are in use today. Digesters in developing countries are fed mostly by animal dung and provide a simple picture of how digesters work. The typical digester consists of a digestion chamber, a metal dome having a gas outlet pipe, an inlet and an outlet for solids, and a mixing agitator. The digestion chamber can be a metal tank or simply a pit in the ground lined with brick or stone. Some sort of stirring or agitation is required. The heavy metal dome traps the biogas above the liquid slurry and maintains a gas pressure for release through the gas-outlet pipe.

Larger and more elaborately controlled agricultural-waste digesters have been built and operated in the industrial countries as well. In industrial countries, however, methane digesters are more apt to use sewage sludge or the organics in industrial waste waters. Food processing, dairy, brewing, pharmaceutical, paper pulp, and alcohol production are among the industries using anaerobic treatments for their waste waters. Municipal landfills are another waste resource for anaerobic methane. In landfills, decomposition is already taking place in the oxygen-deficient atmosphere under the fill dirt. The resulting gas is also a medium-BTU gas of about 50 percent methane. Landfill gas, which otherwise must be vented and flared for reasons of safety, could instead be tapped for use as a fuel.

The output composition of biogas digesters is from 50 percent to 65 percent methane, with the remainder mostly carbon dioxide, depending on the feedstock and the completeness of the process. This puts digester biogas in the class of a medium-BTU gas, since it will not be of pipeline quality unless it is nearly 100 percent methane. Gas production volumes for a 50 percent methane biogas can be in the range of 5–10 ft^3 per pound of feedstock, which would have the equivalent heating value of 2.5–5 ft^3 of natural gas.

Methane has other uses besides heating; it can also be used as the input to fuel cells to generate electricity. In the 1990s, the Environmental Protection Agency (EPA) promoted the demonstration of fuel-cell power plants using landfill methane on site. Combustion turbines and methane-fueled, piston-type engines (see Chapter 3) driving conventional electric generators are also being used on landfill sites. These generating units are typically of moderate output capacity (less than one megawatt). Like other trash-to-energy schemes, revenues from the electricity generated would partially offset the costs of waste disposal for the municipalities. Landfill gas plants are also operating in Austria, France, Finland, United Kingdom, and Tahiti; total annual waste-handling capacities of tens of thousands of tons are typical.

Thermochemical Production

Thermochemical processing holds the most likely prospects for mass production of gaseous biofuels. Biomass feedstocks, such as wood and herbaceous crop residues, may be processed thermochemically to produce gaseous fuels. The products of the thermochemical gasification processes contain, variously, methane (CH_4), hydrogen (H_2), and carbon monoxide (CO) as combustibles. The ultimate gaseous-fuel product is a high-BTU gas, close to pipeline quality, whether it is derived from biomass or coal.

A major thermochemical process is *pyrolysis*, which is a means of breaking down organic compounds, such as carbohydrates, cellulose, and lignin, into their elemental carbon, hydrogen, and oxygen. Pyrolysis is described as the "destructive distillation" of organic material when heated to high temperatures (above 600 °C) in an atmosphere too deficient in oxygen for combustion to take place. Historically, pyrolysis has been utilized in wood kilns for the making of charcoal, but not for production of gaseous fuels. Charcoal kilns are still in use today in the developing countries, but until recently pyrolysis had not been employed in the production of gaseous products.

The processes and gaseous products of pyrolysis with biomass feedstocks are quite similar to those when coal is the feedstock. In that case, coke is the solid product and "town gas" or "producer gas" is the possible gaseous output. The products of pyrolysis vary widely with the temperatures applied to the organics and with the rate at which the temperature is raised above 600 °C. With a wood feedstock, if the temperature is raised gradually from 600 °C to around 1000 °C, charcoal – with a heating value comparable to coal (22–28 MBTU/ton) – and a low–medium-BTU (100–400 BTU/SCF) gas are the results. This gaseous product is composed of combustibles – carbon monoxide (CO), hydrogen (H_2), and small amounts of methane (CH_4) – plus noncombustible carbon dioxide (CO_2) and water vapor (H_2O). Since the gaseous products of pyrolysis have low heating values, they are limited in their uses to industrial heating or as the input for a liquefaction process and cannot substitute for other uses of natural gas.

Another class of thermochemical processes is (high-temperature) *gasification*. As with pyrolysis, biomass gasifiers operate on principles similar to coal gasifiers. Both coal and biomass gasifiers have purely gaseous-fuel output products and both use many of the same thermochemical reactions. A well-known example is the Lurgi process (noted earlier), originally invented for coal but which can be used also with biomass feedstocks, such as wood chips. The steam–carbon and water–gas reactions, familiar historically from coal processing, take place in a reactor in various temperature zones as the processed material falls through the reactor from the top or side (see Fig. C.10).

The reactivities of biomass feedstock are higher than those for coal in the Lurgi process, resulting in more complete reactions and shorter processing times. Products of the various thermochemical reactions vary widely with feedstock and reactor conditions and are still the subject of research. For example, the role of the lignin components of wood and crops in the process must be better understood in order to make best use of all organic components.

There have been a number of simple gasifiers, supplying heat energy only, installed in the United States over recent decades, although several of these have been abandoned. These plants operated mostly in the southeastern states and used wood as feedstock. Thermal gasifiers, supplying low–medium-BTU gas, are in operation elsewhere in the world today, both in industrialized countries (Belgium, France, Finland, Italy, and Sweden) and in developing nations (Brazil, India, Indonesia, and Thailand).

In some developed countries, larger scale, more sophisticated utilization of biomass resources is under way. As an example, the lower BTU output of biomass gasifiers can be used in high-efficiency, combined-cycle (CC) generation (see Chapter 3). The CC operation using biomass gasifiers for input is called a BIG/GT (biomass-integrated-gasifier/gas turbine) system and the biogas output of the gasifier is the fuel input for combustion in the gas turbine (see Fig. C.15). Such plants are in the medium–large range of generating capacities (50–250 MW) for electric power systems, capable of supplying tens of thousands of homes.

Pyrolysis can also be used to create a *bio-oil* fuel (Gupta & Demirbas 2010). A *slow pyrolysis* process creates a fuel having a viscosity comparable to petroleum oils, with a lower heating value than a petroleum-based heating oil. A fast heating process can lead to a product of higher heating value, but only with careful monitoring of the process.

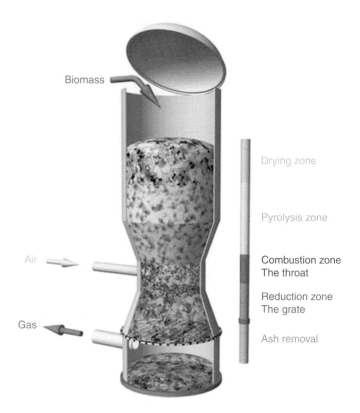

Biomass

Drying zone

Pyrolysis zone

Air

Combustion zone
The throat

Reduction zone
The grate

Gas

Ash removal

Figure C.15 Biomass-integrated-gasifier/gas turbine (BIG/GT) system.

Unconventional Fossil Resources

Oil Shale

As noted in Chapter 10, oil shale represents a natural energy resource comparable to (or larger than) petroleum. It is derived from a sedimentary rock known as marlstone or oil shale found concentrated in particular geological formations on all continents of the world. In the United States the largest formations are in Utah, Wyoming, and Colorado, but there are smaller resources in the Eastern states.

The oil is locked into the rock in a form called *kerogen*, a solid organic substance that is mostly insoluble. Kerogen is a large, complex organic chain which, like other organic compounds, can be broken down by *cracking*. Cracking, in this case, means subjecting pulverized shale to temperatures in the 800–1000 °F range. This produces an oily vapor that condenses into *oil shale* and is then purified by removing sulfur and nitrogen. Raw oil shale is more viscous than crude petroleum, thus presenting some shipping problems until refining techniques can be found that create an oil product more physically similar to high-grade petroleum.

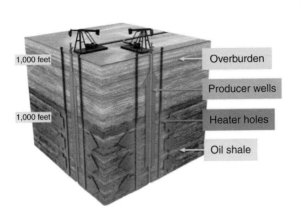

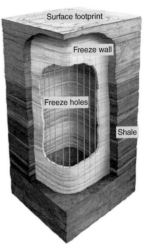

Figure C.16   Graphic representations of Shell's in situ conversion process and freeze wall plans. Electric heaters gradually heat the shale underground until the oil is freed from the rocks and can be pumped to the surface. Around the extraction zone, the underground freeze wall is designed to protect groundwater against contamination. *Source:* Reproduced from http://www.centerwest.org/publications/oilshale/4getitright/1leases.php.

Like coal, shale deposits are mined either underground or at the surface, depending on the depth of the deposits. Both the mining techniques and the problems of coal mining (see Chapter 6) apply to shale as well. Large-scale operations can lead to subsidence, land degradation, and other problems. Shale also creates large volumes of waste material for disposal. In a high-grade formation, about 25 gallons of oil shale can be produced *per ton* of marlstone. About 85 percent of the marlstone is waste. The spent shale is less dense after crushing and thus the volumes of material for disposal are larger than the original deposits by as much as 30 percent. Consequently, spent shale represents a greater disposal problem than coal.

After the shale is mined, it must be pulverized and then fed into a retort or kiln for cracking. In the retort, the crushed shale is heated by direct or indirect means. About 75 percent removal of the kerogen has been obtained at the 800–1000 °F retort temperatures of pilot plants.

Alternatively, cracking can take place *in situ*. The shale is first fractured by explosive charges or high-pressure hydraulic fracturing. The fractured zone in the shale seam underground is then ignited using a combustible gas, and partial combustion of the shale is sustained by compressed air. Then, instead of extracting rock, raw oil shale alone is removed. This technique works when about 30 percent air spaces have been created by fracturing. A cross-sectional conception of *in situ* recovery is shown in Figure C.16. Only about 30 percent kerogen recovery has been obtained with the *in situ* technique.

The *in situ* method has the obvious advantages of eliminating spent shale disposal requirements and, unlike the retorting process, it does not generate air pollution. But there are environmental impacts. One of the major concerns is the pollution of underground aquifers, an especially serious question in regions of scarce water resources in the Western United States. The shale processing itself requires large amounts of water, running close to 0.2 million acre feet annually to produce 1 million barrels per day of shale oil. (An *acre foot* of water is the volume required to fill a basin of 1 acre area to a depth of 1 foot, equal to 325,851 gallons.) The water resources of the arid Western states could be taxed beyond their limits in a multimillion barrel per day oil shale industry.

Oil Sands

Oil sands (often called "tar sands") are natural beds of sand, water, and bitumen. The bitumen is a tar-like hydrocarbon that can be separated from the sand by nonchemical means to give the equivalent of crude oil. Large deposits exist in North America, Russia, and the Middle East. Worldwide, the sands contain the equivalent of two-thirds of the world's petroleum reserves. With high oil prices in the 2000s tar sands have begun to be exploited.

The largest ongoing oil sands operation is the Athabasca sands in Alberta, Canada,[12] which requires the mining of hundreds of thousands of tons of sands each day using giant bucket-shell excavators and conveyor belts (see Fig. C.17). These deposits extend over the border (the "Keystone deposits") into the United States. The highly abrasive sand is mostly composed of quartz particles, leading to excessive wear of shovel bits at more than twice the rate of ordinary use. Rich oil sands contain about 10 percent bitumen by weight. As much as 5 tons of oil sands are needed to extract one barrel of a *crude*, sand oil. As of 2014, this production in Canada was equivalent to 2.3 million bbl/day of oil, which is about 2.5 percent of world crude-oil production. If oil prices rise, it is expected that oil sands production will treble over the next decade.

Oil sands extraction in an Athabasca deposit in Canada.
Source: dan_prat/Getty Images.

Oil sand processing consists basically of heating, sifting, and floating the bitumen tar to separate it from the sand particles. Following separation, the bitumen is upgraded in a process that corresponds roughly to the refining of crude petroleum. Upgrading also includes purification of the product of water, sand, and metals and the creation of useful hydrocarbons such as butane, naphtha, and kerosene. The prime concern in purification is the removal of sulfur, which typically makes up about 4 percent by weight (a level considered unacceptably high for petroleum) of the oil sand. These operations have significant environmental impacts on air, water, and wildlife in the region where these sands are mined. The product has therefore been called "dirty oil" by its detractors.

Geothermal Energy

The core of the Earth is a vast thermal reservoir that is effectively inexhaustible. Yet only a tiny fraction of geothermal energy has been tapped for use. Previously, the principal use of geothermal heat has been with *hydrothermal* techniques, which take advantage of water that has been heated in rock formations deep in the Earth's crust. These rock formations lie just above the *molten magma*, thousands of feet below the surface.

A typical geothermal reservoir is depicted in Figure C.18. In the figure, we are able to follow the path of water flow originating from rain that works its way down through fractures in rock until it reaches the depth at which it is heated by the magma. Once hot, the water rises, due to natural convection. But often its path upward will be blocked by low-permeability strata of rock, trapping it in a reservoir

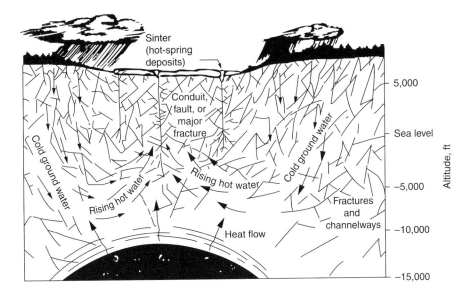

Figure C.18   The structure of a hydrothermal reservoir.
Source: US Geological Survey, 1973, *Geothermal Resources*. US Government Printing Office, Washington, DC.

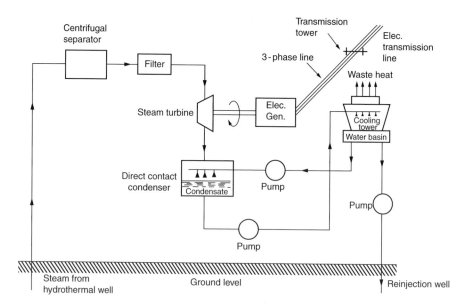

Figure C.19 A hydrothermal electric plant: direct, dry-steam type.
Source: Adapted from DiPippo, 1980, *Geothermal Energy as a Source of Electricity: A Worldwide Survey*, for the US Department of Energy. US Government Printing Office, Washington, DC.

far below the Earth's surface. As a result of heating and confinement, both temperature and pressure in the hydrothermal reservoir build up.

The water temperatures in these strata typically exceed 200 °C and pressures run into the range of 100 psi (about seven times atmospheric pressure). Depending on these temperatures and pressures, the water can be turned to pure or *dry* steam or become a wet mixture of steam and water. The steam or hot water of a hydrothermal reservoir can be brought to the surface through a natural fracture in the overlying rock or through a drilled conduit (Fig. C.18).[13]

The location of sites for drilled wells is a prospecting process similar to that for oil and natural gas. Only particular geological formations are known to be likely sites. When a successful well is found, it can supply steam for a thermal electrical plant, replacing the furnace boiler in a fossil-fired plant. In the case of a well with *dry* steam, the thermal system can have a direct input of the steam (Fig. C.19), and be made to pass through a filter before entering the steam turbine. A typical geothermal electric power plant has a *plant efficiency* around 20 percent, compared with a conventional plant's 35–45 percent. This alone, however, would not be the critical factor in the viability of this technology, because these plants operate on cost-free fuel – once the well is located and successfully drilled. Moreover, capacity factors for geothermal plants are in the range of 60–70 percent, close to that of fossil-fuel generators.

It should be noted, however, that hydrothermal energy is not without possible problems. For example, the steam and water often contain salts and minerals that corrode and pit turbine blades. Further development in corrosion-resistant metals is needed to make the technology less susceptible to breakdown and replacement.

The main barrier to geothermal energy, however, is the limited scope of the resource. The sites with the largest potential generating capacities are located in Central America, Iceland, Italy, New Zealand, the Philippines, and the Western United States. The total world installed capacity is (as of early 2015) 12.8 GW, and there are several gigawatts of additional capacity under development. Nevertheless, geothermal energy depends on natural formations and so will remain a relatively small component of world electric power.

Energy Storage

Conventional Storage

Energy storage is typically taken to mean storage of usable electric power, but should also include *heat storage*. In thermodynamic terms (Appendix A), we consider the storage of *transferred heat* before it is put to useful purposes. A longstanding conventional practice is hot-water storage, in the familiar hot-water tanks in homes and industry. Such heat storage smoothes out in time the delivery of hot water or steam, but it is not usually considered for a term longer than one day.

Electricity can, of course, be stored in batteries. However, the present-day automobile battery, to use an important and common example, is used for starting the (conventional) internal combustion engine and not for locomotion. More powerful batteries used in electric cars are beginning to have some market penetration but do depend heavily on subsidies and mandates.

Advanced storage – meaning a *new* technology not in conventional use – is an integral part of other new technologies that will be made more viable by innovations in storage. Wind and solar technologies especially will benefit from advances in energy storage, because of their intermittent natures.

Advanced Batteries

As noted in Chapter 11, new developments in batteries are important for operation of solar PV and wind generators and also for electric vehicles. As we have noted, batteries require major improvements in technical characteristics and cost reduction to become commercially (and in some cases also technically) viable.

Lead–acid batteries have been used for decades to start the gasoline or diesel engines that power the vehicles thereafter. Advanced battery concepts are typically a variation on the conventional lead–acid storage battery. This operates on the principle of the galvanic cell, whose discovery goes back to the eighteenth century. A single-cell battery consists of two electrodes immersed in an electrolyte (Fig. C.20). Chemical reactions of the electrolyte with the substances at each of the two electrodes release electrons with the potential to do electrical work. In the lead–acid battery, one electrode (the *anode*) is composed of lead (Pb) and is positively charged. The other electrode (the *cathode*) is made of lead oxide (PbO_2). The *electrolyte* in between is sulfuric acid (H_2SO_4). For the battery to deliver electricity, electrons are

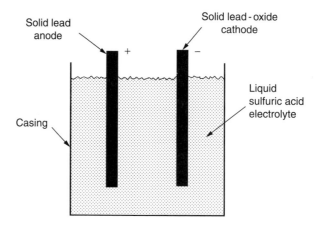

Figure C.20 Conventional lead–acid battery.

supplied to the lead-oxide cathode from the lead anode. But the chemical composition of the electrolyte changes as the battery is discharged. We can see this by writing the electrochemical equation:

$$PbO_2 + Pb + 2H_2SO_4 \rightarrow 2PbSO_2 + 2H_2O$$

We can see the products of the discharge on the right-hand side. The lead sulfate ($PbSO_2$) product is deposited on the electrodes, while water goes into the solution of the electrolyte. These products build up, of course, as the discharge progresses and eventually the battery no longer produces electricity.

However, the reaction is reversible. A storage battery can be recharged by a direct current (d.c.) source (e.g. d.c. generator) to reverse the flow of the charges and compounds. As the charge flow in the cell reverses so do the reactions: from $PbSO_2$ on the cathode back to PbO_2 on the anode; back to lead (Pb) on the anode; and to sulfuric acid in the electrolyte. This *complete* cycle of discharge and recharge can only be repeated, however, a limited number of times due to electrode and electrolyte deterioration. This determines the life cycle of the battery. For the conventional lead–acid batteries presently used in automobiles, this is as few as 300 (complete) cycles. This would mean, if these batteries were used in a solar application with daily (complete) cycling, they would last less than a year. Industrial grade lead–acid batteries have over five times the number of life cycles, but cost more than twice as much and would not be economically viable for such service. The lead–acid storage battery is too short-lived for the new applications envisaged. It is also too heavy and bulk, especially for electric vehicle application.

Although advanced batteries operate on the same galvanic principle as conventional batteries, major innovations have been tried in their construction and operation. In an effort to get away from deposits and corrosion on solid electrodes, liquid electrodes have been designed. An example of this was a sodium–sulfur battery, invented at the Ford Motor Company in the 1960s. In this prototype battery, the two electrodes were made of molten sodium and a molten sulfur–carbon mixture. The use of the liquid electrodes in the sodium–sulfur battery not only reduced corrosion, it extended the life of the electrodes themselves. But this battery had many drawbacks.

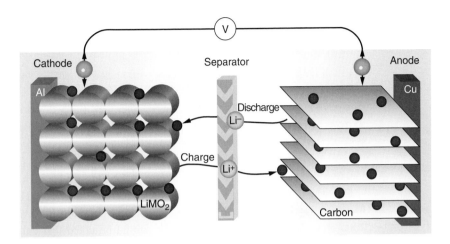

Figure C.21 A schematic of a lithium-ion battery.
Source: http://berc.berkeley.edu/storage-wars-batteries-vs-supercapacitors/.

It required operation at temperatures in the range 300–350 °C, thus presenting a safety risk to passengers in electric vehicles. And it was also too heavy and too costly.

There have been a number of battery designs for electric vehicles. In the 1960s a *nickel–metal hydride* (Ni–MH) battery was developed and this has found application both as a portable rechargeable battery and paired with a gasoline engine in some gasoline–electric hybrid car engines – notably early models of the Toyota Prius.

The main power source for electric vehicles – hybrid cars as well as all-electric vehicles – is a pack of *lithium-ion* (Li-ion) batteries (Fig. C.21). These batteries have longer lifespans (more discharges and recharges), are made from less toxic materials, do not lose their charge when stored, and have greater density (more power in a smaller space) than most other batteries. The positive electrode in a Li-ion battery is typically made from a chemical compound lithium-cobalt oxide ($LiCoO_2$) while the negative electrode is generally made of graphite ("Carbon" in the figure); the electrolyte that carries the ions from the positive electrode to the negative is (typically) a lithium salt in some kind of solvent. These batteries have come down considerably in price over the past decade. As of 2007, an average battery pack of 24 kWh cost about $24,000. By 2015 the price had fallen to under $10,000. Some packs, such as those used on the luxurious Tesla cars, have a range close to double that of other electrics; the estimated price of such battery packs was around $20,000 as of 2015. These are still expensive components and it is not clear as of this writing how durable such battery packs will be, but as the price of the batteries falls, electric cars become more cost competitive.

By 2015, there were over 20 million hybrid (combined gasoline and electric) cars and some 400,000 all-electric cars operating in the United States. The driving range of the all-electric models, however, was in most cases less than 150 miles per engine charge, but a few examples, especially the vehicles made by Tesla Motors, have extended the range to 300 miles – although at very high cost. Such battery-powered vehicles suffer from the lack of infrastructure for recharging and the need for more efficient home-based charging. Charging stations are becoming more common, but it still takes far longer to "fill up" with electricity than it does with gasoline. Therefore, development of the battery technology itself and the expansion of the charging network remain issues for electric car expansion.

Li-ion batteries have had some success with respect to electric vehicles, but do not solve the problem of fixed-location electrical energy storage, that is, the problem of storage of huge quantities of electric power. Other than pumped-hydro (which provides 95 percent of utility storage in the United States), the cost of electric-utility storage systems is currently *uncompetitive*. For example, battery storage may cost up to 50 percent more than pumped hydropower, which is itself expensive (as noted in Chapter 11). Costs for advanced batteries for utilities must come down considerably in the coming years.

One way to make storage batteries more cost competitive would be to extend their lifetimes. Developers of advanced batteries hope to attain life cycles in excess of 2000 and lifetimes of over 10 years. (It should be noted that daily cycling would require 3650 cycles for a 10-year lifetime.) For comparison, the lifetime of a conventional pumped hydroplant can be as high as 50 years and the number of charge–discharge cycles presents no additional limitation to that technology. This has important economic ramifications because the capital cost of equipment is usually amortized over periods no longer than the working life of the equipment. The short lifetimes now expected of batteries means that the projected costs of *delivered* energy will be more than 50 percent higher than those of pumped hydropower. Thus, the cost to the utility ratepayer would be correspondingly higher for that fraction of the electrical energy delivered by batteries. In stand-alone operation, such as with remote wind generation, cost considerations for batteries can be made in a similar manner, but with different operational parameters if high availability is the objective.

It is clear, however, that achieving economic viability for large-scale wind or solar generation will require significantly greater energy storage capacities. That is, the system must be capable of storing several days' equivalent usage of energy, so that the system can deliver the energy at the normal rate of daily consumption during periods of limited sunlight or wind.

Advanced Heat Storage

The conventional technique of storing heat in hot-water tanks is an example of the use of the physical property of *sensible heat* storage (Li 2016). In this form thermal energy is transferred (or "heat is added"; see Appendix A) to a medium and the increase in temperature of the material is proportional to the heat added, as long as the material does not change *state*. Accordingly, water is a material for sensible heat storage as long as it remains in the *state* of a liquid.

Heat can also be stored by hot-air systems by forcing the air through a porous bed of pebbles or crushed stone. Once the stones are warmed to the higher temperature, the heat has been stored – to be extracted later by passing cooler air through the bed.

However, water is the best sensible heat medium per unit of weight of material used (Table C.2). Specific heat is the constant of proportionality for sensible heat, expressed here in units of BTU per pound of material per degree Fahrenheit. With water, it takes 1 BTU per pound to raise the temperature by 1 °F.

Table C.2 Sensible heat materials			Heat capacity (BTU/ft^3 °F)	
Sensible heat material	Specific heat (BTU/lb °F)	True density (lb/ft^3)	No voids	30% voids
Water	1.00	62	62	—
Scrap iron	0.12*	490	59	41
Magnetite (Fe$_3$O$_4$)	0.18*	320	57	40
Scrap aluminum	0.23*	170	39	27
Stone	0.21	~170	36	25
Sodium (to 208 °F)	0.23	59	14	—

* From 77 °F to 600 °F.
Source: Adapted from Hottel and Howard (1971).

Solar-thermal systems probably represent the best examples of advanced heat storage applications. When water is used as the storage medium, heat storage is accomplished by heat exchange coils in the tank through which the hot coolant liquid is circulated from the solar collector. When the rate of heat exchange out of the tank is less than the rate of input from the collector, then the water temperature in the tank rises and thermal energy is stored. Conversely, when the heat output rate is greater than the input rate, energy is supplied from storage and the temperature of the tank water drops. These heat exchanges only work if the coolant from the collector is hotter than the tank water. The exchanges are aided if the input coil is placed in the lower – hence cooler – part of the tank and the output is placed in the hotter top part.

Typical daily residential hot water use is cyclical, rising to a daytime or early evening peak and dropping late at night. Industrial hot water and steam would also have daily cycles, but most often would be confined to daytime work hours and drop to zero on weekends and holidays. The use of short-time storage can extend the heat supply into the evening or possibly to the next morning.

But any attempt to increase short-time storage capacity requires an ever-increasing amount of collector area, which rapidly boosts the capital cost. Because solar energy costs have been at best marginally competitive, where the collector area is sufficient only for one day's supply, any attempt to collect and store heat for several days is currently not cost effective.

Storage tanks for solar-heated water are already quite large. Take the case of residential hot water. A family of four typically might use 75,000 BTU per day for hot water. A typical solar collector would have an area of about 150 square feet and a storage tank (at 2.5 gallons storage per square foot of collector) of 375 gallons. This is a volume of about 2800 cubic feet, which would require a 12-foot-diameter, 25-foot-long cylindrical tank – quite large for a modest sized home. A solar system that also supplies space heat would need a tank over ten times larger. (A still larger space requirement for storage would be required for hot-air solar home space heating.)

The size of storage volumes could be reduced considerably by the use of latent heat materials. High-pressure steam storage, for example, has been considered for large high-temperature industrial or utility operations. For lower-temperature

Table C.3 Latent heat materials			Heat of fusion	
Latent heat material	Melting point (°F)	Density (lb/ft³)	BTU/lb	BTU/ft³
Calcium chloride hexahydrate	84–102	102	75	7,900
Glauber's salt	90	92	105	9,700
Sodium metal	208	59	42	2,500
Lithium nitrate	482	149	158	23,500
Lithium hydride	1,260	51	1,800	92,000

Source: Adapted from Hottel and Howard (1971).

applications, such as residential water or space heating, solid materials that change phase have been considered. Table C.3 shows some common solid latent-heat materials that melt at elevated temperatures and thus absorb latent heat in changing phase from solid to liquid. The heat capacities per unit volume of these materials may be compared with the sensible heat materials.

However, latent heat materials have not been adopted widely for solar and other thermal applications for storage. A technical reason is the long time required for changes of phase and hence long charge–discharge times. Also, materials such as Glauber's salt were found to decompose into a simple sodium sulfate after repeated cycling and then settle out of the liquid solution. Further research on these compounds is being conducted in government laboratories.

Finally, there is a form of long-term heat storage that has found limited adoption. This is called *seasonal storage* and involves the storage of heat, or the *storage of cold*, underground – usually in aquifers. Both heat pumps and air conditioning are *heat exchangers* and, as such, rely on temperature differentials to function efficiently. Heat pumps, for example, can function by exchanging thermal energy from the outside ambient air, but their efficiency drops rapidly as the outside temperature falls below freezing. For subfreezing weather, heat pumps can be made to function by exchanging energy from coils placed underground, where the temperatures typically remain well above freezing throughout winter periods. Summer air conditioning, of course, uses a reservoir of lower-temperature material underground. Seasonal storage has not found wide use, however, mostly because of added investment and operating costs.

Fuel Cells

The construction and operation of a fuel cell appears, at first glance, to be similar to a battery. Both convert chemical energy directly into electricity using a pair of electrodes immersed in an electrolyte. However, the fuel cell is supplied continually by chemical energy, and in the fuel cell, unlike the battery, the electrode material is *not* depleted as it supplies electricity.

The fuel cell could be considered an ideal electric generator. Operation is quiet because the cell has no moving parts, and it is relatively clean. These qualities make it especially promising one day for electric generation in urban areas.

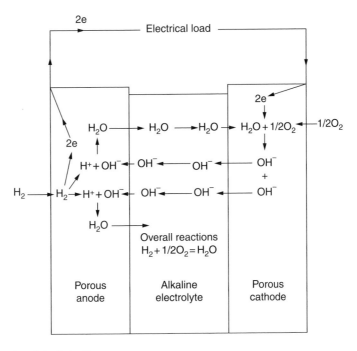

Figure C.22   Basic operation of the fuel cell.

A common fuel cell operates from hydrogen and oxygen and is called the proton-exchange cell. A diagram of the structure of a generic hydrogen–oxygen fuel cell is shown in Figure C.22, while a sketch of the structure is given in Figure C.23. The hydrogen and oxygen are fed into the (porous) anode and the (porous) cathode, respectively, with the hydrogen gas (H_2) flowing in from the left and oxygen (O_2) from the right. These two gases penetrate the porous electrodes, each of which is in contact with the liquid electrolyte. In this example, the electrolyte is alkaline (potassium hydroxide, $KOH \cdot H_2O$), although an acid, such as phosphoric acid, could also be used.

When using an alkaline electrolyte, hydroxyl ions are the *charge carriers* in the solution residing between the electrodes. The hydroxyl ions (OH^-) are created in the porous cathode on the right in Figure C.23 by the reaction of water and oxygen in the presence of excess electrons in the solution. These ions then diffuse in the electrolyte toward the anode. The reaction at the anode liberates electrons (e), which flow through the electrical load and, in turn, feed more reactions at the cathode. Also hydroxyl ions (OH^-) pass from the electrolyte and react with any excess hydrogen there to produce water. An electric potential exists across a fuel cell, just as it does across a battery, and the created electrons can thus do *electrical work* (Appendix A).

Practical operation of a fuel cell requires either a catalyst or elevated temperatures to enhance the electrode reactions. These temperatures typically exceed the boiling point of water and the cells are therefore operated under pressure to prevent the water from turning to steam. Although oxygen can easily be supplied to the cell by air, the supply of hydrogen requires an additional process. The most successful method to date has been steam reforming of either naphtha or natural gas. This reaction is similar to some of those shown earlier in this appendix for coal gasification. If methane (from natural gas) is the input, for example, the reaction is:

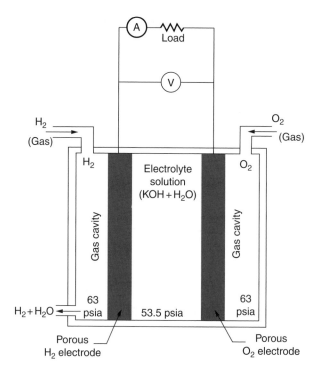

Figure C.23   Construction of the fuel cell.
Source: Crowe (1973).

$$CH_4 + H_2O \rightarrow CO + 3H_2$$

where the water (H_2O) is actually steam at high temperature (1500 °F). In addition to the high temperature, the reaction also requires catalysts, such as nickel oxide or silicon.

Fuel cells have also been considered as the means to power an electric vehicle. As of 2016, Toyota began offering a fuel-cell-powered vehicle called the Mirai. It is a basic mid-size car but costs a minimum of $57,500. At the same time it comes with a $4000 federal government tax credit, and 3 years of free (hydrogen) fuel. California residents receive an additional $5000 rebate. In the past, fuel-cell automobiles had weight concerns. Hydrogen storage was also a problem, but it does not seem so for the Mirai; the hydrogen is stored under high pressure in tanks under the rear seats. Refueling takes much less time than electric car recharging and the range on a full tank is over 300 miles.

Still, the cost of the car is obviously much higher than similar gasoline cars. The DOE has claimed that the cost of the fuel cell itself, formerly more than $1000/kW can now be produced for about $60/kW. Actual production models cost a good deal more. But even at the DOE's figure the cost is still double what it would need to be to be cost competitive with gasoline. And fuel cell vehicles share the same problem as electrics in a lack of infrastructure to provide fuel and servicing. In California, the only state where the Mirai is sold, there are only about 20 hydrogen filling stations.

Cost is also a concern for fuel cells meant to be utilized as backup for electricity-intensive businesses. Web servers, such as Google, and cellular phone operators, such as Sprint, have gigantic electric power needs – needs for power to be available every moment of every day. Most of their electricity comes from the grid, but fuel cells can (and in some cases do) provide instant backup. The capital cost (before installation charges) of these stationary fuel cells was hard to pin down. There were claims of costs as low as $300/kW – or at least claims that that was the expected price in the near future. In general, an electric backup was said to cost anywhere from $800 to $8000/kW; these leave out installation charges, which can be as much as $10,000/kW.[14] But again, there are often incentives that lower the cost to the consumer considerably. Large-scale projects can receive incentives – rebates, grants, etc. – amounting to 50 percent of the initial cost. Ultimately, fuel cells could be cheap enough and powerful enough to serve as primary sources of electricity for communities or for a backup to wind and solar and for quick starting additions to the grid at times of peak demand. But such applications are still well in the future.

With large incentives, fuel cells are penetrating several markets, but there is much development to be done before such technologies as fuel-cell automobiles and backup fuel-cell electricity become commercially viable in their own right.

Nuclear Fusion Power Reactors

Nuclear fusion has been said to represent the ultimate in energy technologies in several different ways. It is a prime example of the so-called *backstop technology* (Nordhaus, 1979), in that its fuels are *superabundant*. Fusion fuels could replace fossil fuels and would essentially never run out. It is potentially a clean technology in that it does not emit great quantities of gaseous or solid wastes and its nuclear by-products are far less hazardous than those of nuclear fission technologies. Because of its enormous potential, it has become by far the largest long-term research and development effort for a new energy-conversion technology. In fact, it has been one of the largest commitments worldwide to scientific development for civilian purposes over the past several decades.

Physical Principles

Nuclear reactions are a source of liberated energy at the two extremes of the nuclear mass scale on the periodic table. At the *high end* (atomic mass number greater than 200) we have noted the nuclear fission reactions of isotopes of uranium and thorium. Nuclear fusion reactions, in contrast, occur at the *low end* of the mass scale, with the isotopes of hydrogen and helium.

It might seem paradoxical that the *splitting* of nuclei leads to the liberation of energy in the case of fission reactors and that the *fusing* together of nuclei in fusion reactions also leads to energy release. The full explanation of this paradox, and why the reactions occur at opposite ends of the nuclear mass scale, lies in the theory of nuclear binding forces (Sproull 1963). We are able to understand the outcomes of the reactions in each case, however, as a consequence of the equivalence of mass and energy – the famous equation of modern physics ($E = mc^2$).

In examining the mass–energy equivalence (Chapter 7) for nuclear fission reactions, we saw that the aggregate mass of the fission products is less than the

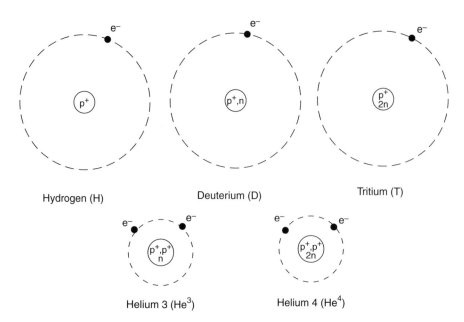

Figure C.24 Hydrogen isotopes and possible fusion reactions. First-generation fusion reactions (DT); $T \geq 4$ keV, 40×10^6 °C; D + T → α + n + 17.6 MeV. Second-generation fusion reactions (DD); $T \geq 34$ keV, 340×10^6 °C; D + D → T + p + 4.03 MeV; D + D → ^{3}He + n + 3.27 MeV. Third-generation fusion reaction (DD); $T \geq 100$ keV, 10^9 °C; D + ^{3}He → α + p + 18.3 MeV.
Source: Adapted from Penner and Icerman (1975).

combined masses of the ^{235}U nucleus and the neutron from which they were created. Similarly, in a typical fusion reaction (of two deuterium nuclei), the combined mass of the helium isotope and neutron fusion products is less than the two deuterium nuclei from which they were created. For both fission and fusion reactions, then, the difference in nuclear masses is manifested as kinetic energy of high-speed motion.

Physicists have considered four different possible nuclear fusion reactions as the basis for fusion power reactors of the future (see Fig. C.24). For each of the four, the liberated energy is shown on the right side of the reaction equation in millions of electron volts (MeV). In the first reaction, deuterium (D) and tritium (T) fuse to yield the helium-4 nucleus ($^4_2$He, or the alpha particle) and a neutron (n); each such reaction also releases 17.6 MeV of energy (see legend, Fig. C.24). The reaction of two deuterium nuclei, on the other hand, can produce either a tritium nucleus and a proton (p) plus 4.03 MeV in energy, or a helium-3 ($^3_2$He) nucleus and a neutron, plus 3.27 MeV in energy. Finally, deuterium and $^3_2$He react to yield $^4_2$He and a proton, plus 18.3 MeV in energy.

Fusion reactions occur only when the light nuclei come sufficiently close together (within 10^{-15} m). The main way that they come into that close proximity, however, is when they collide at high speeds. Otherwise, the repelling force within the nuclei (isolated nuclei are positively charged ions) keeps them separated. The kinetic energy required to overcome those electrostatic repelling forces is shown for each of the four fusion reactions in Figure C.24, in units of keV (thousands of electronvolts) or equivalent temperature (see Appendix A).

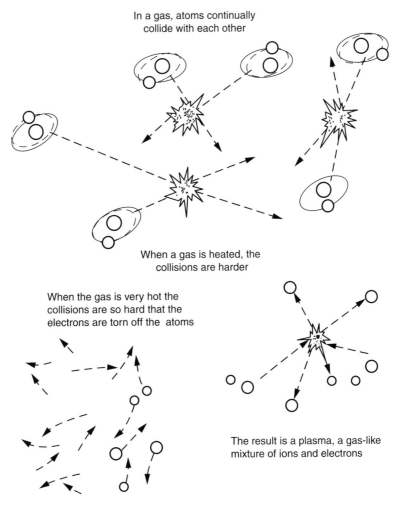

In a gas, atoms continually
collide with each other

When a gas is heated, the
collisions are harder

When the gas is very hot the
collisions are so hard that the
electrons are torn off the atoms

The result is a plasma, a gas-like
mixture of ions and electrons

A hot plasma wants to expand quickly. Also if it is
allowed to hit a wall it will cool quickly and revert
back to a gas again

Figure C.25 Principles of plasmas.
Source: US Energy Research and Development Administration (1976).

Creating and sustaining nuclear fusion reactions is a fundamentally different
problem from that of fission. The fusion reaction is *not a chain reaction*, nor does
it depend on energetic neutrons to initiate it. Rather, the fusion reaction requires the
heating of the isotope constituents, in gaseous form, to extremely high temperatures,
so as to create *high-energy collisions* of the fusible nuclei. For this reason the
reactions are called *thermonuclear* reactions.

The nature of the gas as it is heated is suggested by Figure C.25. It indicates that,
as the temperature of the gas rises, collisions between neutral atoms become sufficiently energetic to strip the electrons from their nuclei, resulting in a gas of
negatively charged electrons and positively charged nuclei (ions) in equal numbers.

This gas, which has a net zero charge over all of its volume, is called a *plasma* and much of the research effort for fusion has to do with its dynamics.

As the gas of hydrogen isotopes is heated, the ionization takes place at temperatures in the range of 100,000 °C or, equivalently, when individual atoms have average thermal velocities over a million meters per second. In order to achieve fusion reactions by collisions of the ions, however, the temperature of the gas must rise to over one hundred million degrees, which corresponds to average thermal velocities about one-tenth the velocity of light (300 million meters per second). Such temperatures create solar-like conditions, and no ordinary material can contain the gas in this state.

Let us assume, however, that containment is possible, and consider an idealized fusion reactor. The ideal fusion reactor should provide a way of transferring the thermal energy generated by the nuclear reactions in the confined plasma to a turbine or other means of conversion to useful work. We can see this idealized concept at the top of Figure C.26. A fusion plasma may be pictured as "*bottled*" and a coolant flows around it, taking the bottled heat out for useful work. In other words, the fusion reactor would be like the nuclear fission reactor in one respect: another extremely *exotic way to boil water*. Although the ideal is unattainable, we can in practice come close enough to contain the plasma and permit controllable fusion reactions.

We must emphasize at the outset that the quest for controlled thermonuclear reactions (CTR) is still in the *demonstration* phase. A successful demonstration fusion reactor must have solved three problems: the temperature must be *hot* enough for the high-energy collisions; the plasma must be confined *long* enough for the nuclear reactions to heat the plasma for further reactions; and the plasma must be *dense* enough for the rate at which reactions take place to overcome the rate at which energy is lost to the outside.

Fusion Research

The first demonstration goal of the fusion programs was to achieve the *breakeven* condition. Breakeven is the point where the energy yield from the fusion reactions equals the energy input required to create the reactions. The required energy input is the amount necessary to confine and heat the plasma to fusion conditions. If the breakeven condition is achieved, the next goal is to realize a *net gain* of energy, that is, where *more* energy is generated by the fusion reactions than has to be invested to initiate them. This condition is also termed *ignition*, with the implication that a thermonuclear *burn* would take place in a manner analogous to combustion in a furnace. Only after ignition is attained could the fusion reactor be developed on an engineering basis and demonstrated for commercial use.

Plasma Confinement

One of the most crucial problems that the researchers had to solve for fusion was how to *confine* the plasma, because there is no known material that could be used for a containment vessel that could withstand the *solar-like* temperatures of the plasma. Confinement of the fusion plasma has been conceived in two major ways: *magnetic*

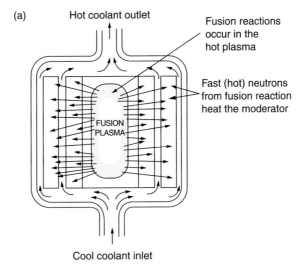

(a)

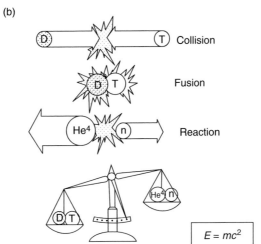

(b)

Figure C.26 Principles of nuclear fusion. (a) Fusion plasma confinement. (b) Nuclear fusion reaction. *Source:* US Energy Research and Development Administration (1976).

confinement and *inertial* confinement. We will focus on the former at this point and will consider inertial confinement later. Magnetic confinement takes advantage of a well-known property of magnetic fields (Fig. C.27). These fields deflect the motions of charged particles that are directed across the lines of force, so that they tend to spiral or gyrate around the lines (Fig. C.27).

The leading design configuration for magnetic confinement has been the *tokamak*, an example of which is shown in Figure C.28. In the tokamak configuration, confining magnetic fields bend the plasma into a torus (a donut-shaped ring). These fields are created by a series of coils looped around the plasma toroid (see the toroidal-field magnet coils in the figure). These magnetic lines of force are concentric with the plasma donut, following its large diameter.

Fusion requirements

Confining force

Energy
investment   Hot
plasma   Energy
output

Confining force

Specifically,

$T_{ION} > 5$ keV (50,000,000°)
$n\tau > 5 \times 10^{13}$ cm^{-9} seconds

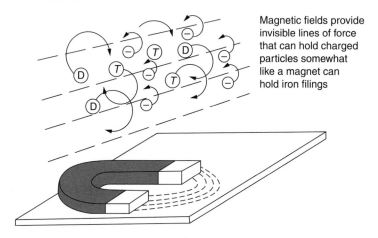

Magnetic fields provide
invisible lines of force
that can hold charged
particles somewhat
like a magnet can
hold iron filings

Figure C.27   Principles of magnetic confinement.
Source: US Energy Research and Development Administration (1976).

The toroidal magnetic fields of the tokamak deflect any radial motion of electrons (or ions) moving out of the toroid and thus confine them within. Thus, the natural force of expansion of the hot plasma is overcome. This confinement field must be maintained to sustain fusion reactions. As we will discuss below, the enormous forces required for a fusion plasma at ignition temperatures make continuous confinement a major engineering challenge.

Magnetic confinement in practice is imperfect, however, due to physical processes, such as *diffusion*, within the plasma. Diffusion is a natural process in any body of particles in thermal motion. Although the magnetic field can contain rapidly *moving* particles, it is ineffective against random motion of a particle momentarily brought to *rest* by a thermal collision. As a result, the objective of magnetic confinement in the face of diffusion and other losses is simply to sustain the density of the plasma for sufficiently long time periods for the nuclear fusion reactions to take place.

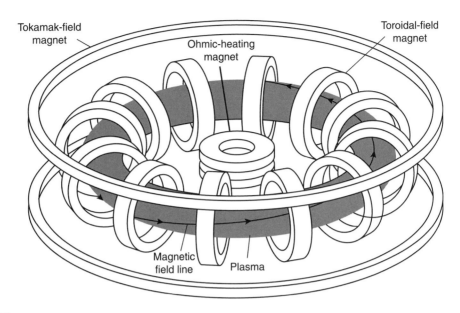

Tokamak-field
magnet

Ohmic-heating
magnet

Toroidal-field
magnet

Magnetic
field line

Plasma

Figure C.28 Idealized tokamak.

After confinement, the next major task in fusion research is heating the plasma. In a tokamak, the principal method has been to induce a strong electric current through the plasma, thus causing electrical losses that in turn heat the gas. This form of heating is called *ohmic* heating, so named after the electrical unit for resistance (ohm). The current is induced magnetically in the plasma torus in the manner of an alternating current (a.c.) electrical transformer (Appendix A). The primary winding of the ohmic heating coil is concentric to the plasma torus (Fig. C.28). The torus is coupled to the primary winding by the magnetic field of the primary coil. According to the principles of transformers, if the secondary coil has fewer turns than the primary, then there is a *step down* in voltage but a *step up* in current. In the original tokamak the plasma torus-shape was, in effect, a single-turn coil, whereas the primary *ohmic-heating* coil was designed with 10 to 20 turns. When 50,000 to 100,000 amperes were driven through the primary coil, 1 million amperes or more were induced in the torus, by transformer action.

Ohmic heating, as powerful as it might seem, was found insufficient for reaching ignition temperatures in the original tokamak. Efforts were made toward *supplementary heating.* One means of supplementary heating is through the injection of a high-energy beam of neutralized hydrogen isotope atoms into the plasma torus, as shown in Figure C.29.

The injected particles have been neutralized – that is, have neither a positive nor a negative charge – in order not to be deflected by the strong magnetic field in the reactor. Once into the plasma, the injected atoms collide with the particles of the plasma and became ionized components of the plasma, adding to the thermal energy.

Supplementary heating was also tried by irradiating the plasma with high-power radiowaves or microwaves, somewhat in the manner of a microwave oven. This

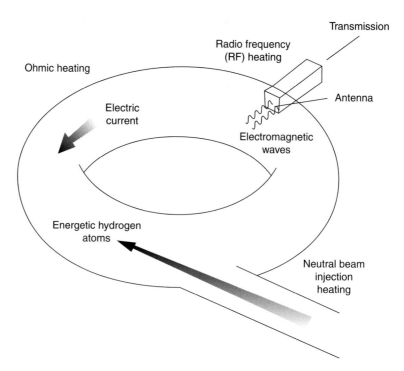

Figure C.29 Neutral beam concept.
Source: Courtesy of the Princeton Plasma Physics Laboratory.

process, called *rf* (for radio frequency) *heating*, was thought to work particularly well when the frequency of the incoming waves is near one that resonates naturally with the magnetized plasma. One such resonance frequency is the *cyclotron frequency*, which is the natural rate of gyration of the electrons around the magnetic field.

In addition to the confining and heating, plasma physicists have faced the challenge of controlling the plasma against *instabilities.* A hot plasma is inherently unstable and can break into violent motion of various types, including strong streaming motions and what are called magnetohydrodynamic (MHD) instabilities. The conditions for controlling plasma are so delicate that some plasma researchers likened the problems to *trying to suspend water in a bottle upside down*, while others said they had to control a stellar Jell-O.

The status of fusion research is summarized in Figure C.30. There have been three measures of the conditions in the hot plasma regarding those required for net energy output: (i) the plasma ion density (n, particles per m^3), (ii) the plasma confinement time (τ, seconds), and (iii) the ion kinetic energy (T, keV). For breakeven reactions to take place, the plasma must have a minimum combination of density (n), confinement time (τ) and heat (T). This is expressed in the figure as the "fusion triple product" ($nT\tau$) on the vertical axis. The horizontal axis depicts the temperature of the plasma (ions). These parameters have been used in measuring the progress toward *ignition* and *sustained fusion* in both the magnetic-confinement and the inertial-confinement fusion programs. The highest triple product to date with magnetic confinement was, as Figure C.30 shows, the Japan Torus 60U.

Table C.4 The experiments reported in Figure C.30	
ALC-C	Alcator-C, Plasma Fusion Center, MIT
ASDEX	Axially Symmetric Divertor Experiment, Max Planck Institute, Garching, Germany
DIII	Doublet III, General Atomics, San Diego
DIII-D	Doublet III-D, General Atomics, San Diego
ITER	International Thermonuclear Experimental Reactor, Cadarache, France
JET	Joint European Torus, Abingdon, UK
JT-60U	Naka Fusion Institute, Japan
ORMAK	Oak Ridge Tokamak, Oak Ridge National Laboratory
PLT	Princeton Large Torus, Princeton Plasma Physics Laboratory
T-3	Kurchatov Institute, Moscow
T-10	Kurchatov Institute, Moscow
TFR	Centre D'Etudes Nucleaire, France
TFTR	Tokamak Fusion Test Reactor, Princeton Plasma Physics Laboratory

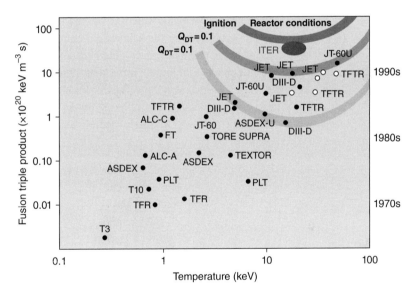

Figure C.30 Tokamak fusion research developments over 40 years (see Table C.4 for abbreviations).
Source: http://forum.tour-magazin.de/

In 1995, the US Congress made major budget cuts in the Department of Energy fusion programs. These cuts were so severe that plans for the next generation of reactor experiments at the Princeton Plasma Physics Laboratory, the leading US fusion research institution, were canceled, including the TFTR (Tokamak Fusion Test Reactor) program, which was closed in 1997. Fusion reactor projects have since been carried on in other countries as cooperative international efforts. The ITER program was formed and funded in 2007 with participation by six countries and the entire European Union. The six countries are France, Japan, China, the Republic of Korea, India, and the United States. The United States has joined this program, having dropped out of earlier phases of international fusion research. The center for the new research is in Cadarache, France. This project will build a new reactor with a capacity of 500 MW and design it to deliver sustained power output for 1000 seconds.[15] This is intended to make gains over the previous international project (JET: Joint European Torus) capacity of 16 MW for less than one second, as a step toward the ultimate reactor, having a GW power output in steady-state operation. The ITER project has high aspirations for bringing nuclear fusion to reality in the coming decades.

Inertial Confinement Fusion

An artist's sketch of laser-driven, inertial-confinement fusion is shown in Figure C.31. The figure shows a small pellet of fusion fuel (such as DT) centered at the point of intersection of several laser beams. The pellet has been timed to drop into this position such that its arrival coincides with a high-power pulse of laser light simultaneously emitted through all laser beam transport tubes (four are shown). The incident laser light heats and compresses the fuel pellet, causing fusion ignition within an extremely short time interval.

When fusion schemes such as this were first proposed in the late 1960s, they were received with great enthusiasm in the scientific community. The concept was elegantly simple compared to the magnetic confinement schemes. Furthermore, with the rapid advances being made with lasers during that period, it was thought that quick advances could be made this way to a *fusion breakeven* experiment, where the amount of energy produced equals the initiating energy input. The new laser-fusion projects appealed to the public as futuristic marvels because of their exotic features.

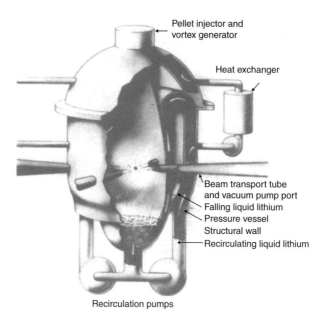

Figure C.31   Inertial confinement fusion reactor concept.
Source: Courtesy of the Lawrence Livermore Laboratory.

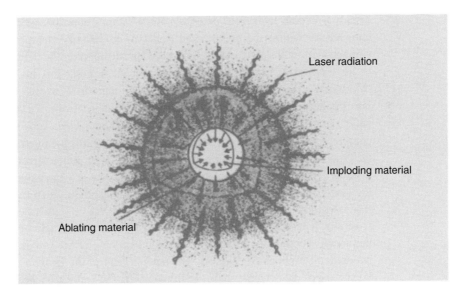

Laser radiation

Imploding material

Ablating material

Figure C.32   Laser fusion: compression stage, implosion of the pellet.
Source: Physics Today (August 1973). Courtesy of the University of California, Lawrence Livermore Laboratory, under the auspices of the US Department of Energy.

Multi-beam laser systems capable of huge power outputs were first built in the 1970s. Early lasers built in the program emitted infrared pulses in eight beams directing 10,000 joules of light energy onto the target in 1 nanosecond (10^{-9} seconds). This rate of energy delivery equaled 1 terawatt (10^{12} watts). The density of the light flux flowing onto the target during the 1-nanosecond interval was huge. For example, if the laser beams were all focused into a 1-mm-diameter spot, the light flux intensity would be over 10^{14} W/cm^2 (compare this with the average solar flux density of 245 W/m^2 = 0.0245 W/cm^2.

Much more powerful lasers have been built subsequently, in attempts to increase this light intensity even further. And there have been efforts to combine the outputs of several hundred such laser beams, thus making the total light energy output of the combination as much as a thousand times that of earlier lasers. This increased laser energy output combined with shortening the laser's pulse length to 0.1 nanoseconds (10^{-10} seconds) would make the power (rate of energy delivery) over 10,000 times higher than that of the early lasers.

With these extremely high concentrations of power in laser light, violent dynamics take place within the pellet. An artist's conception of these dynamics is shown in Figure C.32. They actually evolve in stages. The *first stage* is the formation of a plasma, which is *ablated* directly from the solid pellet material by the rapid laser heating. This plasma surrounding the solid pellet itself gets heated even more and expands violently away from the pellet in a blow-off – the *second stage*. The force of this blow-off is so great that it acts like a rocket thrust, causing a compression wave to propagate *into* the pellet. This compression is calculated to lead to densities of the isotope nuclei of at least 1000 times the density of liquid hydrogen, pressures at 10^{10} times atmospheric pressure, and a temperature within the compressed pellet of millions of degrees Celsius. The compression by the laser can be enhanced by surrounding the pellet with a jacket of an element of higher nuclear weight, such as gold; this jacket is called a "hohlraum" (the German word for *cavity*). The intense deposition of laser energy into the heavier nuclei causes high-energy X-rays to be emitted into the plasma pellet, impinging uniformly from the surrounding jacket, causing tremendous compressive forces to be exerted symmetrically inward. Symmetric compression is important to avoid exciting *plasma instabilities*, which divert energy from heating. Symmetric compression has been difficult to achieve with direct drive by the laser beams themselves and so the hohlraum technique is an important step in this research.

Ideally, a further deposition of heat from incoming laser light accompanies this implosion, serving to raise the temperatures even further within the pellet (to hundreds of millions of degrees). This should lead to the *ignition condition* – the *third stage* of the dynamic sequence (see below). The entire sequence of three stages has been theorized to take place within less than 1 nanosecond. Whereas much of this dynamic sequence has been observed experimentally, the ignition condition has not been readily attained. If it is attained, then and only then will the final stage of thermonuclear burn be observed, which is required to demonstrate the *scientific feasibility* of this version of fusion technology.

Research on inertial confinement (IC) plasmas has the objective of attaining a working, power-producing, fusion reactor. The action of the laser beam on the small pellet plasma, as depicted in Figure C.32, will initiate a miniature nuclear fusion reaction, as the basis for a fusion-reactor power plant. There has been another motivation for IC-fusion projects, however, since laser fusion provides a miniature version of the *nuclear fission* "trigger" of a hydrogen-fusion bomb. This type of experiment thereby was earlier intended to provide the means of investigating the behavior of the hot, dense plasmas of fusion weapons *without* conducting the full scale H-bomb tests (now forbidden by international treaty).

The scientific objective of this new facility, therefore, is to create the fusion *ignition condition* in the DT pellet, using a powerful laser beam. In fact, the new laboratory facility where these exotic conditions are to be created has been named the *National Ignition Facility*. An ignition condition for nuclear fusion is the start of a *sustained* fusion reaction, just as a chain reaction for nuclear fission starts sustained fission reactions. Since ignition is a condition exceeding the minimal-initiation of fusion reactions, this is an even more ambitious objective than the initial objective of the magnetic-confinement reactor program years before. In order to achieve this, temperature conditions in the range of 100 keV (equivalent temperature) and pressures over 10^{12} times atmospheric pressure must be created inside the pellet/plasma for fusion. Figure C.33 is a diagram of the research setup at the National Ignition Facility (NIF) of the Lawrence Livermore National Laboratory in Livermore, CA. The system takes the outputs of (source) lasers and passes the light beams through 192 powerful laser amplifiers to create *beamlets* for a combined optical power output in the terawatt (TW = 10^{12} watt) range, ultimately 500 TW.

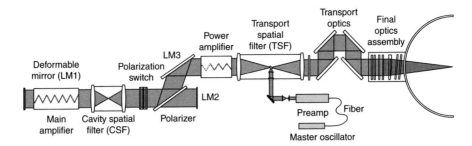

Figure C.33 National Ignition Facility laser system.
Source: Wikipedia – NIF beamline diagram.png-Wikipedia, 2011.

Ignition experiments began in 2009 at the NIF facility but ignition has yet to be attained. Research on inertial fusion continues at NIF according to a research plan entitled the *National Ignition Campaign.*[16]

Ongoing Development

The joke is that fusion is 20 years away from being realized and always will be. But there are a number of projects, both public and private, that might bring the world closer to fusion electric power generation. Defense contractor Lockheed Martin Corporation announced in 2014 the development of a small fusion reactor it hopes to test in the 2020s – a device much smaller than the huge international ITER project and smaller than other fusion efforts. The concept "combines several alternative magnetic confinement approaches, taking the best parts of each," the company said. But it will reveal the elements of its design and operating characteristics only after it is patented and tested. In the meantime, another company, this time a start-up called Tri Alpha Energy, announced a breakthrough in 2015 in controlling hot plasma, though they were also a long way from being able to claim that fusion was at hand.

It is notable that there is an ongoing effort in many quarters, public and private, to achieve the dream of a nuclear fusion reactor and power plant. If such a power plant is successfully developed, it would be a great boon to all humankind.

References

Bryce, R. 2015. "Should the U.S. End the Ethanol Mandate?" *Wall Street Journal,* November 15, online at www.wsj.com/articles/should-the-u-s-end-the-ethanol-mandate-1447643514.

Camatini, E. (ed.). 1976. *Heat Pumps: Their Contribution to Energy Conservation.* Kluwer Academic Publishers, Norwell, MA.

Cassedy, E.S. 2000. *Prospects for Sustainable Energy.* Cambridge University Press, Cambridge, UK.

Cassidy, E. 2013. *"Ethanol's Broken Promise: Using Less Corn Ethanol Reduces Greenhouse Gas Emissions."* Report, Environmental Working Group, Washington, DC.

Crowe, B.J. 1973. *Fuel Cells: A Survey.* National Aeronautics and Space Administration (NASA), US Government Printing Office, Washington, DC.

Gupta, R.B. and A. Demirbas. 2010. *Gasoline, Diesel and Biofuels from Grasses and Plants.* Cambridge University Press, New York.

Hottel, H.C. and J.B. Howard. 1971. *New Energy Technology: Some Facts and Assessments.* MIT Press, Cambridge, MA.

Jaeger, W.K. and T.M. Egelkraut. 2011. "Biofuel Economics in a Setting of Multiple Objectives and Unintended Consequences." *Renewable and Sustainable Energy Reviews*, 15, 4320–4333.

Johansson, T.B., H. Kelly, A.K.N. Reddy, and R.H. Williams (eds.). 1993. *Renewable Energy Sources for Fuels and Electricity.* Island Press, Washington, DC.

Li, G. 2016. "Sensible Heat Thermal Storage Energy and Exergy Performance Evaluations." *Renewable and Sustainable Energy Reviews*, 53, 897–923.

Monteiro, N., I. Altman, and S. Lahiri. 2012. "The Impact of Ethanol Production on Food Prices: The Role of Interplay between the U.S. and Brazil." *Energy Policy*, 41, 193–199.

Nordhaus, W.D. 1979. *The Efficient Use of Energy Resources.* Yale University Press, New Haven, CT.

Probstein, R.F. and R.E. Hicks. 2006. *Synthetic Fuels.* Dover Publications, Mineola, NY.

Sproull, R.L. 1963. *Modern Physics.* Wiley, New York.

NOTES

1 *Water Energy Resources of the United States with Emphasis on Low Head/Low Power Resources*, D.G. Hall et al., April 2004, Idaho National Engineering and Environmental Laboratory, OE/ID-11111.

2 *Municipal Solid Waste: Electricity from Municipal Solid Waste*, US. Environmental Protection Agency, March 17, 2010. http://epa.gov/cleanenergy/energy-and-you/affect/municipal-sw.html.

3 *Integrated Materials and Energy Recovery from Municipal MSW, Microbial Action*, M.S. Finstein & Y. Zadik, BioCycle Energy, November, 2008.

4 Described at http://greatriverenergy.com/we-provide-electricity/making-electricity/biomass/.

5 *Booming European Heat Pumps Market: Latest News*, May 5, 2010. www.robur.com/news/latest-news/booming-european-heat-pumps-market.html.

6 *Geothermal Heat Pumps: Geo-exchanger or Ground-coupled Heat Pump.* NYSERDA – Energy Innovation Solutions, April 11, 2011. www.nyserda.org/programs/geothermal/default.asp.

7 *Lurgi Gasifiers*, US Department of Energy, NETL, April 19, 2011. http://www.netl.doe.gov/File%20Library/Research/Coal/energy%20systems/gasification/gasifipedia/index.html.

8 *Annual Review 2010, Sasol Synfuels,* 2010 Copyright by Sasol Ltd. http://sasol.investoreports.com/sasol_review_2010/operating-reviews/sasol-synfuels/.

9 See Vern W. Weekman, Jr., 2010, "Gazing into an Energy Crystal Ball, *CEP Magazine.* www.aiche.org/resources/publications/cep/2010/june/gazing-energy-crystal-ball.

10 "Why Liquid Coal Is Not a Viable Option to Move America Beyond Oil," Natural Resources Defense Council, 2011. www.nrdc.org/sites/default/files/liquidcoalnotviable_fs.pdf.

11 *H₂ Production via Biomass Gasification,* Advanced Energy Pathways (AEP) Project, Public Interest Research Program, California Energy Commission, July, 2007.

12 "Canada's Tar Sands – Muck and Brass," *The Economist*, 2011. www.economist.com/node/17959688.

13 Natural fractures are the sites of natural geysers and hot springs.

14 From the National Institute of Building Sciences, "Fuel Cells and Renewable Hydrogen." www.wbdg.org/resources/fuelcell.php.

15 *ITER Brings Fusion Into Horizon*, October 31, 2011, Public Service Europe, publicservice.co.uk Ltd, Newcastle-under-Lyme, UK.

16 "Giant Laser Reaches Key Milestone for Fusion," January 28, 2010, Jeff Hecht, *New Scientist*.

Index